Nuclear Bodies

Nuclear Bodies

The Global
Hibakusha

Robert A. Jacobs

Yale UNIVERSITY PRESS
New Haven & London

Published with assistance from the Mary Cady Tew Memorial Fund.

Yale University Press books may be purchased in quantity for educational, business, or promotional use. For information, please e-mail sales.press @yale.edu (U.S. office) or sales@yaleup.co.uk (U.K. office).

Set in Galliard type by IDS Infotech Ltd., Chandigarh, India.
Printed and bound by CPI Group (UK) Ltd, Croydon, CR0 4YY

Library of Congress Control Number: 2021942566
ISBN 978-0-300-23033-8 (hardcover : alk. paper)

A catalogue record for this book is available from the British Library.

10 9 8 7 6 5 4 3 2 1

For Dylan and Max

Contents

Part IV. Heirs

Preface

Millions of people around the world have suffered harm from radiation since the nuclear attacks on Hiroshima and Nagasaki in 1945. Their bodies form part of the fabric of ecosystems where nuclear fallout deposited radioactive particles—whether from nuclear weapon testing, nuclear power plant accidents, or the production of materials used in both technologies. These exposures have led to deaths, illnesses, forced evacuations from homes and communities, continued habitation in radiologically contaminated landscapes, tainted food sources, and endless anxieties and emotional distress. The experiences of these "global hibakusha" have been largely invisible to us because they happened primarily in colonial, post-colonial, or remote parts of our world, or to people with little political recourse. (*Hibakusha* is a Japanese word denoting a survivor of the attacks on Hiroshima and Nagasaki.)

Based on its belief that nuclear weapons were likely to be used in warfare after 1945, the United States conducted large-scale studies to understand the medical consequences of exposure to radiation from the detonations of these weapons. The studies focused on the large bursts of external radiation (primarily gamma waves) that ravaged human bodies in Hiroshima and Nagasaki, but they ignored radioactive particles that cause harm when internalized inside of the body. The radioactive waves remain close to the hypocenter and last less than a minute, while the radioactive particles can travel far downwind and remain dangerous anywhere from a few hours to millions of years later, depending on the specific chemistry of the particle.

World War III never happened, but the detonation of more than two thousand nuclear weapons in "tests" did happen. During these tests, the external waves of radiation were contained to the test sites; however, mushroom clouds heavy with radioactive particles drifted downwind, where those particles could "fall out" and affect the health of living creatures. Long-lived fallout particles embedded into the ecosystem and will continue to pose threats to health: plutonium will remain dangerous for over two hundred thousand years, and uranium particles for more than one million years. Areas where particles fell out in large amounts near Chernobyl and Fukushima are still vexed by ^{137}Cs (cesium-137), a particle that remains dangerous for over three hundred years; easily transports through water, plants, and animals; and has shown up consistently in food produced in downwind zones decades later. Millions of people live in places where these disasters happened; whole communities have been devastated and many abandoned.

The invisibility of these global hibakusha is manufactured in both science and politics. Studies of the hibakusha in Hiroshima and Nagasaki built models of risk on external exposures and ignored the internal exposures that would become far more common. Since the global hibakusha's exposures do not fit our health models, we are unable to see them as enduring risk from radiation. Politically, nuclear weapon states do not want to acknowledge that weapon effects like fallout—which are designed for use in warfare to sicken and kill—constitute actual warfare when inflicted on people during "tests." This is not information that emerged slowly; awareness of the health impacts from fallout is what originally led nations to establish their test sites far from the elite populations in their societies. Some nuclear weapon states never tested in their own countries; others established test sites upwind of ethnic minority populations. Countries like the United States, the United Kingdom, and France tested large thermonuclear weapons on small Pacific islands and atolls that were either colonial or postcolonial trust territories. Awareness of the risks to communities downwind drove those siting choices; little care was given to the actual people living there.

Conversely, nuclear power plant accidents happen in developed nations to communities with some measure of political agency and access to information. The perceptions about radiation of these populations are more actively managed, in part because of their rights to compensation. For these people, the studies of the hibakusha from Hiroshima and Nagasaki are invoked to convey that levels of radiation in zones downwind from explosions, such as at Chernobyl or Fukushima, are too low to be of medical consequence. However, explosions from nuclear power plant accidents also raise clouds of radioactive particles into the air and deposit them downwind. Defenders of nuclear power dismiss downwinders' health concerns and describe people living in contaminated areas as suffering from an irrational fear of radiation, which they pathologize as "radiophobia." When a nuclear disaster happens upwind of your home and fallout clouds deposit radionuclides into your environment, anxiety is a rational response. Communities in need of information and assistance are instead routinely chastised for reacting to the toxic disaster thrust upon them.

The risks from radioactive particles and their behavior in ecosystems were well-known from the start. Senior Manhattan Project scientists had considered using radioactive particles as an offensive weapon against the Germans in World War II long before nuclear weapons were successfully manufactured; they discussed aerosolizing radionuclides so that enemy soldiers would internalize lethal amounts through inhalation. In 1946, after the first American postwar nuclear tests in the Marshall Islands, scientists made extensive studies of the behavior of fallout particles in the waters, soils, and biota of Bikini Atoll and strategized how to best weaponize these effects to both kill and psychologically terrorize an enemy population. Throughout the Cold War, military planners on all sides designed attacks that would weaponize fallout radiation to massacre enemy populations. All of this took place while simultaneously asserting that these same fallout clouds posed no health risk to people living underneath them downwind from test sites. They behaved as though the direct use of fallout radiation in warfare was strategic and calculated, while indirect exposures from testing, production, and accidents were "inconsequential" and "below health concerns." In fact, the effects of fallout do not change depending on the intentions of the party doing the irradiating.

The Cold War was, in part, a limited nuclear war conducted against these communities. We imagine that the nuclear war didn't happen because we had been envisioning the protagonists attacking each other, but the two thousand weapons detonated during the Cold War had profound impacts. The effects on global hibakusha communities—early mortality, disease, displacement, contamination of food sources and ecosystems—constitute a limited nuclear war. The fact that the locations where this happened are on the periphery of our political consciousness is why we are unaware of what happened, and also why they were chosen in the first place.

As the Cold War nuclear arsenals continue to threaten human civilization today, the radiological risks to human beings also extend beyond the twentieth century. Many of the radionuclides produced for our weaponry and electrical generation have deposited all around the world and will continue to migrate through the ecosystem long past our own mortality. We are on the temporal front lines of countless generations of human beings for whom these particles, and millions of tons of radioactive waste, have been stitched into our planet. Much of the waste is classified as high-level and will remain dangerous to living beings for more than one hundred thousand years. It is currently located on every continent (except Antarctica), although primarily in the global north; we plan to bury the most dangerous of it—the spent nuclear fuel—half a kilometer underground in dozens of sites in an attempt to contain the risks. Deep geological storage sites are already under construction in many countries, including Finland and Sweden, and many more are under design. These will present a risk to thousands of generations of human and other beings. We wrestle with what instructions to leave beside the waste—instructions we imagine will help protect people in the future—oblivious to the fact that the presence of our waste in their world is itself the message.

Hundreds of thousands of metric tons of this high-level waste is here now. We cannot keep it out of the future; in a sense, it's already there. But we can do a better or worse job of management. Being responsible about our waste means centuries of funding the facilities necessary to contain it (whatever methods we settle on) and remediating the sites of our production, testing, and accidents. We must compel numerous administrations in multiple governments to fund these remediations and storage sites for centuries. We have already made the central mistake, manufacturing hundreds of thousands of tons of radioactive toxins that will remain harmful for millennia, and so we have to commit our societies to responsible stewardship. The most important thing we must do is to stop making more nuclear waste: we must abolish nuclear weapons and abandon nuclear power.

Our invention of nuclear power plants to produce plutonium and the subsequent invention of nuclear weapons seemed like a powerful path to effectively achieve some immediate outcomes. It turns out we were opening the door to a millennia-long journey, and none of our descendants can opt out: we've made the choice for them. Now we must be more mindful about what we have done and are to do. If we are unwilling to see where we have been, we will have little understanding of where we are going. Has our brutality called us to stewardship? Or has it just extended that brutality into multigenerational-temporal violence? This book seeks to shine light on elements of our recent history that we have buried away, how our actions have already inextricably shaped the future, and what choices remain to us.

Acknowledgments

This work is a synthetic look at the history and legacy of nuclear technologies. It grew out of a lifetime of contemplation on these issues, from childhood terror of nuclear weapons sparked by civil-defense drills in my elementary school in the 1960s, to living most of the last two decades in Hiroshima and being in Japan during the Fukushima nuclear disaster. A great deal of what has been integrated in this book stems out of my work on the Global Hibakusha Project with Mick Broderick, which began in 2010. Research for the project took us to numerous nuclear sites and into communities around the world, providing us countless nights to stew over what we had learned and absorbed, and the stories of the people we interviewed. Mick's intellect and passion is found throughout this book. Many of the oral histories cited in this book were collected by both of us during our years of fieldwork in radiation-affected communities.

While this is a work of global history, it is told from a Western perspective. Its source material is primarily in English, and even oral histories that have been gathered in the field were rendered into English by translators and interpreters. My own training is in both history of science and technology and American history, and so my grounding in the mechanics and particulars of American nuclear history and my proximity to its literature is evident throughout. There are a multitude of pathways into the history considered here, and I hope this work will be a strand in a braid of scholarship that tugs different points of origin and interpretation to provide a more complex story.

Over the dozen years of research and fieldwork that went into this book the number of people who have helped me in various ways is really not calculable. I am most grateful to my family, who have always been the ecosystem where I grow and thrive. My wife Carol helped every day to make this book and the research behind it succeed. Her wisdom and compassion are on every page. My sons and daughters, Kaya, Ocea, Gwynne, and Levi, are ever-present for me. I am lucky, in the way of this modern world, to find myself in the circle of multiple families: the Jacobs, Agrimson, Sala-Garcia, Horner, Haber, Wyatt, and Johnson families. Our cookbooks run deep. My kids by different parents, Yolanda and Dean, keep expanding our family joy.

Professionally, I have benefited from great advice, conversations, and collaborations with many scholars. I am specifically indebted to Susan Lindee, Kate Brown, Jeff Weiss, David Richardson, and Mike Gorman for reading, commenting on, and helping to edit various chapters of this book. The world of nuclear historical scholarship has grown so large and diverse, I find that I work in one *bubble* of that multiverse: those who examine the real-world presence of nuclear technologies and their impact on lives and communities. This is rather separate from the world of nuclear policy analysis and criticism, or pure scientific and technological histories. The complexities on our level are expanding. I have benefited immensely from discussing many of the ideas in this book with Mark Selden, Gabrielle Hecht, Trisha Pritikin, Norma Field, Ran Zwigenberg, Yuki Miyamoto, Shampa Biswas, John O'Brian, Peter van Wyck, Hiroko Takahashi, and Sarah Fox. I have also had many deep exchanges on these issues with artists working on nuclear themes. First is my dear comrade elin o'Hara slavick, with whom I have collaborated elsewhere and who got me writing nuclear poetry (again). The work of Mariko Nagai, Yukiyo Kawano, Paul Miller, Gaku Tsutaja, Jessie Boylan, Linda Dement, Larbi Benchiha, Ian Thomas Ash, M. T. Silva, Julie Salverson, Glenna Cole Allee, Natalia Wehler, John Mandelberg, Peter Blow, Allison Cobb, Shinpei Takeda, Kathy Jetnil-Kijiner, Minoru Maeda, Isao Hashimoto, and Takashi Arai inspire and challenge me. I would also like to thank Ko Sasaki, Volodymyr Shuvayev, Sydney Clark, and David Montgomery for the use of their work in this book.

Important scholars and friends to whom I also want to express gratitude include Joanna Macy, Nico Taylor, Robin Gerster, Kumar Sundaram Pathak,

Kyle Cleveland, Ele Carpenter, Cassandra Atherton, Holly Barker, Dennis Riches, Kathleen Sullivan, Matthew Bolton, Paul Brown, Ken Busseler, Linda Richards, Hugh Gusterson, Cannon Hersey, Taku Nishimae, Jonathan Hogg, Marigold Hughes, Jon Mitchell, Togzhan Kassenova, Jeff Kingston, Kyoko Matsunaga, Ramaswami Mahalingam, Kanade Kurozumi, Ray Matsumiya, Christoph Laucht, Maxime Polleri, Noriko Manabe, Katie Norwell, Tom Engelhardt, Spencer Weart, Jeff Kingston, Mary Palevsky, David Palmer, Jon Mitchell, Bill Geerhart, Joe Copeland, Sabina Puig-Cartes, India Weston, Robert Rand, Takao Takahara, Louise Dunlap, James Thompson, Naoko Wake, Donna Goldstein, Noriko Manabe, Sei Kosugi, Clifton Truman Daniel, Elise Barkley, Sean Kelly, Kenji Ito, Laura Feldman, Yuko Takahashi, Ewa Helena Domanska, Marsha Henry, Jonathan Reinarz, Keiko Nakamura, Tom Le, Satoshi Toyosaki, Gordon Edwards, Paul Josephson, Tatiana Kasperski, Susanne Bauer, Laura Harkewicz, Steve Gilbert, Helena Grinshpun, Wilhelm Hofmeister, Becky Alexis-Martin, and the crew at SimplyInfo.

Doing as much field research as this book has taken, the list of those who supported or helped me there is truly endless; however, I want to especially thank Mariko Ishii, Paul Ah Poy, Hiroko Aihara, Claudia Peterson, Lyolya Filimonova, Aryna Starovojtova, Abacca Anjain-Maddison, Ricardo Garcia, Ursula Gelis, Bill Graham, Roland Oldham, Oishi Matashichi, Kana Miyoshi, Mariko Oda, Nurdana Adylkhanova, Aidana Assykpayeva, Baurzhan Doszhanov, Vercina Kallie, Hertes Biti, Junior Walton, George Masao, Chiho Ikegaya, Giff Johnson, Tom Bailie, Surendra Gadekar, Sanghamitra Gadekar, Yves Kamuronsi, Arman Kozhakhmetov, David Kupferman, Andy Kirk, Brooke Takala, Aitkhazha Bigalievich Bigaliev, Kaisha Atakhanova, Alson Kelen, Lorraine Garay, Andrea Windlass, Haruka Maekawa, Chikako Kamemoto, Kazuho Kubota, Yui Mukoji, Mary Silk, Yann Cambon, Patrice Bouveret, Marylia Kelley, Marie-Jo Floch, Tom Carpenter, Jose Herrera, Jeff Liddiat, Vilya Prokopov, Olexandr Suhetskyy, Zhanna Akushevych, Yevhen Alimov, and the members of the community organization Zemlyaki in Kyiv: Tamara Krasytska, Ludmila Dyatlova, Halyna Dondukova, Tetiana Tsybulska, and Ivan Kuz'min.

Living in Hiroshima has been like being at the center of a wheel whose spokes radiate outward throughout our nuclear inquiries. My academic home at the Hiroshima Peace Institute and Graduate School of Peace

Studies at Hiroshima City University is a wonderful place to be centered. I am fortunate to be paid to do this work by the citizens of Hiroshima, who are my employers. At HPI, my close colleagues who have been there from my start—Narayanan Ganesan, Kazumi Mizumoto, Hitoshi Nagai, Makiko Takemoto, and Sung Chull Kim—welcomed me with open arms and hearts. I am forever indebted to our director, Ryo Oshiba; former directors, Gen Kikkawa, Motofumi Asai, and Haru Fukui; and colleagues, Akihiro Kawakami, Hyun Jin Sun, Xianfen Xu, Tetsuo Sato, Tadashi Okimura, Kyung Jin Ha, Chie Shijo, and Mihoko Kato. Also, every step of the path of this book has been made possible by the administrative staff of HPI, Yukiko Yoshihara, Miki Nomura, Yoshie Yamashite, and Yusuke Akishima, to whom I will be forever indebted. Thanks to my students: Dr. Ágota Duró, Dr. Atsuko Shigesawa, Pinar Temocin, Jerry O'Sullivan, and Ashley Souther. The staff of the Hiroshima Memorial Peace Museum and the Hiroshima Peace Culture Foundation provided ongoing assistance. Thanks also to others among my Hiroshima community, which includes Charlie Worthen, Masae Yuasa, Itsuki Kurashina, Ulrike Woehr, Luke Carson, Yoshi Furuzawa, Brenda and Evan Douple, Flo Smith, Carol Rinnert and Richard Parker, Tatsuya Nishida, Haruka Katarao, Tomoko Nishizaki, Nobuyuki Kakigi, Tomomitsu Miyazaki, Yumi Kanazaki, Ayako Yoshida, Miyuki Kahler, Junko Hatori, Eriko Noguchi, Andrei Mistretu and Eiri and Kaya-chan, Keiko Hiyama, Sane Mizushima at Café Igel, Erika Abiko and Mayu Seto at the Hachidori-sha Social Book Café, Naoko Koizumi, Yasuyoshi Komizu, Ron Kline, Mirei Tashiro and her daughters Leona and Maila, Joy and Paul Walsh, Prakash Lamichhane, Mitomo-san, and, of course, Boku and Yuri at Mac, for making coming to Hiroshima feel like going home from day one.

In all of my work I benefit from training with my mentor, Lillian Hoddeson, who taught me so much about being a scholar and a colleague. And finally, I thank my dear friends Nancy Breslow, Heather Dean, Danny Clark, Jeb Binsted, Takeshi Yodono, Shari Robertson, Mark Menich, Al Rose, Rhonda Welbel, and David Razowsky. For me, writing is intimately tied to music. For keeping my mind and heart moving during the writing of this book, I would like to thank Pandit Hariprasad Chaurasia and Brian Eno for their accompaniment.

I am grateful to the people at Yale University Press, especially Bill Frucht, Karen Olson, and Kate Davis, as well as to Joe Calamia for his early support of this work. The research for this book was funded in part by the Hiroshima Peace Institute, Hiroshima City University, Murdoch University, the Japanese Society for the Promotion of Science, the Nuclear Futures project, and the Australia Council of the Arts. A previous version of chapter 5 was published in Robert Jacobs, "Nuclear Conquistadors: Military Colonialism in Nuclear Test Site Selection During the Cold War," *Asian Journal of Peacebuilding* 1:2 (2013): 157–177, doi:10.18588/201311.000011.

Abbreviations

ABCC	Atomic Bomb Casualty Commission (US)
ACHRE	Advisory Committee on Human Radiation Experiments (US)
AEC	Atomic Energy Commission (US)
AFL	Applied Fisheries Laboratory (University of Washington)
AVEN	Association des Vétérans des Essais Nucléaires (FR)
BNTVA	British Nuclear Test Veterans Association (UK)
BREN	Bare Reactor Experiment, Nevada
CEP	*Centre d'Expérimentation du Pacifique* (France)
CLAB	Central Interim Storage Facility for Spent Nuclear Fuel (Sweden)
DGR	deep geological repository
DNA	Defense Nuclear Agency (US)
DOE	Department of Energy (US)
FBI	Federal Bureau of Investigation (US)
FRN	fallout radionuclides
GTW	global thermonuclear war
GZ	ground zero
HMNZS	Her Majesty's New Zealand Ship
IAEA	International Atomic Energy Agency (UN)
INWORKS	International Nuclear Workers Study

JCS	Joint Chiefs of Staff (US)
km	kilometer(s)
kt	kiloton(s) (of TNT equivalent)
LANL	Los Alamos National Laboratory (US)
LLNL	Lawrence Livermore National Laboratory (US)
LSS	Life Span Study (of the ABCC)
m	meter(s)
METI	Ministry of Economy, Trade and Industry (Japan)
MIRV	multiple independently targetable reentry vehicle
MOX	mixed oxide (nuclear fuel)
MRI	Mitsubishi Research Institute (Japan)
mt	megaton(s) (of TNT equivalent)
MT	metric ton(s)
MTHM	metric ton(s) of heavy metals
NAAV	National Association of Atomic Veterans (US)
NAS	National Academy of Sciences (US)
NATO	North Atlantic Treaty Organization
NGO	non-governmental organization
NPR	National Public Radio (US)
NRC	National Research Council (US)
NSC	National Security Council (US)
NTS	Nevada Test Site (US)
NWS	Nuclear Weapon State
ORNL	Oak Ridge National Laboratory (US)
PPG	Pacific Proving Ground (US)
PRA	probabilistic risk assessment
PRIS	Power Reactor Information System
PTSD	post-traumatic stress disorder
RAI	Rocketdyne/Atomics International
RECA	Radiation Exposure Compensation Act (US)
rem	Roentgen equivalent man
RERF	Radiation Effects Research Foundation (US/ Japan)
RMI	Republic of the Marshall Islands
ROF	Royal Ordnance Factory (UK)
SAC	Strategic Air Command (US)

SDI	Strategic Defense Initiative (US)
SFP	spent fuel pool
SFR	Final Repository for Short-Lived Radioactive Waste (Sweden)
SKB	Swedish Nuclear Fuel and Waste Management Company
SNTS	Semipalatinsk Nuclear Test Site (USSR)
sRAW	solid radioactive waste
TEPCO	Tokyo Electric Power Company (Japan)
TNT	trinitrotoluene
UK	United Kingdom
UN	United Nations
UNESCO	United Nations Educational, Scientific and Cultural Organization
UNSCEAR	United Nations Scientific Committee on the Effects of Atomic Radiation
US	United States
USPHS	United States Public Health Service
USS	United States Ship
USSR	Union of Soviet Socialist Republics
VA	Veterans Administration (US)
WIPP	Waste Isolation Pilot Project (US)
WHO	World Health Organization

Numbers, Chemical Isotopes, and Scientific Symbols

3/11	11 March 2011, the date of the earthquake and tsunami that triggered the Fukushima nuclear disaster
^{14}C	carbon 14
^{90}Sr	strontium 90
^{131}I	iodine 131
^{137}Cs	cesium 137
^{230}Th	thorium 230
^{235}U	uranium 135
^{239}Pu	plutonium 239

GBq	gigabecquerel
Gy	gray(s), unit of absorbed radiation dose
mSv	millisievert
mSv/year	millisieverts per year

Nuclear Bodies

Introduction: Irradiated and Invisible

Bravo

On 1 March 1954, the United States successfully tested its largest nuclear weapon on Bikini Atoll in the Marshall Islands: the first deliverable hydrogen bomb. The Bravo test sparked a revolution in US military capacity and strategy; it was a stunning technological achievement.[1] Since demonstrating two years earlier that a thermonuclear weapon was possible, US weapon labs in Los Alamos in New Mexico and Lawrence Livermore in California had been working to design an H-bomb small enough to be delivered to enemy territory by an airplane. The Bravo test succeeded beyond their wildest expectations and yielded an explosion twice as large as weapon designers had anticipated and a thousand times larger than the weapons used in the nuclear attacks on Japan.

The chief historian of the United States Atomic Energy Commission, Richard Hewlett, and his cowriter, Jack Holl, describe that "from the moment of firing *Bravo* gave every sign of being a spectacular success." The Bravo test was part of a series of six thermonuclear weapon tests in the Marshall Islands in March–April 1954, called the Castle series. Hewlett and Holl write that upon the completion of the series, "the future looked entirely different. It seemed that the American scientists had suddenly found the key to new realms of nuclear weapons. With a few notable exceptions, every new design principle incorporated in the *Castle* series seemed to work, often beyond the hopes of the most optimistic designers. . . . The

1

Atomic Energy Commission, the United States, and the world truly faced a new reality in the technology of war."[2]

Bravo and the other thermonuclear weapons tested in the Marshall Islands in spring of 1954 were imagined to have two profound impacts on the Cold War. The first was the scale of destruction that the weapons could inflict on the Soviet Union: "the hydrogen bomb was so enormous in its destructive power that it defied human description."[3] However, as US military commanders pointed out after the test, beyond the immense explosive power of these weapons was the scale of their radiological fallout. "It is now known that fallout from the larger Castle shots blanketed areas of more than 5,000 square miles with radioactive material that would have been lethal to unprotected personnel. This one result gives a new insight into a method for using high-yield weapons in both strategic and tactical situations."[4] Almost immediately the military began to draw up plans to quickly take advantage of these powerful weapons and their deadly fallout clouds. "In May 1954, Eisenhower was briefed on a paper by the JCS's Advance Study Group which proposed that the U.S. consider 'deliberately precipitating war with the USSR in the near future,' before Soviet thermonuclear capability became a 'real menace.' "[5] Eisenhower rejected such proposals; however, US military strategists immediately integrated the effects of these new weapons into nuclear war–fighting plans.

Eisenhower envisioned that the second impact yielded the true value of the new weapons: their deterrent capacity. The president's Cold War national security policy was formalized in 1953 in the document NSC 162/2. "When both the USSR and United States reach a stage of atomic plenty and ample means of delivery, each will have the probable capacity to inflict critical damage on the other, but it is unlikely to prevent major atomic retaliations. This could create a stalemate, with both sides reluctant to initiate general warfare."[6] Eisenhower believed that the weapons developed and tested during the Castle series would substantiate the threat of "massive retaliation" and effectively deter nuclear war. This is the envisioned logic of a robust and flexible nuclear arsenal during the Cold War: weapons that can almost instantaneously inflict apocalyptic damage deterring war, maintaining peace.

In describing the five-thousand-square-mile radioactive cloud that stretched downwind from the Bravo test, the top-secret 1954 film report

Radiation survey team, 1st Lt. W. J. Larson (USAF) and Ensign R. P. Keiser (USNR), measuring external radiation on unnamed residents of Utirik soon after arrival on 3 March 1954. Thomas Kunkle and Byron Ristvet, *Castle Bravo: Fifty Years of Legend and Lore: A Guide to Off-Site Radiation Exposures* (Kirtland, NM: Defense Threat Reduction Agency, 2013): 123.

of the Castle series mentions that it caused "significant fallout" to occur on Rongelap, Ailinginae, and Utirik Atolls, where "natives" were then evacuated on the "second morning after the shot."[7] Commander George Albin of the USS *Philip,* which evacuated those living on Rongelap and Ailinginae Atolls, later reported to Major General Clarkson, the head of Joint Task Force 7, which conducted the test, that "the Marshallese were excellent passengers, most cooperative, never demanding and exemplary in their conduct. It was a distinct pleasure for the crew of the PHILIP to have been afforded the opportunity to assist these quiet people in the evacuation."[8]

In total, 253 people were evacuated from Rongelap, Ailinginae, and Utirik Atolls.[9] These exposed Marshallese were examined by three radiobiologists assigned to Project 4.1, which was tasked with studying the

Utirik residents awaiting evacuation to USS *Renshaw*, 3 March 1954. Thomas Kunkle and Byron Ristvet, *Castle Bravo: Fifty Years of Legend and Lore: A Guide to Off-Site Radiation Exposures* (Kirtland, NM: Defense Threat Reduction Agency, 2013): 124.

health effects of those exposed to radioactive fallout during the test series. The radiobiologists' initial reports described that "Within hours of exposure to radiation, approximately two-thirds of the Rongelap people felt nauseated and one-tenth of the group had vomiting and diarrhea." These symptoms, which began before the evacuation and were suffered in lesser degree by those on the farther atolls, were followed twelve days later by "lesions of the skin and epilation of the head."[10] Later calculations estimated the average exposure of those from Rongelap (approximately 152 km away from Bikini) was equal to that of people who were 2.4 km from ground zero during the nuclear attack on Hiroshima.[11]

The Project 4.1 researchers continued to study the fallout-exposed Marshallese for decades. Subsequent research by the team, and researchers at the US National Cancer Institute, approximated that 170 additional cancers (beyond what was statistically expected) were produced among the Marshallese because of radiological fallout, predominantly thyroid cancers.[12] The Bravo fallout also contaminated numerous fishing vessels, among them the Japanese tuna boat the *Daigo Fukuryu Maru,* which was over 100 km downwind from Bikini Atoll. When the boat arrived back in Japan two weeks later, the entire crew was hospitalized with radiation sickness.[13] One crew member, Aikichi Kuboyama, died six months later of medical complications from his exposure.[14] The contaminated atolls were considered uninhabitable. Small portions were remediated by the US government during the subsequent years, and some of the displaced Marshallese were returned in 1957. The returnees began to suffer additional illnesses from ongoing exposures on the contaminated atolls and from eating local fish and plants. The Bravo test proved so disruptive and traumatic for the whole of Marshallese society that 1 March is currently a national holiday in the Republic of the Marshall Islands, observed as "Remembrance Day."[15] Thermonuclear weapons can generate five-thousand-square-mile lethal fallout clouds: when that radioactive fallout engulfs inhabited communities, it is no longer a "test." Bravo was experienced by those it irradiated as terror and harm.

The Global Hibakusha

During the Cold War there were more than two thousand nuclear weapon tests. On average there were over forty-five tests per year between 1946 and 1991; statistically, a nuclear weapon exploded every other day during 1962, the year with the most tests.[16] Bravo was the detonation of one weapon on one day. We think of the Cold War as a period of time in which nuclear weapons were not used, but in reality, they were exploding continually. The tests began less than a year after the nuclear attacks on Hiroshima and Nagasaki and have been carried out on every continent except for South America and Antarctica. While never used directly against human beings since Nagasaki, these tests have profoundly affected communities around the world. Weapons have been tested by the United States in several Pacific nations and trust territories, and in Nevada—the location that has seen the most nuclear tests of any single place: 928. The Soviet

Union began testing in East Kazakhstan, then in the European Arctic, as well as in numerous other locations. The British tested in Australia, in Kiribati, and also in Nevada. The French tested in Algeria and then in French Polynesia. The Chinese tested in Lop Nor in Northwest China. India, Pakistan, and North Korea all tested underground and within their national borders. Israel has nuclear weapons but has not tested them.[17]

About one-quarter of the tests were in the atmosphere, facilitating the global distribution of radioactive fallout. Thermonuclear weapon testing created mushroom clouds that carried particles high above the troposphere and into the stratosphere, where they circled the Earth before falling back to the surface, sometimes years later. Subsequently, fallout radiation can be found everywhere around the planet. Its presence has provided scientists tracers with which to observe the fluid dynamics in global systems like the atmosphere and oceans. Their global ubiquity means we can determine whether goods were manufactured before or after 1945 because of the traces of anthropogenic (human-generated) radiation present in anything produced since the advent of nuclear testing. In the soil 3 km from ground zero in Nagasaki, a 2011 study found more radioactive particles deposited by global nuclear weapon testing than from the direct use of a plutonium weapon there in 1945.[18]

When detonated, nuclear weapons exert blast, heat, and radiation. Nuclear weapon states have been largely effective at keeping blast and heat effects contained to the test sites. Radiation can be encountered both as a wave and as a particle. When a nuclear weapon detonates, a burst of gamma and neutron waves spread outward from the reaction. In Hiroshima and Nagasaki these rays were harmful to a range of roughly 3 km from the hypocenter. With thermonuclear weapons, that diameter is larger. These radioactive rays can be understood similarly to X-rays: they are present and then they are gone. The burst and the danger last less than a minute, yet during that minute they can penetrate through most material, including human bodies. A person sufficiently far from the explosion will be unaffected by this burst. As seen at Bravo, there are also radioactive particles that are taken up into the mushroom clouds, travel downwind, and then "fall out" to deposit on the planet's surface, often hundreds or thousands of kilometers downwind.[19] These particles, classified as either alpha-emitting or beta particles (depending on their chemistry), can give off dangerous levels of radioactive waves when concentrated in large amounts, but they are particularly dangerous when they are taken inside the body through inhalation, swallowing, or cuts in the skin. In such cases, single particles can spark deadly diseases. Radioactive particles are not all the same; they are different chemical elements and isotopes of those chemical elements. These differences determine how radioactive a particle is and how long it remains radioactive. Some particles are dangerous for a few hours or days, others remain dangerous for hundreds of years, and some are dangerous for more than a million years. Depending on what and how many radionuclides you internalize into your body, you face differing degrees of risk for developing sickness or experiencing early mortality. When radionuclides deposit far downwind from nuclear test sites, they continue to present health risks to people, sometimes for longer periods of time than human life spans, thus endangering future generations.

While nuclear test sites were imagined to be located away from human habitation, fallout radiation traveled far outside of test site boundaries, affecting millions of people living downwind. Risks from radioactive fallout were well understood before the very first detonation, the Trinity test conducted on 16 July 1945 in New Mexico (three weeks before the attack on Hiroshima). Manhattan Project personnel were stationed in numerous locations to track the fallout and evacuate civilians in high radiation areas if necessary.[20] Fallout from the Trinity test was later found almost 2,000 km away in a field on the Illinois-Indiana border.[21] A 2018 internal report at Los Alamos National Laboratory details, "When Trinity's radioactive debris contaminated the grain fields of the Midwest, the response was to move testing to the Marshall Islands, where the seemingly empty ocean that [sic] would swallow any radioactive fallout. This scheme worked until Bravo demonstrated that the world was not big enough to hide the radioactive fallout from thermonuclear detonations."[22] Almost every atmospheric nuclear test has yielded radioactive fallout outside the boundaries of the military compounds in which they were conducted. Thermonuclear tests have often created "5,000 square mile" zones with lethal levels of fallout. Or larger. Today, substantial amounts of the fallout from Cold War nuclear weapon testing remains embedded in the Earth and in the bodies of creatures living here.

Millions of people have also been exposed to radiation from nuclear power plant accidents. The largest of these, in Chernobyl in 1986 and

Fukushima in 2011, saw explosions at the failed plants carrying radionuclides downwind just like mushroom clouds, depositing fallout over homes, schools, cities, and areas of food production. Radioactive particles cannot be made "un-radioactive"; you can only move them from one place to another, hence the vast areas of northern Japan filled with plastic bags of radioactive soil from "decontaminated" towns.[23] Among the many fallout particles prevalent in soil downwind from nuclear reactor explosions is ^{137}Cs, which will remain dangerous for over three hundred years, far outlasting the plastic bags. ^{137}Cs is particularly good at migrating in an ecosystem once deposited, moving easily from soil into plants into animals and back into soil, and spreading outward via underground or surface water. Decontaminating a town or school in such ecosystems is a temporary measure, as wind and rain will transport more particles from forests and mountainsides back into "decontaminated" spaces. You cannot separate a small zone from its larger ecosystem. We know the names Chernobyl and Fukushima because of the scale of the disasters there, but smaller nuclear reactor accidents have happened regularly since the technology was first invented by the Manhattan Project in 1944.[24]

The production sites of nuclear technologies have also exposed many people to radiation. Since the late 1800s, worldwide uranium mining has seen miners suffering lung diseases from the radon gas given off by uranium ores, and their families suffer from uranium dust tracked back into their homes after work. The processing of uranium both for fuel for nuclear power plants and for weapons left hundreds of sites around the world contaminated with a range of radioactive particles and chemical toxins.[25] The production of plutonium, the reason that nuclear reactors were first invented, has created some of the most radiologically contaminated places in the world.

The people who live in all of these communities are the global hibakusha. *Hibakusha* is a Japanese word that refers to those who survived the nuclear attacks on Hiroshima and Nagasaki; it literally means "explosion-affected person." In recent years "global hibakusha" is being used to signify all who have suffered the radiological effects of nuclear technologies since 1945. We silo off those affected by radiological events into separate historical incident narratives, but many who live through such events understand their bond with those in the other silos. Natalia Manzurova, a liquidator from Chernobyl, was interviewed a week after the initial explosions of the Fukushima disaster. "Their lives will be divided into two parts: before and after Fukushima," she understood about the newly emerging hibakusha. "They'll worry about their health and their children's health. The government will probably say there was not that much radiation and that it didn't harm them. And the government will probably not compensate them for all that they've lost. What they lost can't be calculated."[26] Along with the communities I have described, another cohort of global hibakusha is military personnel who were routinely exposed to radiation during atmospheric nuclear tests. The United States and the Soviet Union, the two most prolific producers and testers of nuclear weapons, each exposed hundreds of thousands of their own troops to nuclear weapons and to radiation. Both conducted single tests in which over forty thousand soldiers participated.[27] The three other NWSs that tested in the atmosphere also exposed troops and test site workers.

As many of these radioactive particles will remain harmful for centuries or millennia, migrating through the ecosystem, more people will encounter and internalize them in the future. Single particles whose half-lives exceed human life spans may pass through several different people. Nuclear waste, laden with a variety of radionuclides, will also far outlast our civilization. Thus, the number of global hibakusha will extend far beyond any current tabulation or imagination. This is the legacy the Cold War wrought.

Defining the Victims

Scientific and medical constructs of the relationship between radiation exposures and subsequent health effects have been utilized to keep the global hibakusha invisible to science, politics, and history. Our models of how radiation impacts human health were built through studies of the hibakusha of Hiroshima and Nagasaki, begun in the late 1940s and early 1950s. The Atomic Bomb Casualty Commission, established by the United States in Hiroshima and Nagasaki during the American occupation of Japan, and its legacy research institution, the Radiation Effects Research Foundation, jointly operated by the US and Japan, have been conducting a large study correlating the disease burdens and early mortality of the survivors of the two nuclear attacks in 1945 with their radiation exposures.

This is the Life Span Study, begun in 1950.[28] In radiation health communities, the LSS is frequently referred to as the "gold standard" database correlating radiation exposures with health outcomes. It is a robust study that has continued over multiple decades and has included the participation of hundreds of thousands of World War II hibakusha.

The LSS is a landmark study, yet there are problems with its design and use. The first weakness is in the ascribing of dose to each participant. For a study to correlate health outcomes to prior radiation exposures, both details must be accurate. Obtaining health outcomes is straightforward: ongoing monitoring of the health and mortality of the participants yields clear data. Determining individual exposures to radiation in Hiroshima and Nagasaki is more problematic. Exposures occurred during a nuclear attack, and doses had to be estimated years later. Each participant has "dose reconstruction" to ascribe a value to their exposure. The key components in reconstructing dosage is distance from ground zero, and shielding—being out in the open, inside a building, or underground. There is functional certainty about the levels of external radiation in the gamma burst from the detonation as it traveled outward, so if location and shielding are ascertained, it's possible to reconstruct an accurate external dose. Interviews were conducted with survivors to determine an individual's location and shielding. Building a statistical database on interviews and memory adds an imprecise variable to the mathematics. Dose reconstruction has been revised multiple times as more-thorough interviews and more-detailed reconstructions of the radiation field and attenuation of shielding are developed.[29] Even with this imprecision, it is widely agreed that the LSS offers a powerful statistical model relating the exposure of human beings to a single, large burst of gamma radiation with possible disease progressions and mortality.

From the outset, it was decided not to include data about internal exposures from particles that were deposited inside the bodies of the participants in the study, even though it had been well understood by US scientists in the Manhattan Project that internalized radionuclides were deleterious and potentially deadly.[30] There were several reasons for this choice. First and foremost, since the participants were close enough to the hypocenter of the detonation to receive high levels of external radiation, any effects of internalized radiation were presumed to have less health impact than the external exposures. Furthermore, it was impossible to survey a large group of people for internalized particles until the manufacture of fully functional whole-body counters in 1964, and even then it would have taken years or decades to examine hundreds of thousands of subjects. Without such instrumentation, you cannot differentiate who in a group of people has internalized a particle and who hasn't. With external exposures, every person in a specific location will have received a similar dose. With internal exposures, any group of twenty people standing in the same place during a radiological event may experience different outcomes. Internalization would depend on the asymmetric distribution of the particles (as opposed to the symmetric distribution of gamma waves), on who inhaled or swallowed one, and on their personal health and metabolism. The science was essentially impossible, and the more pronounced impact of the external exposures took precedence. Excluding incidence of internal exposures made sense; it was the only way the LSS could produce the quality of work that it achieved. Still, this exclusion was to have profound effects on those who ultimately would become exposed to radiation during the Cold War years.

When radioactive particles deposit downwind of a nuclear weapon test from a fallout cloud, or when they scatter throughout a house when a uranium miner hangs up a work jacket, or when large amounts of radio-iodine migrate out from a plutonium production facility, there is an asymmetric distribution: some people will internalize particles and some people will not. It is not possible to predict who will, and almost impossible to know who has. Tools like whole-body counters can assist in analysis, but only very select populations have ever been systematically tested in the decades since their manufacture. When the Bravo fallout cloud descended on the *Daigo Fukuryu Maru,* or on Rongelap, some people internalized particles and some people did not (although all experienced some measure of external exposure from amassed particles). When plumes deposited radionuclides after Chernobyl or Fukushima, they did not uniformly affect the people living below the clouds. After these particles deposit into an ecosystem, some people may escape their effects, some people may encounter (and internalize) them immediately, and others may encounter them years later, not thinking about the presence of radionuclides in the mushrooms they forage. Some will take an indirect route into the human body, such as radioiodine, which is typically internalized by consuming dairy products from farm animals that have eaten contaminated grass or feed. Where in the gallon of milk is the radioactive particle? Who will drink that glass? There is no certainty as to where these particles migrate, where the dangers lie, and who is at risk. It's a risk model directly opposite to that of being exposed to external radiation, in which there can be certainty: being in a specific location at a specific time reveals all. Thus, a primary experience of people in areas of fallout contamination is uncertainty. Is there risk? Are they exposed? Can they protect their children? Any degree of certainty only comes negatively with the presentation of disease, and even then, direct causation is usually indeterminable.

When the United States government authorized the establishment of the ABCC, it seemed common sense that the future of world warfare would include the use of nuclear weapons. The Cold War was a period in which fears of nuclear conflict were endemic. Now, after the Cold War, we can see that didn't happen; there never was a direct use of nuclear weapons between the Cold War protagonists. Populations were not exposed to large bursts of gamma radiation from nuclear detonations. However, millions of people were exposed to radioactive fallout—that actually did happen. And we do not have a medical model to determine what the health outcomes and mortality will be for people whose risks come from internalizing radio-nuclides. We have the LSS. And so, throughout the Cold War and since, governments and health organizations have trotted out the LSS to assess the disease risks of people exposed to radioactive particles in their ecosystem.

The LSS tells us that external exposures to significant levels of radiation may lead to health impacts, and below that level there is less certainty (and much debate).[31] Thus, in communities where fallout has deposited, where radioactive waste has been dumped into trenches or at sea, and where water flows from streams near mines loaded with uranium tailings, we measure the external levels of radiation to determine if people are at risk. Immediately after the deposition of fallout under a cloud from a nuclear test or an explosion at a nuclear power plant, the collection of particles in one location may be high enough to measure as concerning, but after a few days those particles have been dispersed by wind and rain, and thus externally measurable levels steadily decrease. Weeks, months later, the particles will have embedded deeper into the ecosystem and the measurable levels may be very low. To address the worries and care for the health of those in the affected area, we apply the mathematics of the LSS: Are the levels of external radiation high enough to predict future health burdens? That is the wrong question to ask and the wrong tool with which to craft a response. The LSS was developed as a database to inform us about risk and health impacts from large bursts of external radiation; it tells us little about the consequences of large depositions of rapidly dispersing particles, or what the health outcomes are from internalizing one. But it's what we have, so, we use it.

Invoking the LSS invariably tells experts that people exposed to fallout have exposures below levels that correspond to increased risk of disease. It is true that their exposures to external radiation are typically low; however, that's not where danger lies. Radioactive particles are still in the ecosystem around their homes, in their children's schools, and in their gardens, farmlands, and forests. While measures of external radiation can be extrapolated —levels in one part of a schoolyard seeming likely to be close to those in another part of a schoolyard—particles will be asymmetrically distributed, and so low readings in one place tell you little about possible readings just a few meters away. Wind and rain tend to flush particles downward in a landscape. There will typically be more particles collecting in the gutters on the sides of roads than in the middle of the road, or more particles in the streams running off of a field than in the air above the field. Any model of risk from internalizing radionuclides that would be drawn up for a space would have to be specific to that space and the time of its measurement: an afternoon rain could redistribute the risk. The potential for a person to internalize a particle in a contaminated ecosystem also varies. Particles have mass and so typically settle on the ground rather than linger in the air. Children are closer to the ground than adults, tend to sit there more frequently, and often put things from the ground in their mouths. Thus, children are more likely to internalize particles in a setting than adults. Still, one child may and their playmate may not. Who among the radiation-exposed will also become among the radiation-affected? The uncertainty about this progression can in itself become emotionally debilitating, especially for parents.

Utilizing the LSS only occurs in communities where the rights and health of the people are sufficiently respected to actively manage their anxieties. In many nuclear test site regions, especially in colonial or post-colonial spaces, no assessments or information was provided. A quick survey of test locations reveals the structural nuclear colonialism in their siting. Two NWSs did not test even one weapon within their own national borders. Several nations tested weapons, especially thermonuclear weapons, in colonial states or trust territories under their political control in Oceania rather than domestically. Even when testing was done within national borders, sites were located near minority populations with little political power, such as Kazakhstan in the former Soviet Union, Nevada in the United States, and the Xinjiang Uyghur Autonomous Region in China. Test sites were never located near centers of power or concentrations of ethnic, racial, or social elites. No plutonium production facilities were built near capital cities or financial centers. Those irradiated were selected specifically because of their inability to resist. In developed nations where citizens had agency and rights, later distress over the disease load borne by these populations was dismissed by citing the LSS. In colonial or post-colonial spaces like Kiribati, no one bothered to pacify the exposed.

Nuclear War in the Cold War

Our histories of the Cold War fixate on the war that was not fought. Despite the arms race, despite the political tensions, despite the hair-triggered mutually assured destruction, nuclear war was successfully deterred. John Lewis Gaddis called this the "long peace."[32] Recently, scholars have focused on the many regional conflicts and anticolonial wars that were sucked into the Cold War binary, detailing the never-ending military conflicts that, in part, proxied for the superpower face-off. Rarely, if ever, is nuclear weapon testing mentioned in Cold War historiography. When we say the Cold War nuclear conflict never happened, we are really saying that it didn't happen to us. It is a privileged perspective. For people living near atmospheric nuclear testing sites, it did happen. In the lives of those in the Marshall Islands, in East Kazakhstan, in Kiribati, and in French Polynesia, nuclear war was not deterred. For those living in the "5,000 square mile" death zones, this was not an imaginary war—it was a limited nuclear war.

There are multiple reasons that we cannot see that the Cold War was actually a limited nuclear war. We define warfare as the direct use of weapons against an enemy. Cold War nuclear strategists moved past this definition: nuclear attacks were designed to kill and sicken populations far from target locations through the effects of large fallout clouds from high-yield weapons. Tests of these same weapons inflicted these exact effects on downwind populations from test sites, just as they would on downwind populations in warfare. Many people were exposed to and suffered from nuclear weapon effects specifically intended to harm human beings.

Another reason we can't see this as a limited nuclear war is that those harmed were not politically powerful enough for the attacks to bear consequences for the attackers. They were not attacking enemy populations; they were attacking populations under their own care.[33] To minimize their liability, they focused the attacks on colonial, postcolonial, or marginal populations: "nuclear subalterns."[34] The United States tested nuclear weapons in Nevada, upwind of populations made up primarily of Mormons, Native Americans, and Hispanic farmworkers, but they never tested high-yield, thermonuclear weapons there: all of the H-bombs were tested in Pacific territories filled with noncitizens. The Australian government would not allow the British to test thermonuclear weapons in the traditional Aboriginal territories where the first British fission weapons had been detonated, so the larger weapons were tested in Kiribati. The Soviet Union tested all kinds of weapons in Kazakhstan, where, by official definition, only "nomadic" people lived. Each side harmed people under their own care who had no recourse. In the case of the US and USSR, which conducted 84.87 percent of weapon tests between them, nuclear weapon testing became a way of signaling strength and brutality to each other.[35] They engaged in what Liddell Hart would call "indirect warfare," demonstrating the ferocity of their arsenals, delivery systems, and intentions without risking direct conflict, harming only people that were, to them, expendable.[36]

This is an uneven history. In some instances, people were forcibly removed from their homes because of contamination, like the "quiet people" evacuated on US ships after the Bravo test. In such cases, families were forced to abandon their property and belongings, live in temporary housing, be resettled as refugees in new communities, or be returned to their former homes after sloppy remediation. Nuclear scientist Hiroaki Koide has remarked, "Staying in contaminated areas hurts the body, but evacuation crushes the soul."[37] Sickness, death, displacement, loss of property, abandonment to continued habitation in radiologically contaminated homes: the history of nuclear testing is the history of a limited nuclear war.

Killing Our Descendants

Nuclear weapons exert violence across distances. The limited nuclear war that I just described saw violence inflicted on lands distant from the actual location of the weapons' detonations: fallout clouds carried radionuclides to deposit hundreds, thousands, or tens of thousands of kilometers away. A separate nuclear catastrophe threatens to exert violence into distant time and against countless generations of our descendants. This is the disposal of vast measures of highly radioactive waste from nuclear technologies, especially the hundreds of thousands of metric tons of spent nuclear fuel rods from our production of both plutonium and electricity. This waste is laden with the most toxic materials on Earth and will confront future generations with radiological risks beyond ten times longer than human societies have practiced agriculture.

This is not something that might happen; this is something that has already happened. It is a global legacy currently sitting in spent fuel pools and dry storage casks, waiting. We knew throughout the period we generated this waste that there was no method for its disposal—in dozens of countries. We manufacture more every day. The best plan we can devise is to bury it deep, deep, deep underground. We pretend there are places on this planet that can be separated from everywhere else. We intend to build geological repositories 500 m below the Earth's surface, encapsulate the spent fuel in canisters, bury them in holes, and seal the whole mess up with clay. Not once, but dozens of times—in dozens of places—all around the world. We tell ourselves that we can do this perfectly . . . repeatedly. We turn a blind eye to the trail of imperfections that has characterized all human endeavors, that brought us into this very situation.

Spent nuclear fuel consists of uranium fuel rods that have "burned" in nuclear reactors to manufacture either plutonium for nuclear weapons, or electricity. Nuclear reactors were invented to manufacture plutonium, and were operated for over ten years in three countries before contributing electricity to any public grid (done first in the USSR).[38] Once the fuel has been used, or "spent," it becomes waste. The spent fuel rods contain many radioisotopes, including uranium and plutonium, which will remain highly radioactive for over one hundred thousand years and will generate excess heat for the first several thousand years. Fuel rods are typically used for about three years in reactors and then must be contained for longer than human history. There are currently between 300,000 and 400,000 metric tons of this spent fuel on Earth, along with large amounts of other forms of high-level radioactive waste.[39]

None of the thousands of generations that will share their world with this waste will receive any benefit; they will have only burden. Yet we convince ourselves that they will be safe if they just listen to us and follow the instructions we will leave at the repositories: we who generated the waste with no plan for its disposal, we who enjoyed the tiny benefits it provided, we who have put them in this position. If they just listen to us, we tell ourselves, we can protect them. We imagine that they may be irrational and we may have to scare them to keep them from digging the waste up; we may have to devise mythologies or religions that can pass down our instructions to them; we plan to design monuments that will direct them to our information kiosks, where they can receive details from us about how to avoid disaster. We . . . imagine that *they* may be irrational.

We are not seeking the best single place on Earth to bury this waste: we are seeking dozens of places in the specific countries that generated portions of that waste. Each nation that made nuclear weapons or used nuclear power to generate electricity must bury its own waste. Even though we are engaging in millennia-long tasks, seeking to protect thousands of generations who will have no allegiance and perhaps no knowledge of the different nation-states that generated and buried the waste, we are convinced that we must dispose of it within the political units of those nation-states because . . . we are appeasing the necessities of our own socially and politically constructed world. It's who we are. We can't conceive of acting outside of our constructs, but we can imagine it will go perfectly—as long as humans in the future pay attention to our instructions.

The disposal of this waste, in the most effective and responsible way possible, is perhaps the largest challenge to ever face the human race. Failure to meet this challenge will subject thousands of generations of human beings and other creatures to extreme dangers. The manufacture of this waste is the most ecologically significant event caused by living creatures in the long history of the planet. Unlike asteroid impacts that have led to extinctions, the risk from this waste will be ongoing, enduring long past any specific disaster affecting one generation along the way. It is, by definition, temporal violence, striking at the future bodies that will grow on the Earth.

De-National History

There have been many books exploring the history of nuclear weapon testing, radiation exposures, and the legacies of Chernobyl, Fukushima, and other nuclear disasters. Based on how we narrate war and politics, those histories have often focused on the radiological legacies of specific nations. They are histories of US nuclear testing, British nuclear testing, Soviet or Japanese power plant disasters. However, the nature of these events, the distances their effects travel, the penetration of their legacy radionuclides into deep time all make this a global history. When we look at the victims of Soviet testing in Kazakhstan, we can reduce their number to a tragic but understandable level. When we talk about those impacted by British nuclear testing, the locations seem contained: a small area of the desert in Australia and some islands. It is when we de-nationalize and survey this as a global history that we can begin to see the limited nuclear war hiding behind the long peace. As a group of national histories, it has been dismissed; as a global history, it is profound. When we focus on the experiences of human beings exposed to radiation from fallout particles rather than on elite discourse about deterrence and geopolitics, we grasp the actual role of nuclear weapons in the Cold War: the embodied history.

Kate Brown, the scholar of Cold War plutonium production and the human impact of the Chernobyl disaster, has written that "There ought to be a new frontier of scholarly inquiry, one that learns to read bodies as historical texts so as to re-create historically voided bodies living on contaminated landscapes in a way that does not dismiss bodies in pain."[40] This new frontier must not be distracted or bogged down by political ideologies and national histories. Human bodies are of a species, and this history is a species history whose political location is Earth. When it is seen as global history, the full impact becomes visible—the radiological harm no longer appears episodic, but systemic. We must hold nations responsible and detail their actions: as the fallout particles are transported around the world, so too should be our gaze.

PART

Technicalities

1

Hypocenter

Nuclear Weapon Effects

When we speak of the effects of nuclear weapon detonations, we talk about three primary effects: blast, heat, and radiation.[1] Each is a form of energy moving through space.

Blast refers to the force, or pressure, that is exerted by the detonation on buildings and other matter. When looking at photographs taken of Hiroshima and Nagasaki after the nuclear attacks of 1945, this is what is most striking about the images: the absence of the structures that previously stood in the cities. In those two places, only reinforced-concrete buildings survived the blast near the hypocenters, although they too were damaged. We refer to the size of a nuclear detonation strictly by a measure of this energy: its blast force—using the metric of how much force would be released by an equivalent blast from the detonation of TNT. Hiroshima-sized fission weapons are typically in the thousands of tons of TNT range (kt, kilotons), while H-bombs (fusion weapons) are in the range of millions of tons of TNT (mt, megatons).

Heat refers to the astonishingly high temperatures produced by the fire-balls of nuclear detonations. The temperatures of this fireball can exceed tens of millions of degrees Celsius.[2] Many types of physical matter close to the hypocenters dissolved instantly, such as biota (biological life-forms), including human beings. Moving away from the center, these high temperatures still caused combustible materials to burst into flames. In Hiroshima and Nagasaki, where the majority of buildings were made of wood, these flames engulfed most of the structures within 3 km of ground zero, producing massive, city-wide fires that burned for hours. Much of the debris left behind by the blasts was turned to ash by the fires that followed the intense burst of heat from the weapon, leaving a charred scar where the city center had stood and where its population had lived, worked, and attended school.

The final effect, radiation (henceforth using this term to refer specifically to ionizing radiation), is more complicated and not so easily understood. Humans have experiences of force and heat in their daily lives and have a visceral, bodily understanding of them. Our senses are not attuned to detecting radiation. Additionally, complicating our understanding is that radiation can affect us in multiple ways. For our purposes, the main thing to understand about radiation is that there is an important difference between radiation experienced from a source external to the body, primarily in the form of waves, and radiation coming from particles that have been internalized into the body; in the case of a nuclear weapon detonation humans can be affected by both.

When a nuclear weapon detonates, it emits a massive burst of gamma and neutron radiation. In Hiroshima that burst is estimated to have delivered 35 Gy of gamma radiation and 6.04 Gy of neutron radiation at 500 km. In Nagasaki it is estimated that the burst delivered 78.5 Gy of gamma radiation and 3.31 of neutron radiation at 500 km.[3] (Note: it is not necessary to understand the specifics of this science to understand this chapter.) Gamma radiation is in the form of waves (as opposed to particles), and as waves they can be understood to function similar to X-rays: the energy penetrates through most materials, including through human bodies. The penetration of the gamma wave is moderated by the material it passes through; its energy is lessened by materials that provide shielding, such as lead or soil, but only slightly by materials like wood and flesh. Therefore, a burst of gamma radiation passes through the entire body of a person close to the detonation of a weapon and can affect cells and organs throughout the body. Those within 1.5 km of the detonations experienced massive doses of gamma radiation to their whole bodies. This dose lessened away from the hypocenter, being significantly lower at about 3 km than at 1 km. Physicist Jeffrey Weiss explains that "the intensity drops smoothly as the square of one over the distance. The 3 km is roughly a threshold for human effects, not a sharp change in the dose."[4]

Many people who experienced radiation sickness in the hours, days, and months after the detonations were suffering from this massive external exposure to gamma rays, although some early radiation sickness likely came from ingestion, walking into ground zero, and suffering from induced radiation (groundshine).[5] Such gamma exposures can damage internal organs and cells, as well as harming DNA. The reason that workers can still not enter any of the buildings housing the nuclear reactors that experienced full meltdowns at Fukushima Daiichi is because of the risk of their exposure to massive amounts of gamma radiation coming from the melted nuclear fuel inside or underneath the buildings (this melted fuel amalgam is referred to as "corium"). The gamma levels are high enough to kill a person within minutes of entering the buildings. In Hiroshima and Nagasaki this initial burst was short-lived, lasting about a minute. However, any living thing that experienced this burst would suffer from damages that would endure long after the burst diminished.[6] The affected cells and organs of those who were exposed to the massive burst were often fatally damaged, and the organism as a whole typically experienced steadily degrading health and organ failure over the ensuing days and weeks.

Internal radiation is more popularly understood as "radioactive fallout." The fireball of a nuclear detonation draws material upward as it rises higher into the sky. "Regarding fallout, there are many components rendered radioactive in the dust drawn up from the earth by the mushroom effect of the explosion," explained Councilman Morgan, the executive director of the Council of Life Sciences of the National Research Council (NRC) in 1980.[7] The detonation ionizes many of these particles, making them radioactive. "*Ionizing radiation* refers to radiation that is electromagnetic or particulate in nature and that has sufficient energy to cause ejection of an electron from an orbital shell of a target atom. This ejection creates charged particles or ions," explains a guidebook on the medical management of radiation injuries. "Ionization of biologically important molecules (eg, DNA) can cause cellular death by means of apoptosis and mitotic catastrophe. Ionizing radiation at higher doses can cause damage to actively dividing and undifferentiated cell types (eg, stem and progenitor cells in the bone marrow, gastrointestinal system, or skin)."[8] The fallout also includes material from the weapon itself, such as ^{239}Pu and ^{235}U that didn't fission during the detonation. A 2005 study conducted by the US Department of Health and Human Services, the Centers for Disease Control and Prevention, and the National Cancer Institute explains, "All nuclear weapons use some combination of uranium-235 (235U), uranium-238 (238U), and plutonium-239 (239Pu) as the source of fission energy. Even in the most efficient modern weapons, some of the fissionable material in the bomb does not fission. A typical nuclear weapon will use both plutonium and uranium as the source of fission energy, so every nuclear weapon detonation scatters large quantities of uranium, and most of them also scatter plutonium."[9] These scattered particles become an additional radioactive hazard and can then be internalized into the bodies of those breathing, drinking, and living in the affected area.

As the cloud cools, these particles, and the many other radiological particles and nonradiological particles in which radiation was induced, *fall out* of the cloud and drift back down to Earth. Lawrence Livermore National Laboratory health physicist Brooke Buddemeier described: "A nuclear explosion can produce fallout, which is generated when dust and debris created by the explosion are combined with radioactive fission products and drawn upward into the cloud produced by the detonation. Due to the heat of the explosion, the cloud rapidly climbs through the atmosphere, potentially reaching heights of 5 miles (8 km) for a 10-kt explosion, and forming a mushroom cloud under ideal conditions. Highly radioactive particles drop back down to earth as the cloud cools."[10] Precipitation brings this fallout down in larger quantities as raindrops condense around the particles.[11] In Hiroshima and Nagasaki, since the cities consisted primarily of wooden buildings that were set ablaze, large quantities of the materials from the fires filled the clouds with additional particulates from the rising smoke. When the rain fell after the Hiroshima attack, the density of the particles from the fires darkened the rain, and it was called "black rain."[12]

These individual particles, beta particles or alpha-emitting particles, depending on their specific chemistry, can be internalized into the body primarily through inhalation or swallowing, but also through cuts in the skin. Once internalized, a particle may pass through the body and be excreted, but it may also lodge in the body for a period of time: the body may determine that the particle is a useful chemical and retain it. For example, the body normally puts iodine into the thyroid gland. If a person internalizes ^{131}I, which is a radioisotope or radioactive form of iodine, the body may put it in the thyroid gland, as it would any isotope of iodine

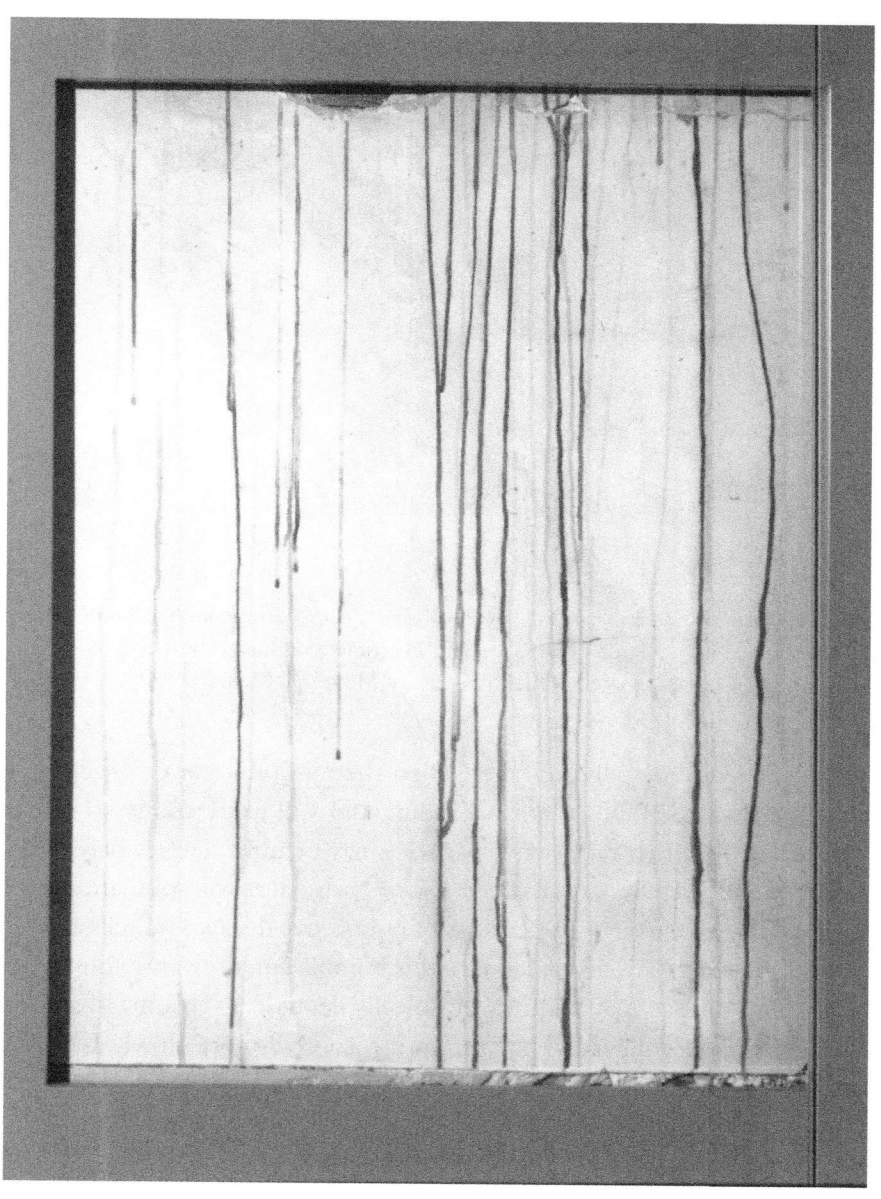

Traces of radioactive fallout–laden black rain that dripped down the white plaster
wall of a damaged house about 3,700 m from the hypocenter in Hiroshima.
Donated by Akijiro Yajima, Collection of the Hiroshima Peace Memorial Museum.

Gym uniform of sixteen-year-old Toyoko Kubata, with stains from black rain that fell as she fled her collapsed high school in Hiroshima. Donated by Toyoko Matsumiya, Collection of the Hiroshima Peace Memorial Museum.

(this is why iodine pills are taken when there is a nuclear emergency, so that the thyroid is full of normal iodine and will excrete any ^{131}I that is internalized). Other radioactive particles may be utilized elsewhere in the body. When particles are retained in the body, they will irradiate nearby cells as they go through the process of radioactive decay. The cells around the internalized particle will endure this bombardment twenty-four hours a day, and disease may form as those cells degrade.[13] This may result in cancers or other illnesses, but they will progress more slowly than the disease presentation of a body that experiences a large dose of external gamma radiation. Thus, both manners of experiencing radiation are harmful, but in different ways and in different time frames.

Sickness and Mortality in Hiroshima and Nagasaki

On 6 August 1945, approximately 68,000 people were killed instantly during the nuclear attack conducted by the United States, and 76,000 people were injured (this does not include the indeterminate number who internalized radionuclides).[14] Three days later in Nagasaki, approximately 38,000 people were killed and 21,000 were injured by the US nuclear attack.[15] Within a year after the bombings, approximately 114,000 people had died as a result of the nuclear attack on Hiroshima.[16] By the end of 1945, 73,884 people had died in Nagasaki.[17] In each city, among the affected were both those who had visible injuries, such as burns and wounds, and those whose harm from radiation had not yet presented or been diagnosed as disease. Many of those who died during the first months and year after the attacks were victims of severe burns and also of high doses of external radiation; victims suffered harm from all of the effects of the explosion simultaneously.

At the time of the nuclear attacks on Hiroshima and Nagasaki, death tolls in the tens of thousands and more had become normalized during World War II. On the night of 9–10 March, the United States conducted its first incendiary attack (firebombing) on Japan, in Tokyo: an attack that killed between 90,000 and 100,000 people.[18] The United States would ultimately firebomb sixty-five cities in Japan, and massive civilian casualties from single nights of attack had become common by the summer of 1945.[19] However, there were many shocking things contained in the news on 6 August about the nuclear attack on Hiroshima. The first shocking thing was that the destruction was accomplished by a single bomb from a single plane in a single second. The second thing was the realization that so much energy was held inside the nuclei of atoms, and that humans had figured out how to release this energy as a weapon. The third and most abstract thing was the use of radiation as a weapon. This was hard for people to understand, and in the days and weeks after the nuclear attacks, global newspapers, magazines, and radio shows were full of basic primers on the nature of radiation.[20]

The United States actively worked to blunt the notion that there was any danger from radiation besides that of the initial burst of gamma radiation. "Destructiveness of the atomic bomb in Hiroshima exceeded scientists' expectations, but there is no evidence that a single person that entered the area after the bombing died from radioactivity, Brig. Gen. T. F. Farrell, chief of an American mission studying effects of the bomb, said today," assured a September 1945 article in the *Chicago Tribune*. " 'No measurable radio activity was found under the point of detonation or elsewhere,' he said. In other words, it would be safe now to live in the bombed area. The largest number of casualties in Hiroshima probably resulted from blasts, missiles, and fires, he added."[21] Of course that statement was not accurate. Measurable radiation was found under the point of detonation and elsewhere after the bombings, and ultimately those who entered the city for several weeks after the attacks would obtain legal status in Japan as hibakusha, in recognition of the radiological dangers that they endured.[22]

Scientists had known since the 1920s that exposure to radiation could cause mutations, but people in Hiroshima and Nagasaki had no awareness of the illnesses that would slowly unfold, and they could not have anticipated the legacy that this single second of destruction would yield over the coming decades. As this disease burden slowly came into view, the fears of death and sickness that resulted from nuclear detonations and exposures to radiation would mark how people around the world felt about nuclear weapons and the impacts they have on communities. Nine years after the nuclear attack on Hiroshima, twelve-year-old Sadako Sasaki would develop swelling in her neck and ears and a fever and was soon diagnosed with leukemia. Sadako, who was two years old at the time of the bombing, had not been injured on the day of the nuclear attack, and such hidden and unexpected illnesses would come to typify the deepest fears of many who had received exposures to radiation but who had appeared to remain healthy.[23] The association of nuclear weapons with mass casualties and slowly emerging cancers grew in the nuclear imaginary.

The Opportunity

Within days of the nuclear attacks, Japanese military physicians arrived in Hiroshima and Nagasaki to assess the weapons' impact and their effects on those who had survived. "Military physicians from the Army Medical College and the Tokyo First Army Hospital arrived in Hiroshima on 8 August, two days after the bombing," writes historian Susan Lindee.[24] Soon after the American occupation officially began in September, multiple teams of Americans began inquiries into the medical effects of the attacks, and before long the teams were unified into a "Joint Commission." These researchers sought to access the work already being conducted by the Japanese Special Committee for the Investigation of the Effects of the Atomic Bomb, as well as to carry out their own inquiries and interviews with survivors. Much of this effort was haphazard, and it suffered from both a lack of planning and a lack of effective execution due to problems with translation and cooperation between the two teams unfolding in a form that Lindee describes as "colonial science."[25]

The Joint Commission released its report in 1946, recommending that a " 'permanent American Control Commission' supervise studies at Hiroshima and Nagasaki of possible hematological effects, genetic and fertility effects, and carcinogenic effects."[26] This recommendation was soon followed by a more authoritative directive: "During this period . . . the National Academy of Sciences—National Research Council received a Presidential Directive instructing the National Research Council to undertake a long-range, continuing study of the biological and medical effects of the atomic bomb on man."[27] The directive stated, "Preliminary surveys include about 14,000 Japanese who were exposed to the radiation of atomic fission. It is considered that the group, and others yet identified offer a unique opportunity for the study of the medical and biological effects of radiation which is of utmost importance to the United States. Such a study should continue for a span of time as yet undeterminable. However, the study is beyond the scope of military and naval affairs, involving as it does, humanity in general, not only in war, but in anticipated problems of peaceful industry and agriculture."[28] The Atomic Bomb Casualty Commission (ABCC) was formally established by presidential mandate in November of 1946, and "The first representatives entered the devastated cities in 1947 and began to build an extensive system for the collection and evaluation of medical and statistical data."[29]

Aside from the moral and political dimensions of the nuclear attacks on Hiroshima and Nagasaki, the reality of the medical situation there offered a unique research opportunity to scientists. While numerous people had been exposed to radiation during the less than fifty years since the discovery of radioactivity, the exposure of more than a hundred thousand human beings to varying doses of gamma radiation offered the potential to follow the subsequent health outcomes of that cohort and establish a robust database correlating radiation exposure with later disease presentation, life span, and mortality.

In the immediate aftermath of World War II, the future of human society seemed wedded to the development and destiny of nuclear weaponry.

Having endured two world wars in a single generation, the second being marked by the strategic implementation of "total war" and the use of atomic bombs, it seemed inevitable that these weapons would be used again in future wars.[30] Chemist George Kistiakowsky, the leader of X Division (explosives) at Los Alamos during the Manhattan Project, responded to the Trinity test, the first detonation of a nuclear weapon on Earth in New Mexico on 16 July 1945, remarking, "at the end of the world—in the last millisecond of the earth's existence—the last man will see what we have just seen."[31] Working to generate the most detailed analysis of the effects of ionizing radiation on human beings seemed to be essential information to prepare for the future, and the research opportunity provided by the existence of this cohort of radiation-exposed survivors at Hiroshima and Nagasaki provided an opportunity to gather such data. Lindee cites a former director of the ABCC, Robert Holmes, as proclaiming, for just this reason, that the Hiroshima and Nagasaki hibakusha were "the most important people living."[32]

The ABCC and the Life Span Study

Scientists became aware that exposure to radiation was deleterious to health soon after the discovery of radiation in 1896 by Henri Becquerel and the subsequent discovery of radium by Marie Skłodowska Curie and her husband Pierre Curie in 1898. "Pierre Curie and other radium researchers found that if they carried the substance around in a pocket, it would burn their skin."[33] In the late 1920s Americans became horrified by the illnesses suffered by the so-called Radium Girls, who had developed a variety of radiation-related diseases from working with radium-infused paint making glow-in-the-dark hands and numbers on watches and clocks. Their ultimate legal victory established the right of workers in the United States to sue for health damages incurred in the workplace.[34] Around that same time, biological researchers learned that exposures to radiation produced genetic mutations. Radiation mutagenesis was discovered by Hermann Muller, then a graduate student working in the drosophila (fruit fly) lab of Thomas Morgan at Columbia University.[35] Muller would receive the first Nobel Prize in Physiology or Medicine awarded after the nuclear attacks on Hiroshima and Nagasaki (1946), in part to highlight the importance of genetic dangers that the new weapons posed to living creatures.[36]

The ABCC sought to study the impact of the irradiation of the people of Hiroshima and Nagasaki on several tracks. In March 1948 they began a genetic study of pregnant women and the children of hibakusha. Pediatric examinations began in both cities later that same year. In 1950 a systematic study of leukemia was begun, and in 1951 a study of children who were exposed in utero.[37] At the behest of the chairman of the NRC in 1955, a committee was formed under the leadership of University of Michigan epidemiologist Thomas Francis and was sent to the ABCC to assess the progress and organization of work being done there. The Francis Committee made several recommendations, including that a long-term study be initiated with a set population whose health could be tracked for decades.[38] This formalized previous work into the Life Span Study, the most profound and impactful of the studies conducted by the ABCC and its legacy organization, the Radiation Effects Research Foundation.[39] "The broad program of study by Atomic Bomb Casualty Commission (ABCC) concerning A-bomb survivors in Japan includes, as a major component, a general mortality investigation known as the Life Span Study. This large-scale statistical study represents a systematic search for mortality differentials associated with radiation," explained the first report of the LSS.[40] However, it cautioned, "A specific weakness of the Life Span Study which should be recognized at the outset is the present unavailability of accurate knowledge of radiation doses received by individual survivors."[41]

Researchers relied on the 1950 national census, and supplementary censuses carried out by the ABCC, and identified 284,000 people alive in 1950 who reported residing in Hiroshima or Nagasaki at the time of the nuclear attacks in August 1945. From that number, 98,000 were currently living in Hiroshima in October 1950, and 97,000 were currently living in Nagasaki.[42] According to initial plans, "The sample for the Life Span Study will include around 100,000 persons."[43] As for the "unavailability of accurate knowledge," a program to estimate the radiation doses received by the participants in the study was outlined: "The basis for a dosimetry program is provided by air-dose estimates together with information on the attenuation of dose by common shielding materials, and detailed information on the location and shielding of each individual near enough to the hypocenter to have had potential radiation exposure. Under such a dosimetry program ultimately it will be possible to assign a numerical dose estimate to most individuals with significant exposure who have continued to reside in or near the two cities."[44] While an initial sample of about 100,000 persons was included in the LSS, the sample population was expanded in 1967 and again in 1979 to about 120,000 persons.[45]

The LSS has been one of the most robust databases created to track the mortality of a large population in response to an environmental exposure and is regularly referred to as the "gold standard" of data on the health outcomes and mortality that can follow exposures to radiation. A 2017 article in the *Journal of Nuclear Medicine* asserted that "The atomic-bomb survivor cohort of the Life Span Study (LSS) is the single most important dataset—the gold standard—for estimating radiation effects in humans."[46] The LSS is an ongoing study and is the fruit of the labor of thousands of researchers, technicians, and participants, primarily from Japan and the United States.

Internal Problems with the Life Span Study

The LSS database was designed with a heavy reliance on the reconstruction of radiation doses. For a study that correlates radiation dosage with later health outcomes, both data points need to be precise. Doses are primarily reconstructed by ascribing location and attenuating shielding to each participant in the study. "Although there is a lack of knowledge regarding the amount of radiation released by either of the two weapons, distance has been used as a surrogate measurement for radiation exposure," explains John Phair, an American researcher who worked at the ABCC in the early 1960s.[47] After the initial reconstructions were formalized in 1965,[48] individual dosage for LSS participants was recalibrated in 1986 and in 2002.[49]

There have been numerous problems with the LSS, both in terms of how the study has been structured and also in terms of what has been left out. The problems of its structure are less important for this book than the issue of what was excluded. The LSS remains both historically and scientifically valuable regardless of its flaws. However, like any scientific inquiry, it was constructed by human beings and reflects the strengths and weaknesses of a collective human intellectual inquiry.[50]

Regarding structural problems, these have mainly to do with the ascribing of dose. As I detailed earlier, the method of determining dose is based on two things, one general to all hibakusha, and one specific to each person. The nature of the radiation released by the two detonations informs us of what measure of gamma and neutron radiation was present at any distance from the point of detonation. The level of external radiation decreases in a quantifiable manner as one moves farther from the hypocenter. The specifics of each person's location and shielding is then situated within this model of the radioactive field. Emergent from these data points, dose is then assigned to every participant in the LSS.[51] Each of these has subjective qualities that slightly diminish the veracity of the eventual dose reconstruction, not significantly enough to invalidate the data, but nonetheless to some small degree.

The primary method for detailing distance and shielding for hibakusha who participate in the LSS is through oral interview. Participants were asked exactly where they were on a map at the moment of the explosion, and then where they were in terms of shielding: inside, outside, upstairs, downstairs. Thus, at the outset, people were asked to accurately remember the most traumatic moment of their lives, often years later, to determine a statistical data point. For many people this was a moment of extreme psychological disorientation. Many believed that they would die, many watched loved ones and strangers die; few understood why the entire observable world had so completely transformed into a hellscape in less than a second. Many had shame over their actions and choices in that hellscape, and many others wanted to hide the fact that they were hibakusha. Memory and testimony are inherently problematic sources for statistical data.

Of these early efforts, as Lindee has noted, bad translations during the interviews were a frequent problem, and many interviews consisted of only two or three questions. Interviewers sometimes doubted the testimony that they received and altered the ascribed dosage to reflect the symptoms they observed in the interviewee. "For the present," wrote the authors of the first LSS report, "it is necessary to rely entirely upon distance from hypocenter, a rough shielding classification, and reports of symptoms of acute radiation illness. . . . Symptoms are useful but must be used cautiously because varying individual symptom-response to radiation dose must be presumed. Presence of symptoms is used merely to sort out a presumed more heavily irradiated group but the measure of radiation is usually taken as distance, perhaps with shielding controlled."[52]

The two weapons used in Hiroshima and in Nagasaki, while both rely-ing on the release of energy from nuclear fission, were of entirely different designs and utilized different fissile materials: ^{235}U and a "gun" design for the Hiroshima weapon, ^{239}Pu and an "implosion" design for the Nagasaki weapon. Scientists understood the general nature of the radiation burst of the weapons but did not have certainty about the exact levels of radiation, and thus the effects of shielding. "Neither bomb has been monitored for its radiation potential and it is necessary to rely on estimates of air dose derived from similar, but not identical, nuclear weapons," cautioned the first LSS report.[53] John A. Auxier, director of the Health Physics Division of Oak Ridge National Laboratory, explains, "In the Nagasaki type of bomb, the surrounding material contained tons of high explosive whose hydrogen and nitrogen slowed down and absorbed most of the neutrons; thus detonation produced an intense source of high-energy gamma rays. In the Hiroshima bomb the iron slowed down, but did not capture many neutrons, but it did absorb gamma radiation substantially."[54] It should be pointed out that there was never another weapon like that used in Hiro-shima, a gun-design weapon utilizing ^{235}U as the fissile material, ever built or detonated, denying the utility of a comparable explosion being used as a model. There was also uncertainty about the scattering effects of particles in the atmosphere and of the ground itself, but most importantly about the effects of the structural shielding provided by buildings and furniture.

As unforeseen health problems emerged among the hibakusha, the need for a more precise dosimetry of the detonations was clearly required. In 1956 the Atomic Energy Commission (AEC) commissioned ORNL to continue research on this issue, and it initiated Project Ichiban (*ichiban* is Japanese for "first" or "number one") under the direction of Auxier. "Most of the survivors in the two cities were exposed in residential wood-frame structures, and the highly modular and uniform construction of Japanese houses at that time made a definitive dosimetry study feasible. Hence the shielding factors for typical Japanese houses were another principal prod-uct of the ICHIBAN project," explain the authors of the "Historical Re-view" that introduces the publication of the 1986 dose-reconstruction report. "Several Japanese-type houses were constructed at the Nevada Test Site (NTS) and used to make radiation-shielding measurements at atmo-spheric weapon tests prior to the Limited Test Ban Treaty of 1962. After

Project Ichiban "Japanese-type house" at the Nevada Test Site (1957). John A. Auxier, *Ichiban: Radiation Dosimetry for the Survivors of the Bombings of Hiroshima and Nagasaki* (Springfield, VA: National Technical Information Service, 1977): 32.

1962, the house-shielding measurements were conducted using other neutron and gamma-ray sources mounted on a 500 m tower at the NTS. These sources included a small unshielded reactor."[55] These *Japanese-type houses* were built on skids and rearranged at varying distances, filled with arrays of radiation-measuring equipment and subjected to neutron and gamma radiation from the unshielded reactor on the BREN tower for several years—in a macabre reenactment of the nuclear attacks themselves at the Nevada Test Site—to gather data on shielding for use in dose reconstruction efforts at the ABCC.

Last standing Ichiban house with BREN tower in background at the renamed Nevada National Security Site (2012). The BREN tower was torn down four months after this photograph was taken. Photo by author.

While both the subjective element to the determination of the location of participants and the uncertainty about the shielding effects of both buildings and atmospheric factors have necessitated recalibrations of dose reconstruction, neither is seen as significantly compromising the veracity of the LSS. What was left out of the LSS is the more concerning issue, and while it doesn't challenge the value of the LSS as a database, it does call into question the political deployment of the study.

External Problems with the Life Span Study

While the data yielded from the Life Span Study is highly relevant for external exposures to radiation, it is virtually irrelevant for internal exposures. During the subsequent decades of the Cold War, far more people would be exposed to fallout radiation than to the burst of gamma radiation from weapon detonations; internal exposures would become far more common than the external exposures assessed in the LSS. While the LSS has given us a map for understanding how exposures of various levels of external radiation may eventuate in illness or mortality, it does nothing to clarify the health impacts of internal radiation. Yet when people have been exposed to fallout, we have primarily used the LSS to assess their risk. This inaccurate presentation of LSS data in relationship to health risks faced by those exposed to fallout has contributed to their structural invisibility.

Serious risk from internal exposures was understood at the time.[56] From the initial speculations about nuclear materials as weapons for use in World War II, the potential of internalized radionuclides as a means of sickening and killing enemy troops and civilians was considered alongside the use of nuclear materials as explosive agents. Imagining that the Nazis might use radiological agents against Allied troops (specifically during an Allied invasion of Europe), Manhattan Project scientists reverse-engineered this fear to consider how the US could similarly use them against the Germans.[57] In a meeting of the Subcommittee of the S-1 Executive Committee (formerly the Uranium Committee of the US government's Office of Scientific Research and Development) in October 1943, several of the top scientists overseeing the Manhattan Project considered methods of weaponizing radioactive particles in warfare. One method would be to spread large quantities of radionuclides peppered in a "chemical carrier," to expose troops to high levels of gamma radiation. Because of concerns that wind might disperse the material before a fatal or harmful dose could be assured, the committee considered methods of compelling internalization of radioactive particles: "A somewhat different use or [*sic*] the radioactive material would depend on the fact that extremely small quantities of certain of the radioactive elements appear to be absorbed in the lungs of animals and produce fatal effects after a period of some weeks. The amounts necessary to produce eventual death under such conditions are extraordinarily small." This section of the report, titled "Radioactive Gas Warfare," detailed how to facilitate the inhalation of these microparticles: "This means that if such materials could be kept in the air in the form of a fine dust or smoke in concentrations as low as two thousandths of a microgram per liter, inhalation of such atmosphere for one hour would be sufficient to establish a lethal concentration; the material would accumulate in the lungs and the individual would eventually die of the radioactive poison. There is no known way in which such a contaminated individual could be successfully treated once he had thus' [*sic*] been exposed."[58] While these pathways for radiation and dire health effects were known early in the Manhattan Project, they were considered irrelevant when assessing the effects of radiation on the hibakusha in Hiroshima and Nagasaki, who had clearly spent days and weeks immersed in atmospheres containing aerosolized and deposited radionuclides.[59]

The decision to exclude data on internalized radiation was made at the outset of the LSS. As explained in the 1961 LSS Technical Report 05-61: "Although the facts as to fallout and induced radiation are not incontrovertible, in the judgment of the authors these potential sources of radiation are, at most, of secondary importance. Comparison of the mortality of early entrants with that of other persons not directly exposed revealed no differences between these two groups (Appendix I). Accordingly, in the present report exposure is classified on the basic direct radiation from the burst."[60] This early decision to exclude data on the health effects of internalized radioactive particles would have a profound effect on how risk and damage from exposure to radiation would be understood and legitimated throughout the Cold War and beyond. In this statement, it was assumed that the health effects of internalized radiation among the hibakusha of Hiroshima and Nagasaki were dwarfed in significance compared to the massive external exposures, and not significant enough of a detriment to health to warrant the difficulty of including it in the data sets of the LSS. Clearly the number of hibakusha in Hiroshima and Nagasaki that were exposed to high levels of external radiation precipitated an overwhelming medical crisis, and the internal exposures, both of those in the cities at the time of the attacks and of the "early entrants" who came into the cities soon afterward (to look for relatives or to assist in cleaning up) presented a less visible and urgent health concern.[61]

These concerns, and the internal logic of this decision, shaped the subsequent study and also the radiation paradigm internalized by ABCC researchers, as well as by scholars in the field more broadly since then. James V. Neel and William J. Schull, two senior researchers at the ABCC and later RERF, presented this logic in their 1991 assessment of the genetic study of the hibakusha of Hiroshima and Nagasaki:

> The estimation of the amount of residual radiation at Hiroshima and Nagasaki presents many difficulties. The distribution of residual radiation around the hypocenter in the two cities was asymmetric, the exact pattern depending on local meteorological details. . . . There are persistent Japanese reports of members of rescue parties and others not actually in the two cities at the time of the bombing later developing symptoms of radiation sickness. . . . We shall adhere for the time being to the more conservative view concerning residual radioactivity. By and large, individuals exposed to the effects of the atomic bombs tended to leave the area as rapidly as possible. Accordingly, for the purposes of this study it has been felt that although exposed individuals and also those entering the city immediately after the bombings may have been subjected to some residual radiation, by comparison with the amount absorbed by those present at the moment of the explosion, this was on the average small and could be disregarded.[62]

Built into this original choice in the 1950s and 1960s was the concern, widespread at the time, that wars fought with nuclear weapons would be likely in the future, and that the most valuable medical research would be on the health outcomes of those exposed to the massive bursts of gamma radiation that would accompany such a catastrophic war.[63] This would aid battlefield commanders in assessing the health of troops exposed to nuclear detonations, in strategies for the use of nuclear weapons against enemy troops, and in the design of buildings and shelters that could provide safety under nuclear attack.

But this was not how the Cold War actually unfolded. In the more than seven decades since those first nuclear attacks, it would be internalized radioactive particles that would most plague the health of those who suffered from nuclear explosions. Many who would be exposed to dangerous amounts of radiation in the period of atmospheric nuclear weapon testing (1945–1963) would have their primary exposures delivered through internalized radionuclides deposited from the fallout clouds of those tests. These people did not experience direct attack with nuclear weapons, and so were not exposed to the massive bursts of gamma and neutron radiation that accompanied them (nor the blast and heat). But they were often exposed to large amounts of radioactive fallout, far more than were the populations of Hiroshima and Nagasaki, especially after 1952 in the era of atmospheric testing of thermonuclear weapons. The health effects that individuals in these communities experienced were routinely dismissed because they were judged below levels of medical concern *according to the Life Span Study.*

Similarly, most of those whose health has been affected by living near nuclear production sites or the aftermath of nuclear power plant accidents, such as at Mayak (USSR, 1957), Windscale (UK, 1957), Chernobyl (1986), and Fukushima (2011), would suffer from living in areas with massive distributions of radionuclides resulting from the fires and explosions of those accidents. They would grapple with the illnesses that manifest from internalizing these particles and living in ecosystems rich in contaminated particles. Their health problems would also be trivialized as unrelated to these exposures because they did not fit the profile of radiation-induced illnesses delineated in the LSS. Shizuyo Sutou of the School of Pharmacy of Shujitsu University in Okayama, Japan, would frame his explanation exactly so in his 2016 article "A Message to Fukushima: Nothing to Fear but Fear Itself." His examination uses the LSS to document to his readers that there is no danger to people living in Fukushima from radiation, and that their problems are caused by *incorrect understanding.* He concludes by advising, "When people return to the evacuation zones in Fukushima now and in the future, they will be exposed to such low radiation doses as to cause no physical effects. The most threatening public health issue is the adverse effect on mental health caused by undue fear of radiation."[64] Sutou's dismissal of the fears of the residents of Fukushima characterizes the radiological threat to their well-being as coming exclusively from the levels of external radiation in the evacuated communities they might reinhabit. In actuality, the primary danger to the people he is discussing is from internalizing the widely distributed radionuclides that contaminate the forests surrounding these communities. What presents risk to the residents of Fukushima is not the doses they will receive from returning (although there is evidence that even such low doses received externally can be dangerous; see the INWORKS study),[65] but the proliferation of radionuclides throughout the ecosystems to which they return, and the dangers that they or their children might internalize such particles. Those pose very real threats, and the data in the LSS does not address dangers. I will return to the discourse pathologizing the "mental health" of those anxious about living with radiation in chapter 3.

Conclusion

Some have ascribed malicious intent to ABCC scientists for excluding the study of internalized radiation from the LSS, especially since internal effects were well-known before the nuclear attacks in 1945. I would argue that the truth is more mundane: the impact of external radiation was more pressing in Hiroshima and Nagasaki, and importantly, quantifying internal exposures was simply hard science to do. Determining the dosage received externally can be straightforward: calculate the level of the burst from a central point and then the distance and shielding of those close enough to fall within this radiation field. Conversely, determining who in a large cohort of people has internalized individual radionuclides may not have been possible. The risk is stochastic: two people may live in the same house, eat meals together, and walk the same streets, yet only one of them happens to internalize a particle. It is difficult to determine which of the two that was before disease presentation, and even afterward. One tool to determine if a person has internalized a beta particle or alpha-emitting particle is the whole-body counter. Whole-body counters were not yet invented when the LSS began, and not generally available until 1964 when they began to be manufactured at ORNL, each weighing about sixty tons (more about whole-body counters in the next chapter).[66] After the nuclear accident at Three Mile Island in 1979, which resulted in a partial meltdown of the nuclear fuel, it was decided to administer whole-body counting to a selected cohort of the exposed population, and it took seventy-four days to test 769 people.[67]

In retrospect, it was functionally impossible to have taken this research task up at that time, given the current technology, and it is understandable that it was not then pursued. In recent years, researchers have been able to verify the presence of internalized radiation in the remains of hibakusha from Nagasaki that had been preserved in paraffin. In a 2018 study, "Autoradiography was carried out with the 70-year-old paraffin-embedded specimens taken from Nagasaki atomic bomb victims who died within 5 months after the bombing." Researchers determined that "Pu was deposited in the bodies of the Nagasaki A-bomb victims presumably via various routes."[68]

While understandable, the decision to study the health impacts of internalized radionuclides becomes problematic when it is left out of descriptions about just what the LSS does tell us about the health risks from radiation exposures. The LSS is globally invoked as the gold standard of explaining the health effects of exposures to radiation, when it should be presented as the gold standard for describing the health effects of *external* exposures to radiation. It shouldn't be used to dismiss the risk of health effects from internalized radionuclides, as those outcomes are outside the purview of the study. However, this is precisely what has been done, and this has resulted in institutionalizing the invisibility of most of those who have actually suffered radiation exposures and dire health effects since Hiroshima and Nagasaki.

2

The Particles That Remain

Bone Seekers

In 1956, AEC commissioner Willard Libby released restricted information about radioactive fallout into the wild by publishing an article in *Science* titled "Radioactive Fallout and Radioactive Strontium."[1] He opened with a definition: "The radioactivity that falls out of the atmosphere after the explosion of a nuclear weapon is called the radioactive fallout. In the ordinary atomic bomb, for example, for each 20,000 tons of TNT equivalent of explosive energy, about 2 pounds of radioactive materials are produced. In these 2 pounds are some 90 different radioactive species varying in lifetime from a fraction of a second to many years." His article focused particularly on the risks to human beings from radiostrontium (^{90}Sr). As radiostrontium is chemically similar to calcium, the body treats it as though it is calcium after it has been internalized and moves it to the teeth or bones. Libby described it as "bone seeking." He explained its means of transportation into the ecosystem: "The radiostrontium comes down mainly in raindrops although morning mists and fogs may be particularly effective in this regard also, as well as surface contact and direct falling. It descends on the foliage and on the soil. That fraction of it which falls upon plant leaves has a good chance of being absorbed directly into the plant—much in the way that modern leaf fertilizers operate."[2]

Libby described the dynamic process by which the radioactive materials fell out of the mushroom clouds of these explosions: "A bomb fired on

the surface of the earth . . . may have an appreciable portion of its radioactivity reprecipitated within relatively short distances." However, by 1956
the United States and the Soviet Union were both aggressively detonating
extraordinary nuclear weapons, H-bombs—as much as one thousand times
larger than the "ordinary atomic bomb" Libby utilized as a baseline. Atmospheric detonations of thermonuclear weapons sent radioactive "debris"
far higher, raising particles through the troposphere and into the stratosphere, where they circled the Earth for years before deposition: "As a
result of its residence time in the highest layers of the atmosphere, the
winds mix and distribute the radioactive materials broadly over the earth
and one finds, when the fallout does finally find its way down into the
troposphere were the rain and snow wash it out, that the rates of precipitation are relatively uniform over the entire earth's surface."[3]

This global deposition of radioactive fallout particles from nuclear weapon
testing would link all living creatures as participants in nuclear weapon testing. The most immediate and deadly of the effects of nuclear detonations—
the blast, heat, and external burst of radiation—would be largely contained
at nuclear weapon test sites. Risks from internalizing radionuclides from
fallout clouds would escape containment and circle the Earth: finding their
way to us all through rain, mist, dust, and food. Ultimately, there would be
520 atmospheric nuclear weapon tests, spreading radionuclides around the
planet: "World-wide deposition of long-lived fission products (e.g. ^{137}Cs,
^{90}Sr) has largely resulted from stratospheric fallout."[4] Further distribution
of radionuclides would occur through explosions at nuclear accident sites
and decades of waste leaking from nuclear production sites. The whole world
remains touched by nuclear technologies.

Sickness and Mortality at Nuclear Test Sites

While cases of mortality from the blast and heat of the actual
detonations of nuclear weapons have occurred, typically only military and
test site personnel were close enough to have suffered from these effects.
Nuclear tests are sited away from permanent human habitation; as the
weapons became larger, their effects reached out farther from the test sites
and closer to those inhabited spaces. A 2002 study of residents living within
100 km of the epicenters of Soviet tests in the Polygon—the nuclear

weapon testing site of the former Soviet Union near Semey (formerly Semipalatinsk) in Kazakhstan—was conducted by Japanese researchers who found that "546 of 606 respondents (90%) saw the flash of the nuclear explosions," while 70 percent felt the blast waves and 18 percent felt heat from the nearby tests.[5]

Soviet nuclear weapon designer Andrei Sakharov describes several deaths and injuries resulting specifically from the blast wave of a thermonuclear test conducted there on 22 November 1955: "the force of the explosion had collapsed a nearby trench sheltering a platoon of soldiers, and one, a young boy in his first year of service, had been killed." In a nearby village estimated to be beyond the reach of the shock wave, "inhabitants had been ordered to take refuge in a primitive bomb shelter. After they saw the flash, they decided it was safe to emerge. They left behind a two-year-old girl who was playing with blocks; the shock wave demolished the shelter, and the girl was killed." Meanwhile, in another village, the ceiling collapsed in a local hospital ward "seriously injuring half a dozen people."[6] Later histories reveal a more lethal toll than the girl left behind with her blocks. Togzhan Kassenova, a fellow in the Nuclear Policy Program at the Carnegie Endowment, writes, "A three-year-old girl died under a collapsed ceiling in the village of Malye Akzhary. Six soldiers from a security battalion died. They had waited in trenches, thirty-six kilometers away from ground zero. Dirt buried all of them. Other local people sustained multiple injuries including broken bones."[7] These deaths and injuries, including accounts of forty-two people being injured by flying glass, all resulted from the blast wave of the weapon.[8]

There have been cases of deaths and illness from the acute radiation experienced externally in the immediate aftermath of nuclear weapon tests.[9] During the detonation of the first Soviet nuclear weapon at the Polygon in 1949, residents of the small village of Dolon received doses of 160 rem, and several villagers who were working outdoors at the time of the test additionally suffered radiation burns to their skin.[10] Four years later the Soviet Union tested its first thermonuclear weapon (the 1955 test mentioned in the previous paragraph) at the Polygon; several thousand local villagers (and tens of thousands of livestock) were evacuated before the test. But some villagers were intentionally left behind: "The residents of Kara-Aul were hurriedly evacuated a few hours after the arrival of the radioactive cloud in their settlement. However, out of the 1,620 residents of Kara-Aul, 191 adults were left behind in order to guard the properties of the residents."[11] One of those left behind, Nurzhan Mukhanov, recounted in an interview years later:

> On the day of the explosion, military personnel took all forty men who were left behind to a beautiful area a few kilometers from the village. They gave them food and alcohol and instructed them to relax and enjoy the day. Before leaving, they provided Mukhanov and his co-worker with some radio equipment and told them that further instructions regarding the exercise would be sent via radio. He and his friend waited for this news, but before it arrived, there was a massive explosion, followed by a mushroom cloud. The other men were very worried that something might be terribly wrong, and they were angry that Mukhanov did not provide them with any information. Mukhanov and his friend insisted that he had not received any radio transmissions. A few hours later, military personnel, covered in white protective clothing from head to toe, returned to the scene to see how they were doing. The men were still angry. They felt that they were in danger, and they demanded to know why they had never been contacted with further instructions. The soldiers responded by saying that the radio operator on the other end was drunk and failed to do his job properly. Mukhanov now feels that he and the others were intentionally used as guinea pigs. He remembers how the soldiers took blood and urine samples from him and the others on a regular basis.[12]

For those who evacuated, when they returned, "they found unusual things: many of their chickens were dead, some of the baby chicks were deformed, their dogs and cats were losing their fur and some were covered with scabs. Some people described how their dogs and cats died before their eyes in the first days after their return."[13] Villagers kilometers away reported a deafening explosion, followed by fires and violent shock waves rumpling the earth, shattering windows, and collapsing houses. A seventy-four-year-old woman who would not give her name said blood ran from her ear after the explosion as she hid from it in the military garrison about 6.5 km away.[14]

French electrician Jean-Claude Hervieux had worked at nuclear weapon testing sites in Algeria and French Polynesia. In a 2020 interview he describes "visiting a village in French Polynesia where high radiation levels had been detected. 'A local teacher said children were sick and vomiting,' he recalled. 'Mothers were asking why their children's hair was falling out.'"[15] While the sickness was distressing for these mothers, the uncertainty about what the symptoms signified and what they might portend was more destabilizing.

All nuclear states that tested weapons in the atmosphere compelled military personnel to participate in nuclear tests. For those positioned close to hypocenters, there are frequent reports of soldiers seeing their bones through their closed eyelids. Private John Hall of the British Royal Air Force (RAF) was stationed on Christmas Island in the Pacific when he participated in nuclear tests in 1958. Hall and his fellow servicemen "had been ordered to turn away from the mushroom cloud and put their hands in front of their faces. He later said that as he did so, his hands 'lit up like an X-ray,' and he saw his bones outlined through the flesh."[16] Derek Allen, another British serviceman, recalls his participation in a nuclear test in 1962: "It was so bright that he saw all the bones in his hands as if he were holding up an X-ray."[17] For those close to larger, thermonuclear detonations, such occurrences intensified. Robert MacKenzie, a US Marine who was stationed on the USS *Curtiss,* recalls the detonation of the Bravo weapon: "there were only two of us that had goggles, so it was the lieutenant and myself. All the other guys didn't have goggles. And the goggles are absolutely black. They're like welder's glasses. *Real* dark. So anyway, so we all sat on the deck, faced away from the blast. And we put our heads between our knees and put our hands in the back of our heads. . . . Then all of a sudden, it starts getting bright. *Real* bright. And brighter. And brighter. And I've got these welder's goggles on and my arm in front of it like this [demonstrating] and I can see my own bones. . . . Several of the guys looked up and they could see a guy right through a guy standing in front of them."[18] These reports reveal the gamma and neutron radiation resulting from the nuclear detonations and are sensorial indicators of the high radiation levels that were being experienced by the whole bodies of these soldiers simultaneous to their skeletal awareness.

Residual Radiation Harm

There has been no more consequential impact from nuclear weapon testing than that of residual radiation. "Human bodies—porous, renewing, and transforming—are as much a repository, a dump of man-made waste products, as are rivers, ground water, soils, plants, and animals," notes Kate Brown.[19] Millions of people living downwind have been exposed to radiation from the fallout clouds of nuclear weapon tests, and the deposition of immense loads of radionuclides into various levels of the atmosphere transported those particles throughout the entire world. "Recent estimates suggest that over 73 million cubic meters of soils and sediments in the U.S. alone have been contaminated with actinides and fission products by Department of Energy defense nuclear activities, with even larger volumes of Pu contaminated soils and sediments generated in other countries such as the former Soviet Union."[20]

Anticipation of radioactive fallout was structured into preparations for and operations during the Trinity test in New Mexico on 16 July 1945, three weeks ahead of the nuclear attack on Hiroshima. Australian physicist Mark Oliphant wrote to his friend, biochemist Hedley Marston, that while working on the Manhattan Project, "In 1943, when the first reasonable large-scale fission yields were examined by us in Berkeley (Seaborg and MacMillan), and it was clear that fission products would be a very great hazard if we ever succeeded in obtaining chain reactions of military significance, the report was labelled with the reddest of classifications."[21] During a "rehearsal" shot for Trinity conducted on 7 May 1945, radionuclides were placed in one hundred tons of TNT and detonated to simulate fallout and prepare downwind radiological monitoring teams to assess its intensity.[22] Wright H. Langham, an associate division leader of the Biomedical Research Division of Los Alamos National Laboratory recalled of the Trinity test at a conference in 1969: "Stafford Warren [chief of the Manhattan Project's Medical section] mounted evacuation teams and monitoring teams to cover the potential fallout area. We didn't have to evacuate anybody; we almost did. The arbitrary limit chosen for evacuation was an infinite life-time dose of 50 rems. One family approached this limit, and there was much debate as to whether we should evacuate them or not. We did not evacuate them. . . . Cattle were burned by fallout at

Trinity, and we had experience with fallout at Bikini where there was fallout on ships. I can't imagine anyone thinking that there wouldn't be fallout involved with weapons tests."[23]

The Bikini test that Langham referred to was the Baker test of Operation Crossroads in 1946, an underwater test that had the effect of suspending radionuclides that would have dispersed in a fallout cloud within the waters of the lagoon of Bikini Atoll, where it had been detonated. Crossroads was the first nuclear test series after the war, conducted less than a year after the attacks on Hiroshima and Nagasaki, and included the participation of approximately forty-two thousand troops.[24] The focus of the tests at Crossroads was to gauge the effects of nuclear weapons on naval ships. The US Army and Navy had been grappling (in part through the Strategic Bombing Survey) as to who could take credit for defeating Japan, the army, which had destroyed Japanese cities through aerial bombing, including the two nuclear attacks, or the navy, which had destroyed the Japanese Navy and was preventing food and fuel from reaching Japan.[25] Both service branches realized that nuclear weapons would be an important part of the future of warfare (and thus of ongoing funding), and Operation Crossroads designed tests to determine if military ships could survive a nuclear attack. The tests were conducted at Bikini Atoll, in part because atolls provided a better location to both stage and test the weapons. Atolls are coral remnants that had developed around undersea volcanoes, and when the volcanoes eroded, the coral atolls left roughly circular series of islands that contained lagoons in their center. The lagoons had still water, compared to the turbulent ocean water outside of the atoll, a property that also made them desirable places for habitation by Polynesian and Melanesian populations across the Pacific.[26] For Crossroads the US towed 70 American, Japanese, and German naval vessels (surplus after the war's end) inside of the lagoon, and staged 5,404 animals and other materials destined to endure the blasts on the ships. The 149 support vessels and crew were then pulled out of the lagoon to the open ocean for the actual tests.[27] In the first test, Able, the weapon was air-dropped from an airplane like the two used in the attacks on Japan. US sailor Nuell Paschal remembers that after the Able test, "We swam in the radioactive water, and we drank the radioactive water. We walked barefoot on the radioactive sand and we slept on the radioactive ships."[28]

The second test, Baker, involved detonating the weapon 27 m beneath the surface in the lagoon of Bikini Atoll.[29] Approximately forty of the target vessels (including submarines) were arrayed within one mile of the weapon, while the other vessels were slightly more distant. Members of the Radsafe (radiation safety) team had anticipated that an underwater detonation might create radiological problems, yet the test proceeded as planned. "The amount of radioactive material that collapsed back into Bikini's lagoon moments after the Baker shot was simply staggering. Unlike the Able blast, the fission products at Baker did not dissipate in the atmosphere. The water surrounding the bomb trapped most of the radioactive material and rained it down over the target vessels. As much as half the bomb's fission products remained in the lagoon's water or in the mist remaining in the air after the surge of spray fell back in to the lagoon."[30] Hours later the support ships and tens of thousands of service personnel began to reenter the lagoon. "Radioactivity in the lagoon's contaminated waters quickly spread to the support ships . . . every nontarget vessel became contaminated just as the planners had feared, as fission products became concentrated on underwater hulls and in condensers, evaporators and saltwater pipes. . . . For all its thousands of pages of detailed plans, the U.S. Navy managed to expose tens of thousands of men and more than 200 ships to radioactive contamination more than 2,000 miles from decent port facilities."[31] According to the *Official Report of Operation Crossroads,* continued presence on these irradiated ships would have been lethal to service personnel: "These contaminated ships became radioactive stoves, and would have burned all living things aboard them with invisible and painless but deadly radiation."[32]

Charlie, the planned third test of Operation Crossroads, was canceled soon after Baker. The cancellation was attributed to several factors; however, "What the Navy did not say was that the fallout from the Baker test had left such high levels of radioactivity in the lagoon, in debris washed up on the beach, and in the atoll's water and fish that it would have been impossible to stage the Charlie test from Bikini."[33] The US government estimated the test "left 500,000 tons of radioactive mud in the atoll's lagoon."[34] This scale of radiological contamination also forced the US to abandon Bikini as a test site and develop a second atoll, Enewetak, for nuclear weapon testing from 1946 until the Bravo test (back at Bikini) in 1954.

While the US military had not taken steps to sufficiently protect its personnel during Crossroads, the crisis of their irradiation was quickly integrated into official strategy. The 1947 report of the Joint Chiefs of Staff on the tests explains, "When a bomb is exploded underwater, lethal residual radioactivity assumes an importance greater than the physical damage caused by the explosion. Vast quantities of water falling from the explosion column and travelling outward in the base surge and, also falling as 'rain' from the cauliflower cloud, carry, not only highly radioactive fission products, but unfissioned material as well."[35] At the same time that US scientists at the ABCC were excluding consideration of internalized radiation as a significant health concern to the survivors in Hiroshima and Nagasaki, the US military was strategizing how to use it as a weapon against enemy populations.[36]

Military service personnel who participated in or observed atmospheric nuclear tests, in all nuclear weapon states, comprise a specific population cohort heavily impacted by the residual radiation of nuclear weapon testing from the tests throughout the atmospheric testing era. Although the United States did not keep adequate records, the government puts the figure of troop participation in US nuclear tests at 220,000.[37] While many observed tests at the Pacific Proving Ground in the Marshall Islands, other troops conducted actual military maneuvers with live nuclear weapons at the Nevada Test Site. The largest single maneuver was the Marine Brigade Exercise, involving 2,025 troops during Shot Hood of the Plumbbob series (74 kt) conducted on 5 July 1957.[38]

Many Soviet troops were exposed to high levels of radioactive fallout. On 14 September 1954, the USSR conducted the Totskoye nuclear test (40 kt) in the Southern Urals, in part as an exercise modeled on a conflict with NATO countries at the Fulda Gap in Germany, in a location where the landscape was imagined to align closely with that found at the gap. Envisioning the use of a nuclear weapon on NATO forces, the design was: "To learn if troops could deploy and wage war in a zone over which a nuclear weapon had just exploded, the high command selected a spot in eastern Russia nearly 1,000 kilometers from Moscow for a test that took place at 9:33 hours on 14 September 1954. The area chosen was near a village called Totskoye, between Kuibyshev (today's Samara) and Orenburg, where at least 45,000 personnel of the Soviet Army 'and many thousands of civilians' were dispersed among temporary shelters situated 3 km from a blast that took place 350 m

Mushroom fallout cloud drifts downwind in Nevada from the 12 kt Boltzman nuclear test, Operation Plumbbob, 28 May 1957. Photo by Dave Cicero. "Cloud from A-Bomb Test," *New York Journal American,* 31 May 1957. Held in Harry Ransom Center, University of Texas, Box B028, Folder "Atomic bomb—Tests—Yucca Flat, Nev., 17 Mar 1953 [3 folders]," Record Number: NYJA000294.

Mushroom fallout cloud drifts downwind in Nevada from the 29 kt Apple-2 nuclear test, Operation Teapot, 5 May 1955. Terrence N. Fehner and F. G. Gosling, *Battlefield of the Cold War: The Nevada Test Site*, vol. 1 (Washington, DC: US Department of Energy, 2006): 140.

above ground."[39] One veteran later recalled, "Some, the majority even, had no protective clothing, and besides it was impossible to use gas masks."[40]

Detonating the weapon at 350 m meant that the fireball was low enough to touch the ground, which significantly increased the resulting radioactive fallout. While the weapon was detonated at 9:33 a.m., "At approximately noon, the 'Eastern' troops, having begun to enter the tactical defense zone and working their way through the fires, entered the area of the nuclear explosion. The area was unrecognizable: trees had been cleared, only splinters remained from the enormous oaks that had stood there moments before, all of the grass had been 'combed' to one side, as if after a flood. During the exercise, nuclear explosions were simulated twice more and were psychologically perceived as nuclear assaults."[41]

Former Soviet colonel I. Krivoy wrote in 1991 that "Beyond the firing range several villages burned down, but the people in them were evacuated in good time with all their property to other settled areas. The villages had natural, non-dangerous radiation levels. After the test, the burned-out villages were completely rebuilt—at state expense—at a considerable distance from the test site. I saw the rebuilt villages with my own eyes."[42] However, Nikolay Leonov, one of those villagers, remembers, "When we returned the village was still burning. There was military equipment ablaze. The fire engines were putting the fires out, bulldozers were working away. But they actually allowed us to eat everything right away. We've got cucumbers, tomatoes and melons in our vegetable garden, and when we got back all of this vegetable crop was ripened. The tomatoes and such, they were all red. And they said, 'go ahead, you can eat everything, it's not dangerous.' Of course, we and the children began eating."[43] The Totskoye test was the largest direct troop participation in a nuclear weapon test in history, and subsequent analysis of the health impact on those who were exposed to radiation by a team of Russian researchers concluded that "As a result of the Totskoye nuclear test, 45,000 soldiers and approximately 60,000 civil residents of roughly 20 villages were exposed to direct radiation from the blast, or delayed radiation from fallout."[44]

Populations downwind from major test sites were significantly impacted by the fallout radiation that tracked across their lands from nuclear weapon testing. Estimates of the radiation-exposed population downwind from the Polygon is about 1.6 million people.[45] A 2015 article in *The Lancet* asserts that "atomic weapons testing at the US Government's Nevada test site exposed the public across the entire country, although most people were exposed to only fairly low doses of ionising radiation." Nonetheless: "Risk modelling studies of exposure to ionising radiation from the Nevada Test Site in the US suggest that an extra 49 000 . . . cases of thyroid cancer would be expected to occur among US residents alive at the time of the testing—an excess of about 12% over the 400000 cases of thyroid cancer expected to develop in the absence of fallout."[46] For those living close to the test site, the health consequences were more pronounced. Commenting on the fallout cloud from the 1954 Bravo test that irradiated the crew of a Japanese tuna trawler, the *Daigo Fukuryu Maru* (*Lucky Dragon No. 5*), a few hours after the test, Ralph Lapp cautioned, "Had the Japanese fishermen been on dry land instead of on a small ship they would have received several times the lethal dose of radiation in the first day," since the fallout would not disperse as quickly as it did at sea.[47]

Radionuclides Embed in the Ecosystem

When fallout plumes from nuclear explosions deposit large amounts of radiological particles into ecosystems, or nuclear production sites leak streams of them, the particles begin a process shaped by both the nature of the particle (its particular half-life and weight) and the dynamics of the ecosystem into which they deposit: "The overall dispersion of a contaminant through the environment is the result of the transport by different kinds of more or less mobile abiotic and biotic carriers such as masses of air or water, particulate matter and biota."[48] This was modeled soon after the Baker test by researchers working under the AEC at the Applied Fisheries Laboratory of the University of Washington studying the radiological consequences of the tests on marine biota. "Now all we need is time, time to trace the intensity and effect of radioactivity through every type of life at Bikini," mused a 1949 AEC film, *Bikini: Radiobiological Laboratory*:

> Bikini is quiet again. But under the silence of the atoll's waters something is still happening, something that must be studied and measured. The lagoon still contains traces of energy released there on Baker Day: radioactive energy. For us, the biologists, the story of Bikini merely began in 1946; for us the far more significant story has been unfolding year after year since Operation Crossroads. The long-range problem of whether these traces of radioactivity have observable effects on various forms of life. The problem of how radioactive materials persist and circulate in this completely natural environment. Fortunately, we can study the patterns and courses of radioactivity because certain forms of it prevail for many years.[49]

The presence of these long-lived radioactive tracers in our ecosystem continues to enable scientific research. These particles will move through the ecosystem, transported by "abiotic and biotic carriers" such as wind, rain, and living creatures, and will remain traceable until their radioactivity has decayed to that of background radiation.

A particle may deposit on the soil from a fallout cloud, be taken up by a plant, be internalized into a bird or animal that feeds on the plant, be redeposited into the soil when the animal defecates or dies, be taken up by a different plant, be put back into the soil when that plant decays, be carried by rain or underground water into a reservoir, be taken into a fish, be taken into a bird when it eats the fish, be transported a far distance, and be redeposited into soil in a new place when the bird defecates or dies and decays. The particle will cycle through the ecosystem just as chemicals do, and as long as it remains radioactive (generally assumed to be the period of time equivalent to ten of the particle's half-lives), it will pose a danger to living beings. For some particles that period will be short: some decay in a matter of days or weeks or even less, but for other particles this may be far longer than the life spans of the creatures that internalize them. ^{137}Cs, which has a half-life of 30.2 years, will remain dangerous for over 300 years, or roughly ten half-lives. ^{239}Pu, with a half-life of over 24,000 years, and uranium-235, with its half-life of over 700 million years, will both cycle through the ecosystem for far longer than most species have existed. ^{137}Cs is widely dispersed in the areas downwind from Chernobyl and Fukushima, creating an ongoing, asymmetric, and ill-defined challenge to safe habitation that will last beyond the lives of the grandchildren of those living there at the time of contamination.

Radionuclides are carried high into the atmosphere by nuclear explosions. Smaller nuclear explosions may carry particles into the troposphere (which extends to 12 km above the Earth's surface), after which most will "fall out" of the cloud within a few hours, while mushroom clouds from large thermonuclear explosions may carry fallout up past the troposphere and into the stratosphere, where particles may circle the Earth for years, facilitating global distribution, as Libby described in 1956. When they descend, close to or far from the test site, particles can condense water vapor, creating raindrops that fall in more-substantial amounts with heavy precipitation.[50] "After entering the troposphere from above, the fission products are transported to the level of rain-bearing clouds by turbulent mixing, and below this level the particles are rapidly deposited by precipitation."[51] A 2004 study by IAEA scientists on the radiological legacy of the nuclear tests at the former Soviet test site in Novaya Zemlya found that "variations in the radioactivity levels from region to region are due not to the distance from Novaya Zemlya, but to the amount of atmospheric precipitation."[52] They are carried by water down to our level of the biosphere and begin their cycles of transport within its dynamic systems. Many will pass through human beings and other living creatures: "Severe radio-active contamination of land might result from the deposition of fallout originating in the detonation of nuclear weapons or nuclear reactor accidents. The deposition of fallout on soil or plants would introduce radioactive isotopes into the food chains of animals and man."[53]

Uptake routes into plants can be immediate: "There are two important ways in which plants may become contaminated by fission products from the atmosphere: (a) indirect contamination, which occurs when radioactive material enters the soil and passes into the plant through the roots as do soil nutrients; (b) direct contamination, in which passage through the soil is by-passed." As one moves higher up the food chain these materials can concentrate in animals from contaminated food sources in a process known as bioaccumulation. "Radionuclides in or on plants may reach man directly by his consumption of foods of plant origin or indirectly by his consumption of animal products. The grazing animal effectively collects contamination from plant material and concentrates it in animal products."[54] As each animal is eaten by one farther up the food chain, the increasing load of these radionuclides deepens the health risks to the specific animal and renders its eventual remains more dangerous when it is eaten or decomposes.

^{137}Cs presents a persistent problem once it has been deposited on forests: "Forest soils are rich in organic matter, and contrary to agricultural soils generally have a low clay content. This leads to a higher transfer of radio-caesium to most forest products, such as berries or mushrooms compared with agricultural products. As radiocaesium is accumulated in the organic horizon, and also remains plant available, this transfer persists over years."[55] Soil scientists studying cesium contamination from the Chernobyl Nuclear Power Plant disaster have found that this transfer makes "decontamination" increasingly difficult. "In the radiobiological context, radioactive isotopes of caesium are some of the most hazardous radionuclides which were released into [sic] biosphere due to the Chernobyl Nuclear Power Plant accident as they are chemical analogues of potassium. They are included in the biological cycle and cause long-term irradiation to the forest biota and man. Their capability to enter the internal plant tissues by roots and other ways limits the possibility of the decontamination of the forest products."[56]

A 2016 RAND Corporation study explains this further, using the Fukushima contamination zone as a model:

At Fukushima, the character of the environment exacerbated the difficulty of the decontamination effort. As several of our sources indicated, forest and mountain environments retain radioactive materials much better than do most urban environments due to cesium capture attributes of the natural clay in soil. Cesium binds tightly to the clay in Fukushima's soil, making it difficult to extract using decontamination methods but allowing it to slowly percolate through the ecosystem and hydrological cycle. The dynamic forest environment shuffles radioactive material around from soil to the tree canopy, and then back into the forest floor or creeks as the leaves fall late in the year. Rain also causes radioactive material to shift from elevated areas to lower ones, and Japan's abundance of underground water helps to redistribute radiation.[57]

As a result of this specific ecology, even when towns or schools are "de-contaminated," residents may find that radionuclides have remobilized back into these sites once storms, high winds, groundwater movement, and plant life cycles have transported additional cesium down from the forests and mountains surrounding the towns. A 2019 study published in the journal *Soil* points out that while decontamination has reduced the levels of radiation at specific points where decontamination efforts were carried out, "Further research is required to investigate the perennial contribution of radiocesium from forest sources. In addition, the re-cultivation of farm-land after decontamination raises additional questions associated with the fertility of remediated soils and the potential transfer of residual radiocesium to the plants."[58]

The fact that radioactive fallout clouds from high-yield nuclear weapon tests carried these radionuclides into the stratosphere means that their subsequent distribution, as the particles descend atmospheric levels, resulted in their widespread dispersal around the planet, often far from the sites of the originating detonations. Thus, the problem of radiological contamination from nuclear testing is ubiquitous in spite of distance from test sites or the nuclear status of host governments. A 1993 report from the United Nations Scientific Committee on the Effects of Atomic Radiation to the UN General Assembly explains why the slow distribution of these particles from the upper atmosphere did little to mitigate the radiological dangers:

"Since the residence time of particulate debris injected into the stratosphere is of the order of one to a few years, most of the longer-lived radionuclides are deposited without appreciable decay."[59]

The US, USSR, and UK banned atmospheric testing in 1963, while France and China continued to test in the atmosphere as late as 1980. Most of the particles carried high into the stratosphere have by now returned to the Earth. "The fallout from the atmospheric nuclear weapon's tests in the mid-20th century has by now largely been transferred from the atmosphere to the surface of Earth. Since about 70% of the surface of the planet is ocean, much of this mixture of artificial radionuclides has ended up there," point out Hugh Livingston and Pavel Povinec. "Following entry to the ocean, the behavior of these radio-elements was determined by their physical and chemical properties and their fate by oceanic physical and biogeochemical processes."[60] As many of these particles are heavy, many will eventually sink to the bottom of these bodies of water. "Some radionuclides will behave conservatively and stay in the water in soluble form, whereas others will be insoluble or adhere to particles and thus, sooner or later, be transferred to marine sediments."[61] However, this deposition in marine sediments does not conclusively remove the radionuclides from posing a risk to local biota. In 2007, researchers documented "the resuspension of radioactive material from the Irish Sea into the atmosphere," and subsequently into the food chain of human beings in North Wales.[62]

As the radionuclides slowly decay, they continue to cycle through the environment, mobilizing from where they originally deposited, presenting ever-fluctuating and dynamic risks to unsuspecting living creatures for whom they are imperceptible. "There has been some redistribution of the radionuclides in the ocean due to both advection and mixing processes. Waterborne releases from reprocessing in western Europe and global fallout have thus been dispersed to the Arctic Ocean and contamination from the Baltic Sea, due to the Chernobyl accident, has reached the North Atlantic and Arctic Oceans."[63] In 2019, Chinese scientists found ^{14}C from nuclear weapon testing in hadal amphipods (tiny shrimplike crustaceans) at the bottom of the Mariana Trench in the Pacific, the deepest ocean location on Earth.[64]

Since radionuclides cannot be made nonradioactive beyond their natural decay rate, once deposited into the ecosystem they can only be moved from one place to another: as the Canadian physicist Gordon Edwards has remarked, "when you hear the word 'decontaminate' you should think of the word 'distribute.'"[65] As of September 2015, the government of Fukushima Prefecture and the Japanese Environment Ministry calculated that there were over 9.1 million one-cubic-meter plastic bags filled with radioactive soil that had been removed from towns, homes, and schools in Fukushima, then being stored at 114,700 locations. One cannot speak of the "decontamination" of Fukushima without also speaking about the establishment of these 114,700 new nuclear waste storage sites.[66]

Short- and Long-Term Disease Presentation

Exposures to high levels of external radiation can rapidly cause illness and death, sometimes in days or hours. Exposures to lower levels of external radiation can also cause illness and death, but the diseases associated with such exposures typically develop more slowly. The LSS does a good job of tracking and analyzing the relationship between such exposures and later disease presentation, although research such as the recent INWORKS study have provided more nuanced analysis of risks of lower exposures sustained over long periods of time (which is what faces those being encouraged to return to their homes in evacuated villages in Fukushima). Over the more than seven decades that the ABCC and its successor institute, the RERF, have been collecting information, it has determined that there was an increase in solid cancers, leukemias, and also noncancer early mortality among the survivors of the nuclear attacks on Hiroshima and Nagasaki.[67]

There are no comparable studies of the health effects of internalized alpha-emitting radionuclides and beta particles on large populations, and no databases like the LSS to predict the relationships between exposures and health outcomes.[68] While it is possible to extrapolate the external dose of radiation for someone close to a nuclear detonation based on distance and shielding, the specifics of who internalizes radionuclides when they are dispersed into an ecosystem and how many radionuclides an individual internalizes are problematic to determine. Furthermore, the effects are stochastic: "Radiation effects depend not only on the physical properties of emitted radiation, but also on the physiology and biochemistry of the exposed person and the physical and chemical characteristics of the radio-nuclides, which control their deposition, transport, metabolism, excretion, and reuse in the body." These health effects can be quite serious, and include "cancer induction, genetic disease, teratogenesis (induction of developmental abnormalities), and degenerative changes. The most important target tissues for cancer induction are the respiratory tract, bone, liver and the reticuloendothelium system. . . . The human data and the alpha-radiation dosimetry alone are, at present, inadequate to provide direct calculation of cancer-risk coefficients in the radiosensitive organs and tissues."[69]

Different radionuclides present different health risks to those who internalize them, are dependent on the prior health of the organism, the specific radioactivity and chemical toxicity of the particle, the number of particles deposited, and a host of other factors. Since many of these particles deposit into the body through ingestion routes, the health problems they spark may be difficult to specifically attribute to radioactive exposures, unlike the disease legacies of external radiation. Lung cancers produced by smoking or pollutants in the air can be impossible to differentiate from lung cancers produced from inhaled radionuclides. And, unlike significant external exposures, the disease presentation from internalized particles can take longer to manifest. This lack of clear causation facilitates the dismissal of any health impacts at all from internalized particles, even though their toxicity is well established, and has emboldened nuclear apologists to boldly claim things like that fewer than fifty people died because of radiation from Chernobyl.[70]

Radioactive Tracers Worldwide

While the United States government had been aware of fallout from nuclear detonations before the Trinity test, it had been assumed that the radionuclides raised into the atmosphere and then deposited by fallout clouds remained "close-in" to the detonation sites. This presumption was proved fallacious after fallout from the Trinity test was found to have fogged X-ray film produced by the Eastman Kodak company thousands of kilometers from the test site. The X-ray film had been shipped in boxes using strawboard produced at a Kodak-owned mill in Vincennes, Indiana, which had been contaminated by Trinity fallout, revealing that the fallout had spread across the continent.[71] As nuclear testing accelerated, fallout was being found farther from nuclear tests, and concerns over potential thresholds that might endanger significant portions of the global human population found voice. In 1948, David Bradley, a medical doctor and a radiation monitor during Operation Crossroads, published a book about his experiences titled *No Place to Hide*.[72] The book alerted readers to the radiological disaster that followed the Baker test, from which there was "no place to hide." The dangers presented in Bradley's book described radiation that had found its way onto every boat and threatened every person working in Bikini Lagoon. Yet, "Internal radiation exposure was not measured at Operation Crossroads," states a US government study of internalized radiation at the test series, describing how the Defense Nuclear Agency (DNA) "has estimated the exposure from inhaling radioactive materials but used a constant ratio between alpha, beta, and gamma radiation that may have underestimated alpha radiation by a factor of from 5 to 10. Moreover, DNA has not evaluated internal radiation exposure from ingestion and open wounds. DNA believed, incorrectly, a prohibition against food consumption aboard target ships effectively precluded ingestion and did not know how to calculate for open wounds."[73] While the tens of thousands of service personnel were evacuated because of fears about their external exposures on ships that were rapidly becoming more contaminated, testimonies like that of Nuell Paschal illustrate that few precautions had effectively been in place to guard against internalizing particles.

Awareness of the capacity of radionuclides to transport far from the sites of nuclear detonations would unfold in the years that followed Crossroads. After the Soviet Union acquired nuclear weapons in late 1949, nuclear testing by both the Soviet Union and the United States accelerated. While the epidemiological impacts of radiation were being delineated during this time through animal and human experimentation, much more was scientifically understood about the damage that exposures to radiation could cause to genes as a result of Muller's work on radiation mutagenesis.[74] As nuclear testing increased, initial worries about the fallout load being deposited into the environment from nuclear testing focused on the genetic risks they posed to the whole human species.

Project GABRIEL was a study undertaken by the RAND Corporation in 1949 under contract to the AEC, "to evaluate the radiological hazard from the fallout of debris from nuclear weapons detonated in warfare," motivated in part by this concern over the long-term genetic toll that fallout might exact.[75] Fears about the increasing deposition of fallout and the rapidly escalating number of tests led scientists to try to "determine how many atomic weapons could be detonated before radioactive contamination of air water and soil would have long range effect on crops, animals and humans. The AEC created a worldwide network for the collection and measurement of fallout."[76] The development and testing of thermonuclear weapons by both the United States and the Soviet Union substantially increased fallout levels beyond that of fission weapons and carried that fallout higher into the atmosphere, resulting in increasing depositions of fallout in locations far across the globe from the test sites. Official AEC historian Richard G. Hewlett and coauthor Jack M. Holl explained the shift of focus to radiostrontium:

> In the original *Gabriel* studies the principal concern had been the potential toxicity of plutonium disbursed as particles in the radioactive cloud. But since 1950 scientists had become more concerned about the possible effects of strontium-90, which behaved much like calcium in plant and animal chemistry; hence it tended to concentrate in the bone, where, with its twenty-eight year half-life, it could cause bone cancer. Later *Gabriel* studies had used strontium-90 as the critical factor in determining the number of weapon detonations that constituted a radiological hazard. Not until the *Upshot-Knothole* tests in 1953, however, was it evident that strontium could be widely distributed over the northern hemisphere, not only by nuclear war but also by fallout from testing.[77]

It was this realization that led to the large research project known as Project SUNSHINE. As the GABRIEL report explained, "In the summer of 1953 RAND held a short conference of selected consultants to make an over-all review of GABRIEL. The conference recommended that studies then current be supplemented by a world-wide assay of the distribution of Sr–90 from the nuclear detonations which have occurred. This assay has been designated Project SUNSHINE." The new project was to collect a wide range of materials: "Samples for assay have included soil, alfalfa, animals, dairy products, human bones, rain and other water, etc. Samples of one or more of these materials have been obtained from each of some 20 foreign countries. Many of the samples have been obtained through the Department of Agriculture and directly by the participating laboratories; others have been obtained through miscellaneous contacts."[78] At the end of the Cold War these methods of sample collecting would receive harsh criticism since they included obtaining human ashes, bones, and teeth taken from cadavers (including of babies) without the disclosure or consent of the next of kin of those whose remains were collected.[79]

Scientists working on Project SUNSHINE in the 1950s became aware that human beings had all likely amassed traces of radiation in their bodies from the frenetic nuclear testing of the atmospheric period, writing in the original 1953 RAND report that "The release in the world of several kilograms (kg) of Sr^{90} within less than a decade has probably disseminated enough of the contaminant to provide amounts that are probably now detectable in samples of inert and biological materials throughout the world."[80] During the Cold War, the United States government continued to assure Americans that the radiation from these tests did not pose a threat beyond the test site borders. However, "In August of 1958, *Nature* published a brief article by Danish biochemist Herman M. Kalckar. Though scarcely more than a page long, the article's innovative thesis sparked a vast research project that defied official public safety claims with evidence taken from the mouths of St. Louis's child population. Scientists suspected that radiation from the above-ground atomic weapons tests taking place from 1945 onward could pose a significant health risk to the populace, but no one knew how much radiation human bodies were absorbing."[81] In 1958 the Greater St. Louis Citizens Committee for Nuclear Information began its own research in an early iteration of citizen-driven science. They solicited baby teeth that fell out naturally from local residents and dentists to survey levels of strontium 90. The first results of the "baby tooth survey" were published in *Science* in 1961, and showed elevated levels of the radionuclide in the teeth.[82] This inquiry was followed by similar, public studies, of both babies and adults, in the UK and other countries.[83]

The realization that had been dawning on scientists since Operation Crossroads deposited such large amounts of radionuclides into the lagoon of Bikini Atoll, and which was to be investigated in Project SUNSHINE, was that once released into the atmosphere from nuclear testing, radionuclides could actually be traced as they were transported through the global ecosystem, including in biota: "Project *Sunshine* had analyzed the physical and biological behavior of strontium-90 as it traveled from the nuclear fireball through the atmosphere into the soil, up through the food chain, and finally via human metabolism into bone."[84]

An early clue to this system dynamic came after the Baker test of Crossroads. Members of the Radiobiology Division of Joint Task Force One, which administered the two nuclear tests at Operation Crossroads, began to notice unexpected fluctuations in readings of the radioactivity in the lagoon of Bikini Atoll in the days after the underwater Baker test. "The levels of radioactivity in the lagoon were found to be rising at night and falling in daylight hours, a circumstance later realized as attributable to the vertical movement of lagoonal plankton in response to light," wrote Neal Hines of the Applied Fisheries Laboratory (AFL) of the University of Washington.[85] As they began to examine fish in the lagoon, they realized that "Radioactivity was concentrated not in the skin of the fish, which had come into contact with the radioactive water, but in their digestive tracts."[86] This fundamental realization, that there was uptake of the radionuclides into the fish not through their immersion into radioactive water, but through the natural biodynamic process of feeding, suggested to scientists that once deposited into the ecosystem, these particles would not simply rest where they fell, but, as with any chemical, would begin a journey of transport through the ecosystem. This fundamental insight would lead to profound transformations of our understandings of those mechanisms, especially for the study of both atmospheric and oceanic fluid dynamics. As Hines points out, "The addition of radioactivity makes possible broad-gauged studies to trace the movement and concentration of radionuclides in the environment. These studies, in turn, can disclose new information on biological complexes and mechanisms."[87]

In fact, it was this entry of large quantities of radionuclides into the ecosystem that facilitated modern understandings of critical dynamics of atmospheric and oceanic systems.[88] Just as they do when you take a radioactive tracer for medical doctors to see the functioning of systems inside of your body, the entry of the vast load of radionuclides into the air and

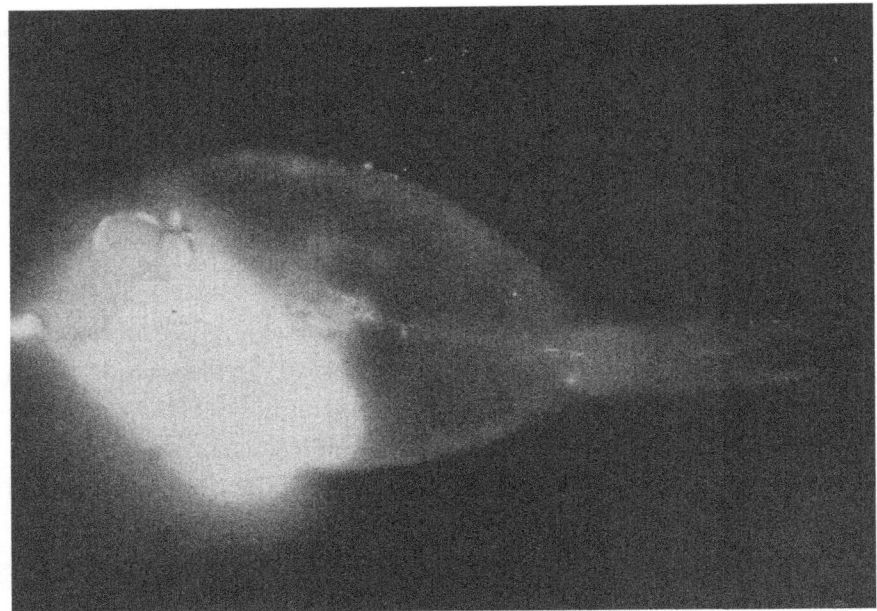

Autoradiograph of a surgeonfish caught in Bikini Lagoon after the Baker nuclear test of Operation Crossroads (1946). Scientists had expected fish swimming the contaminated waters of Bikini Atoll to have radiation on their skin but were shocked to see that the bulk of the radiation the fish received was from radionuclides internalized as they ate. The Office of the Historian Joint Task Force One, *Operation Crossroads: The Official Pictorial Record* (New York: William H. Wise, 1946): 216.

sea provided researchers with radioactive tracers that brought the underlying structure and dynamics of atmospheric and oceanic systems into view. "The presence in the atmosphere of radioactive debris from nuclear tests provided a unique opportunity to study atmospheric motions and transfers of trace substances. . . . The worldwide network of fallout measurements and all the other measurements at the surface, by aircraft and with balloons, set a precedent for the monitoring of other air qualities on a global as well as local scale."[89] Down below that, for oceanographers, "Radionuclides are now found in all of the oceans; they present a potential health hazard and thus are of concern to the people of many nations. However, radionuclides as tags for chemical elements are a valuable tool for the study of biological, chemical and physical processes of the ocean."[90]

Since most nuclear weapon tests were then being conducted in the atmosphere, and an imagined nuclear war would largely involve atmospheric detonations rather than underwater detonations, the movement of these radionuclides through the atmosphere became a key focus of research for AEC scientists. Concerns over the transport and deposition of radionuclides around the world through atmospheric detonations fueled the work done on both Project GABRIEL and Project SUNSHINE. Willard Libby, who was a physical chemist by training, argued that fallout that reached the troposphere would likely remain there and not fall out back to the surface of the Earth, becoming a de facto "reservoir." Historian of science E. Jerry Jessee describes how a group of atmospheric scientists countered this view: "Spearheaded by US Weather Bureau scientist Lester Machta, this group centered their objections largely on newly developed atmospheric circulation theories and firmly grounded empirical data derived, ironically, from the use of radioactive fallout as a tracer to study stratospheric air motions." Jessee recounts that "The model that Machta advanced to account for this new information demonstrated that stratospherically injected radioactive debris was falling faster and less uniformly than Libby's model supposed. . . . As Machta's model of vertical transboundary movement gained favor among AEC and independent scientists, it contributed significantly to the scientific rationale for ending aboveground testing in 1963."[91] This use of "radioactive fallout as a tracer" would prove revolutionary in numerous fields studying environmental mechanisms in the atmosphere, the oceans, the soil, and the biota of the planet following the massive deposition of these radionuclides into the ecosystem during the years of atmospheric nuclear weapon testing.

Swiss scientists asserted that "Anthropogenic radionuclides have been injected into the atmosphere by nuclear weapon tests in the fifties, sixties and seventies, the burn-up of the SNAP-9A satellite, and accidents in nuclear power plants such as Chernobyl (1986) and Fukushima (2011). Radioactive debris became rapidly attached to ambient aerosol particles, which determined the mechanisms governing their atmospheric transport."[92] Japanese scientists describe their utility for inquiries into oceanic mechanisms: "The presence of artificial radionuclides in the ocean surface is due to global fallout from atmospheric nuclear weapons tests and nuclear reactor accidents. The distribution of these radionuclides in the ocean has been investigated in order to understand the transport of materials in the ocean."[93] Similarly, in the study of the chemical and geological dynamics of rivers, "Caesium-137, a radionuclide originating as fallout generated by the atmospheric testing of nuclear weapons, can provide the geomorphologist with a valuable tracer for investigating the movement of sediment through the fluvial system."[94] IAEA scientists found that "soil erosion rates can be estimated using 137Cs, a human-induced radionuclide of caesium released into the atmosphere during nuclear weapon tests more than half a century ago. The 137Cs method for soil erosion assessment effectively provides long-term mean soil redistribution rates, representing the period since its release (mid-1950s) until the time of the 137Cs sampling."[95] The introduction of anthropogenic radionuclides into the ecosystem in large quantities as a result of atmospheric nuclear weapon testing and nuclear accidents, while disastrous for the health and well-being of global biota, facilitated a revolution in the holistic inquiry into dynamic global mechanisms that had previously been difficult to study on a systems-wide basis in numerous scientific fields. Fallout radiation made the underlying dynamics of the global ecosystem visible to human inquiry.

A case in point that has had enormous impacts, in terms of both radiological damage and the capacity of scientists to learn about the movement of those radionuclides through various layers of the ecosystem, was the Chernobyl nuclear disaster in 1986. While the Chernobyl Nuclear Power Plant accident, like the Fukushima nuclear accident that would follow twenty-five years later, deposited an array of radionuclides into the ecosystem, it was cesium 137 that would offer insights into the slow migration of the radiological danger from its initial deposition to the particles' ongoing uptake in multiple nodes of chemical and biological entry. "Forest ecosystems can retain 137Cs for a long time due to continuous cyclic transfer between the upper organic layer, bacteria, micro-fauna, microflora, and vegetation,"[96] explain scientists from the Institute of Radiology, Toxicology and Civil Protection in the Czech Republic. Thus, the initial entry is followed by ongoing chemical and biological processes that facilitate migration and ongoing contaminations of the ecosystem and creatures living therein.

"In forest ecosystems, two stages of radionuclide migration after the accident were identified. In the first stage (2–4 y) contamination of forest stand components (the aboveground parts) was mainly caused by airborne deposition of radioactive aerosols. In the second stage, the radionuclide content in forest compartments was caused by the root uptake of radio-nuclides by plants," explained Russian health physicists in 2007. "Cher-nobyl radionuclides were the source of irradiation of plant and animal communities and, consequently, their radiation damage. On the other hand, when released into the environment and then included into the trophic chains of biological cycling, the radionuclides became the source of contamination of environmental resources (soil, water, agricultural products, forests), the consumption of which, or contact with which, led to irradiation of humans."[97]

Scientists have traced these radiological migrations through the ecosys-tem and upward through the food chain ever since. "The Chernobyl ac-cident showed how a released radioactive or otherwise hazardous plume can, under certain meteorological conditions, rapidly move over long distances," describe Finnish scientists. "At first the emissions were trans-ported north-westwards over Poland, the Baltic States, Finland, Sweden and Norway. During 27 April 1986 emissions were spreading to eastern-central Europe, southern Germany, Italy and Yugoslavia. Radioactivity mapping over Finland between 29 April and 16 May 1986 showed that the ground deposition in Finland covered southern and central parts of the country but had an irregular distribution."[98] In the United Kingdom, "Areas receiving high levels of deposition tended to be upland areas in the west. The grazing of free-ranging sheep is the predominant land use in many of these areas which are characterised by nutrient-poor, highly organic acidic soils. This resulted in radiocaesium in the tissues of sheep that ex-ceeded 1000 Bq kg–1 fresh weight."[99]

These initial depositions were just the beginning of the journey of the ^{137}Cs particles throughout the regional biospheres. Eventually they would work their way further up the food chain.[100] In Scandinavia this would have a devastating impact on reindeer herding and consumption, and in central Europe the cesium would bioaccumulate in wild boar. Scientists in the Czech Republic sought to determine the source of the cesium 137 observed in the bodies of wild boar killed in the republic: "The aim of the study was to find the sources of 137Cs in wild boar food in the natural ecosystem. The main emphasis is focused on the analyses of wild boar muscles and the content of wild boar stomach." Their findings did not bode well for the reduction of these "radioactive boars" in the near future: "Underground mushrooms probably represent the main source of radio-cesium in the food chain of boars. A remarkable reduction of 137Cs specific activities in boar muscles is not expected at the post-Chernobyl radiocesium contaminated locations with the occurrence of *Elaphomyces granulatus* within next [sic] two decades."[101] A key reason for this is that "edible mushrooms harvested from forests . . . are known as accumulators of 137Cs."[102] Radioactive wild boars have similarly been found in significant quantities in Germany and Sweden.

Radioecologists comparing the 137Cs levels in wild boar found in Europe, whose contamination originated from the Chernobyl nuclear accident, with those in Japan, whose contamination originated in the Fukushima nuclear accident, found that while the contamination and numbers of radioactive boars in Japan appeared lower than those in Europe, "The levels of radio-cesium in boar appear to be more persistent than would be indicated by the constantly decreasing 137Cs inventory observed in the soil which points to a food source that is highly retentive to 137Cs contamination or to other radioecological anomalies that are not yet fully understood."[103]

After the Chernobyl accident, which was first detected outside the Soviet Union by radiation monitors at the Forsmark Nuclear Power Plant in Sweden detecting fallout on the shoes of an employee entering the plant, risks to the health of those consuming reindeer meat, and also the reindeer industry in the region, were immediately apparent.[104] "During the summer of 1986 it was realized that the Chernobyl consequences had dramatic effects for reindeer husbandry in central and southern Norway, and trade of reindeer meat from these areas were [sic] banned in autumn 1986. Totally 545 tons of reindeer meat was condemned, and it was realized that reindeer husbandry in the most contaminated areas could be affected for decades."[105] Researchers determined the pathway for radionuclides into the reindeer was lichen, a primary food source: "Following the atmospheric nuclear tests in the 1950s and early 1960s several radioecological research projects, focused on the (sub)arctic food chain lichen–reindeer/caribou–man, were initiated in Scandinavia and North America. Lichen collects deposited radionuclides efficiently and is the main fodder for reindeer during the winter season. The enrichment of radionuclides in this food-chain can lead to exceptionally high body burdens among the indigenous Sami and Inuit (Eskimo) populations consuming large quantities of the meat and edible organs of reindeer and caribou."[106]

Strict guidelines for allowable levels of radiation in slaughtered reindeer remain in Norway, Sweden, and Finland to reduce exposures to humans, as the annual levels fluctuate depending on various local climate and ecological conditions. "Measurements of caesium-137 in reindeer meat reveal large variations between different areas. The radioactive deposition from the Chernobyl accident was unevenly distributed across the country, causing the large differences in caesium-137 levels between different reindeer populations. There can also be large annual variations within the same area, which are caused by local differences in mushroom abundance. Because mushrooms can absorb more radioactive caesium than green plants, reindeer may contain higher levels in years in which there is a lot of mushrooms available."[107] The bioaccumulation of radiation in reindeer and boar will remain problematic for decades in Europe.

The Ubiquity of Anthropogenic Fallout in the Global Ecosystem

Beyond tracing the specific migration of radionuclides in local ecosystems, the entry of quantities of fallout radionuclides into the troposphere, especially thermonuclear weapon testing in the atmosphere, has resulted in the global distribution of many of these particles, as they take years or even decades to fall out of the upper atmosphere and deposit back on Earth. One ironic place we can see the ubiquity of radionuclides in the global ecosystem is through the rigorous temporal requirement on steel used in the construction of whole-body counters. As mentioned previously, whole-body counters are used to test for the presence of individual radionuclides internalized within the human body. Whole-body counters were first developed by Leonidas D. Marinelli in 1950 as a means of analyzing the radiological condition of people who had internalized radium in the 1920s and 1930s.[108] More-comprehensive counters were developed at Oak Ridge National Laboratory in Tennessee in 1964: "A 60-ton steel 'box' with walls five inches thick has arrived in Oak Ridge to become part of an extremely sensitive medical-research device to measure low-level radiation from the human body. . . . The steel chamber was fabricated from steel manufactured before 1945, because steel manufactured after that year contains traces of radioactive fallout that would impair the low-background characteristics of the counter. The concrete housing around the steel box is also formed from material with low-activity constituents."[109] The preferred shielding came from the "Walls of battleship steel 6 to 8 inches thick," but ideally, even 1945 was too late for the manufacture of the desired steel: "Pre-World War Two surplus armor plate became the preferred shielding material. Thick slabs of battleship steel were available after the war at low cost. . . . Old warship armor is used when possible."[110] A source for many of the early counters, and also for steel used in some brands of Geiger counter, is salvaged from the World War I–era German naval fleet scuttled north of Scapa Flow off the coast of Scotland in 1919.[111]

In the last few decades, the ubiquity of fallout radiation throughout the biosphere has aided experts in detecting forgeries in a number of different fields. "Elena Basner, former curator at the Museum of Fine Art in St Petersburg, and subsequently consultant in Russian avant-garde art at Bukowski's auction house, thought of using such isotopes to identify forged paintings purported to have been created before 1945. Her idea was that the isotopes (^{137}Cs) and (^{90}Sr) would have made their way into soil and water throughout the world, and ended up in all post-1945 art, via the natural oils extracted from plants which were used as binding agents in paints." Basner's hunch proved true: "Geochemist Andrey Krusanov and several other researchers of the Russian Academy of Sciences in St Petersburg validated Basner's hypothesis by using mass spectrometry to identify traces of these two isotopes in paintings created in the 1950s, but not in paintings created in the first half of the twentieth century."[112]

The technique also proved useful in detecting fake high-end wines. While tests of the chemical composition of wine bottles was already a technique used to spot fake wines, there remained the problem that old bottles could be filled with new, fake wines. Physicist Philippe Hubert of the University of Bordeaux developed a technique to determine if wine being sold as having been produced prior to 1945 was actually produced later, no matter what bottle contained it and without opening the bottle: "It uses a gamma ray detector to study the levels of radioactive particles in the wine, in this case caesium-137, that have been present in the atmosphere since the era of atomic weapons testing began after World War II. 'The main advantage of this technique is we don't need to open the bottle to do these kinds of measurements,' Professor Hubert relates. 'We just have to put the bottle close to or on top of the detector.' Using bottles donated from the chateaux, Philippe Hubert has built up a record of caesium-137 levels in wine across the second half of the 20th Century. 'In the wine,' he says, 'is the story of the atomic age.'"[113] In 2020, analysis of ^{14}C from fallout began being used in determining the precise age of Scotch whiskys.[114]

The age of deep-sea turtles can be determined because of the ubiquity of globally distributed radioactive fallout: "Bomb radiocarbon dating is widely used for age-validation in marine organisms. Environmental signatures of radiocarbon (14C) shifted abruptly after the proliferation of thermonuclear testing in the mid-twentieth century. . . . Recent applications of bomb 14C dating to fish otoliths (ear stones) in Hawaii are an important precedent in applying this method to sea turtles."[115] Atmospheric testing fallout was also used to determine that the vertebrae rings in whale sharks, the largest fish species, are added annually and not biannually.[116]

The global distribution and ubiquity of transuranics is elemental to our definition of the Anthropocene. Simon Lewis and Mark Maslin of University College London remark on the utility of anthropogenic radionuclides when considering what "signal" to use as a "golden spike" by which to reference other Anthropogenic changes: "This signal is found within lake and ocean sediments, glacier ice, tree rings, cave stalactites and other archives, and is usually very clear."[117] We have efficiently distributed radionuclides from our nuclear technological endeavors around the world and throughout the ecosystem: they have marked our historical moment for millennia.

Conclusion

After twelve years of field research on radiological impacts to small mammals in the Chernobyl Exclusion Zone, biologists Ronald Chesser and Robert Baker realized that the prior assumptions that they had brought to their work had not served them well. Among the presumptions they caution against is the belief that measuring radioactivity in a system can produce certainty about the exposures of individual creatures in that system. Chesser and Baker realized that "Radioactive fallout from the Chernobyl accident was not deposited uniformly around the reactor. Distinct 'excursions,' known as the northern and western traces, carried the ash in plumes across the countryside and through the city of Pripyat, a mere 3 kilometers from the power plant. This produced a mosaic of radioactive habitats that are separated by relatively unaffected areas. Such heterogeneity makes it difficult to evaluate the effects on animal populations because animals from 'clean' habitats might migrate into the contaminated areas."[118] Thus, there are dangers when extrapolating measured radiation levels in ecosystems into assumed dosages for individuals in those ecosystems: "When we entered Chernobyl, most animal studies assumed that individuals from the same location would have similar dose rates and that the rates were proportional to the animal's distance from the source, in this case a damaged reactor. Neither of these assumptions proved to be correct. Mice living in the same habitat vary considerably in how much radiation they are exposed to externally from the soils and the vegetation, and internally from things they have ingested. Our analyses showed that we must examine the internal and external doses for each individual, rather than relying on population averages or the animal's proximity to the reactor."[119]

What Chesser and Baker realized was that no amount of data on external levels of radiation in the Red Forest, or other areas of the Zone of Exclusion, could determine the level of risks that creatures found there faced. The situation was filled with uncertainty, which could only be dispelled by specificity: What is the health risk for this individual? This understanding affords the animals being assessed far more individual integrity than do most studies and discussions of the human beings living downwind from nuclear test sites, nuclear production sites, and nuclear accident sites. For these populations, the asymmetric distribution of risk, and the uncertainty of their exposures—the stochastic nature of the dangers ahead—can be profoundly destabilizing. Referring to those living in the tracks of the fallout clouds from the NTS and the streams of radiological waste from production sites like Hanford, Trisha Pritikin writes, "For the downwinders, life often hinges on the cataclysmic unknown—will my cancer return; what else will happen to me?"[120] Their anxieties being dismissed beneath invocations of the LSS renders them invisible to their doctors, researchers, governments, and our histories of the Cold War.

PART

People

3

Falling Apart Inside

Parenting amid Particles

It was a sunny April day in 2015 in Fukushima City, about 60 km northwest of the Fukushima Daiichi Nuclear Power Plants. Four years after the nuclear meltdowns, residents were becoming accustomed to living with the radiative fallout carried by clouds from the four explosions that had deposited in the region.[1] The removal of soil contaminated above legally allowable levels into plastic bags, which were stacked in fields and along roads in the region, had become routine. For many, there was a hope that a measure of normalcy could be carved out in narrow slices around homes, schools, and shops. Five-year-old Haru was playing with his friends in the small fenced yard attached to his day-care center when he tripped and tumbled, scrapping his knee and drawing blood. It was a minor injury; the day-care attendant disinfected the cut and put a bandage over it. For his mother, Nozomi, this would mark the beginning of an anxious journey. Concerns that Haru may have internalized a radioactive "hot particle" through the cut in his skin when he fell nagged at her.[2] The next day, Nozomi and several other mothers came to the yard at the day-care center with a small doll they had built, roughly the height of Haru. They covered their test child with two-sided tape, stood it up at the spot where Haru had fallen, and knocked it over. They took a Geiger counter and measured the exterior of the tape, especially where the doll's "knee" would be. It did not click above background levels. They stood the doll back up and pushed it over a second time, and again measured the tape with the Geiger counter. They repeated this ritual several times before Nozomi could relax and tell herself that it appeared unlikely there were "hot particles" present at the spot where Haru's skin had torn.[3]

It may be that Nozomi forgets this episode and that Haru will never hear this story about his childhood in later years. It will become part of the fabric of stress that accompanies living in an area with ill-defined but widely distributed radioactive fallout. Haru may never suffer physical illness related to the Fukushima disaster, but he will have lived his early life in an emotional landscape laden with fear, worry, and trauma. His mother will have worked hard to protect him, but uncertainty about the location and scale of the risks from dangerous radionuclides will have braided that vigilance with anxiety about what may unfold in time. Dwelling in a world in which our homes, streets, parks, and schools are littered with migrating, dangerous particles that may attach to our clothes or shoes or sneak into our bodies via our breath or abrasions embeds us into a hostile ecosystem in which there is only a fleeting sense of refuge, comfort, or safety.

In a 2002 study of the psychological distresses that blew into Northern Europe along with the radioactive clouds from Chernobyl, Arnfinn Tønnessen, Bertil Mårdberg, and Lars Weisæth (psychiatrists and psychologists working in both academia and military institutes) describe the dynamics of this stochastic radiological risk: "Individuals will want urgently clear answers to the questions: 'Is the situation dangerous to my family and me, and what should I do about it?' However, there are no satisfactory answers to give at the individual level; one may provide only generalizations. Even for a large population exposed at a particular level there will be uncertainties about the long-term impact." They elaborate:

> Toxic disasters constitute a special class of man-made disasters characterized by the exposure of a large number of people to hazardous substances. A nuclear disaster, such as Chernobyl, differs from other man-made disasters in that it lacks clear boundaries. Rather than being an isolated event, it triggers a sequence of events that will continue to unfold over several years, thereby creating a situation of chronic stress for many people. Such toxic disasters may also lack a clearly defined low point from which improvement may

be judged. . . . An essential feature of ionizing radiation is the uncertainty over its health effects. Higher doses that yield deterministic effects, such as acute radiation syndrome, could be attributed to discrete events; but for the major class of effects inherent in far field radioactive fall-out, namely, stochastic effects, it is impossible to identify individual victims even after the effects have occurred. Anxiety provoking features of stochastic health effects are their unpredictable nature, their long incubation or latency period, and scientific uncertainties about the whole issue.[4]

Exposures to radiation, both externally and internally, are accompanied by spectrums of potential outcomes. How the exposure may later present as illness is varied and uncertain: time may pass and a person may develop one or more of a variety of illnesses, but the relationship between exposure and outcome is not linear. Being free of illnesses for years or decades may give way to the grave and sudden onset of a disease that had been slowly incubating. No period of healthy survival can be assumed to be complete; illness may lurk just ahead, hidden within one's body, a radiogenic time bomb, ticking.[5] Most people who internalize a radionuclide have no idea this has occurred. Dwelling in a space where fallout has deposited "hot particles," some of which may be dangerous longer than a human life span, means risk will endure as people go about their daily lives. In a radiologically contaminated zone, the occurrence of radiogenic illnesses among neighbors would only heighten the sense of risk and fatalism. The uncertainty of living one's life, of raising children in such a landscape of invisible risk, takes a toll even on those who will ultimately avoid becoming sick.

The Chernobyl fallout that deposited in Scandinavia continues to have devastating effects on the indigenous Sámi communities in Norway, Sweden, Finland, and Russia.[6] Lichen, which is a primary food for reindeer in the region, absorbs much of its nutrients directly from the atmosphere and thus concentrates the radiation deposited from Chernobyl's fallout clouds. This has resulted in a significant contamination of the reindeer herds that are the economic anchor of the region and fulcrum of the Sámi diet.[7] The bioaccumulation of radiological contaminants in the reindeer and the lack of alternative sources of protein forced local communities to adopt strategies to minimize the impact of the radiation on humans: "Some families

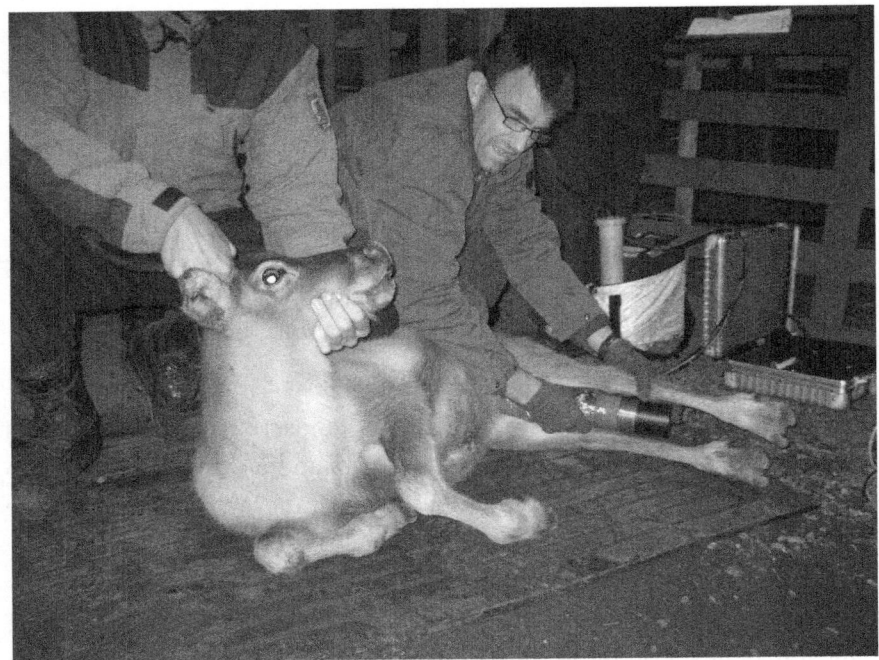

Lavrans Skuterud, a senior researcher at the Norwegian Radiation Protection Authority, is surprised by the high levels of radiation found in reindeer and sheep in 2014. Photo: DSA.

kept three separate food compartments in their freezer sheds—relatively uncontaminated reindeer meat bought from outside for the children, moderately contaminated meat for the middle-aged, and the most radioactive deer for the old people."

The impact of the reindeer's irradiation penetrates deeper than simple changes to diet: "This contamination poses serious threats not only to the health and economic well-being of Sámi people, but to their political strength and to a Sámi cultural identity bound by material and symbolic connections between humans and reindeer." Gerd Persson, a South Sámi reindeer owner in Sweden who was interviewed by anthropologist Sharon Stephens, explained that the community did not simply consume the deer meat: "Our men care for the deer and know them. When deer are slaughtered, it is done with respect. We women know how to care for the meat, to use every bit, the blood, the head, even the feet in soup. We know how

to make thread from sinew and how to prepare the skin and furs for clothing and shoes. The work of our hands puts food on our tables and clothing on our backs."[8] The connection of the people to the animal that
provides the community with sustenance has evolved over millennia and
has served to structure the activities and sensibilities of today's Sámi. The
people can be understood as being sustained by the practices taught by
their ancestors and in turn passed on to their descendants.

Over time, the community began to grapple with the more-abstract
impacts contamination was having on the family structures. "A mother of
two children, Britt-Inger Gustaffson-Blind of Sansa, Norway, recalled the
initial confusion and fear. 'I remember lying in bed, unable to get any kind
of clear picture of what was happening. I began to think of our deer and
who would be affected by their loss—not just my man and my children,
but our old parents and other relatives dependent on people in the reindeer
work. Then the Sámi doing handicrafts and those in the reindeer administration and the teachers at the Sámi school and the people at the Sámi
cultural center. The webs of people I knew just got wider and more tangled,
and my heart grew heavier and heavier.' "[9]

Anthropologist Robert Paine wrote that "The invisibility of 'Chernobyl'
plagues the Snåsa Saami of Norway. 'If only it left an algae behind . . .'
said one housewife," wishing for a sensorial cue to aid in determining radiation risks.[10] Additionally, the differences in individual physiological
response—when two people have the same exposure, one may develop
illness while the other does not—compounds uncertainty for those who
have been exposed. Before disease arrives, and even if disease never arrives,
this uncertainty undergirds the epidemiology of radiological exposure. As
one interviewee told Svetlana Alexievich about the aftermath of Chernobyl,
"Something occurred for which we do not yet have a conceptualization,
or analogies, or experience, something to which our vision and hearing,
even our vocabulary, is not adapted."[11] Stephens talks of how people she
interviewed among the Sámi "frequently expressed to me their sense of a
jarring post-Chernobyl dislocation—of a world split apart into visible realities and invisible dangers."[12] One mother recounted, "I think most
people wanted to believe the official reassurances because the alternatives
were so horrifying. I became pregnant shortly after Chernobyl, and I remember myself at that time as almost having a wall around me. I didn't

want to read about the fallout or talk to people about it because then my fears would have to be specific. If I kept them general enough, I could push them away."[13] She was seeking a sense of control over this risk, and if not, then over her anxieties.

Tønnessen, Mårdberg, and Weisæth caution that a radiological disaster "lacks clear boundaries. Rather than being an isolated event, it triggers a sequence of events that will continue to unfold over several years, thereby creating a situation of chronic stress for many people."[14] Humans cannot physically perceive the presence of radiation, when either receiving a lethal dose of gamma waves or internalizing a particle.[15] This imperceptibility is not unique to radiation; there are many toxins in the modern world whose wound is unfelt, heightening a sense of uncertainty and lack of agency in the face of industrial/technological society generally. Sociologist Kai Erikson points out that radiation encounters "contaminate rather than merely damage; they pollute, befoul, taint, rather than just create wreckage; they can penetrate human tissue indirectly rather than wound the surfaces."[16] For those who live in areas that become contaminated with fallout radiation, the disaster lingers: it is the beginning of a potentially lifelong struggle to step between the raindrops of risk left behind by the passing clouds. Daily life can feel like an obstacle course run through a minefield of particles, ever shifting in their location, stalking our families, hiding in our food. While the eventual disease load that is borne by a community will present in time, the terror of incubating a radiogenic cancer can cripple us day after day regardless of actual disease presentation. This psychological burden can degrade and deteriorate our quality of life. Uncertainty about eating from one's own garden or fields, anxiety about children walking to school, perceptions of nature as hostile and perhaps deadly—all drag life into the dystopic. You may find yourself building small dolls the size of your own children in a desperate attempt to cleanse your anxieties.

Fragmenting Families

Claudia Peterson grew up in southern Utah and then raised her own family in communities that were particularly hard hit by fallout from the nuclear tests at the nearby Nevada Test Site. "Over the years . . . she watched nuclear fallout slowly kill friends and family around her."[17] In

2001, she told a reporter from her local paper, the *Deseret News,* "I was barely holding it together. I had a sick child, my father had just died, my grandparents had died and I had a sister with six kids who was dying too. I was trying to be so together, and inside I was just screaming."[18] After the death of her six-year-old daughter, Bethany, from cancer, Peterson found that her reception in the small town of St. George, Utah, went from compassion to disdain: "I think for [a] long time the community was nice to me and accepted me because I was this crazy woman that lost a child. Everybody in the town went through Bethany's illness, the mayor sent her gifts, I mean everybody knew. The paper did big articles on her, that she was off treatment, doing well, riding her bike, and her hair had grown back. And then she got sick and died. . . . It was like, okay, when is she going to get over this? We're gonna tolerate it for a little while but she's gotta get over this. Well, this isn't something you're going to get over, you've got loved ones, you're losing family, you don't get over it."[19]

Communities downwind from nuclear test sites suffered ongoing radio-logical contamination from nuclear fallout during the years of atmospheric testing (1945–1980).[20] Diane Nielson remembered radioactive fallout from nuclear tests at the NTS that fell all around her childhood home in the hills outside of St. George during the 1950s: "After a bomb, there it would be, the fallout, fine like flour, kind of grayish white. We would play like that was our snow. We never had snow there because it was a warm climate. Then we would go out and write our names in it. It would be thick enough you could write your name in it and see it. It would burn your fingers, it would irritate you, and you would have to wash your hands."[21] In these cases, the scale, density, and immediacy of the radioactive fallout from nuclear weapon tests that had occurred hours earlier brought visible fallout, and some of the harm it did to the body (burns to the skin inflicted by beta particles) was directly experienced as painful. This was only true for communities close in the sites.

Scientists estimate the number of people exposed to fallout radiation from atmospheric nuclear testing downwind from the Semipalatinsk Nu-clear Test Site (known as the Polygon) in Kazakhstan is approximately one million people.[22] "The total yield of nuclear devices tested in atmospheric detonations and conducted at the SNTS is reported to be 6.58 megatons of TNT equivalent, which corresponds to approximately 2/3 of the total estimated yield of Soviet atmospheric bomb tests."[23] In the villages near the Polygon, where fallout clouds from both fission and fusion weapons deposited immense waves of radionuclides, the populations have been left in place to grapple with this radiological and toxic contamination over several generations. These villages are located on the western edge of Siberia, with soil and climate not suited to large-scale agricultural production. Most families have small gardens to produce vegetables, fruit, and grains, and raise horses for both meat and milk.

While residents have advocated for the remediation of their villages and schools, the political entity responsible for this contamination, the Soviet Union, no longer exists. Thus, there is no governmental body that considers itself responsible for this contamination, for medical harm to people living there, or for any cleanup. Those living in the villages near the Polygon find themselves burdened with risks that will last hundreds or even tens of thousands of years, having been inflicted on them by a political entity that has itself already passed. This demonstrates the political, economic, and social disconnect of governments, largely focused on what they perceive as immediate or near-term security threats engaging in actions that have millennia-long consequences. In this case, people find themselves grappling with the legacies as individuals, families, and communities.

Uranium mining is the first waste stream of the nuclear fuel cycle.[24] The Navajo Nation, which contains over two hundred former uranium mines, continues to grapple with the legacy of this mining in the form of ongoing contaminations that will long outlive current members of the tribe. A 2019 study conducted by researchers at the University of New Mexico found that "About a quarter of Navajo women and some infants . . . had high levels of the radioactive metal in their systems, decades after mining for Cold War weaponry ended on their reservation." The study shows that those being born into this toxic legacy are as likely to suffer from it as those who have already lived with the particles for decades: "Among them, 26% had concentrations of uranium that exceeded levels found in the highest 5% of the U.S. population, and newborns with equally high concentrations continued to be exposed to uranium during their first year."[25] Uranium particles, specifically, will remain dangerous for millions of years; untold generations of babies born into this ecosystem will have to grapple with the legacy of our mining.

Downwind from some test sites, people have been forcibly removed from their land after contamination. For many poor communities, this loss of inheritable wealth perpetuates dependency and poverty. It is hard to catalog the less material benefits to family members that can slip away following such a separation; generations provide a deep and organic context to human behavior. Recipes are lost, as the younger generation doesn't eat or cook with their grandparents; stories are lost—the events that shaped a family. When families live in large groups, the range of available perspectives, inter-personal familial roles, and emotional resources are enhanced and expanded. In many societies multigenerational homes are traditional and common. There are tangible daily benefits to such a living arrangement. Elderly grandparents and relatives are cared for and may enjoy a higher quality of life, even after they are bedridden. They have companionship—people who love them to help them through illness, during decreasing capacity, and in grappling with death. Grandparents can provide childcare for working parents and for parents to engage in the household work that supports the entire family. All family members gain from the intimacy and interactions of daily living, adding a depth to the quality of life that would be unavailable if the generations lived separately. Additionally, family resources such as food and money can be better utilized to support more family members. Only a single household need be maintained, food can be cooked in quantities, support can be given for family members in need. Living among those who know and love you increases the quality of life, enhances personal security, and contributes to the collective well-being of the family.

When people are displaced from their homes because of radiological contamination, either forcibly or by choice, some of the support structures of their lives can be rebuilt, but others cannot. What a child loses when he or she can't eat fruit from a tree that was planted by a great-grandparent can't be measured. What's lost when children can't live in a house built by their parents, grandparents, or more-distant relatives is not replaceable simply by occupying a new house. The community is sustained by a web of relationships: to place, to culture, and to ancestors. Speaking at a Havasupai Prayer Gathering of the Havasupai Tribe in Arizona in 2017, Krysta Manakaja described her opposition to the enlargement of uranium mining of the Canyon Mine onto traditional tribal territory: "I am standing here in front of you as an ambassador of the Havasupai, to protect my home land and the waters of the Grand Canyon. Red Butte behind me was our first home, our first land, this was where we first lived. There are a lot of our ancestors buried out here and we are here to protect them."[26] Tony Hood, a Navajo former uranium miner from Gallup, New Mexico, spoke of the sense of loss when contemplating the necessity for his community to abandon their homes because of radiological contamination: "Our umbilical cords are buried here, our children's umbilical cords are buried here. It's like a homing device."[27] Ancestral graves can function like spiritual roots that bind communities to a specific place, a location where the lives of the ancestors created the world in which one dwells. Separation from or abandonment of them may feel disrespectful.

The first people forcibly removed from their homes because of nuclear weapon testing were the people of Bikini Atoll. In preparation for postwar nuclear tests conducted in July 1946, the United States chose Bikini Atoll as the location for these tests.[28] The US held the Marshall Islands as a trust territory under UN auspices for less than a year when it made this choice.[29] The Bikinians were fortunate in that they were not exposed to radiation from the nuclear tests, because they were displaced before the tests, but they lost their homes and their homeland to radioactive contamination. An American newsreel short produced to promote the Crossroads nuclear weapon tests, and shown in American movie theaters in 1946, informed audience members that the Bikinians actually liked leaving their homes: "The islanders are a nomadic group and are well pleased that the Yanks are going to add a little variety to their lives."[30] The Bikinians were made to perform the ritual of willfully giving up their homes to nuclear testing, a process in which they had no choice or control, by reenacting their consent to allow the American military to test this "new weapon" on their atoll in front of US military cameras. Afterward, off camera, they "decorated the community cemetery with flowers and held a ceremony to bid farewell to their ancestors. . . . That afternoon the last of the Bikinians left their atoll aboard a Navy landing craft. A few of them lining the rails sang songs of farewell; some wept. Most were silent."[31]

The Bikini evacuations were to be the first of multiple forced evacuations of the Marshallese because of American nuclear testing. When the United States decided to prepare a second atoll in the Marshall Islands for nuclear testing in 1947, Enewetak, the entire population of 145 individuals was moved to Ujelang Atoll. "Ujelang was chosen because it was uninhabited and geographically close to Enewetak. Located approximately 130 miles to the southwest, Ujelang was much smaller, with one-third the land surface of Enewetak and a tiny lagoon."[32] The people of Enewetak experienced the same lack of agency in their own destiny as had their neighbors on Bikini. "Four days before Christmas, 1947, a U.S. military landing craft drove up on the beach of Aomon island at Enewetak Atoll. Everyone was told to gather their personal belongings and get aboard. One of the translators warned us: 'You cannot protest or fight. You are like a rabbit fish wriggling on the end of a spear. You can struggle all you want, but there is nothing you can do to escape,'" recalls one of the evacuees. The future looked sparse: "Ujelang atoll was uninhabited at the time and no one had been living there for many years. Useless brush had grown up everywhere and there were no breadfruit trees. Navy personnel hurriedly put up tents and gave us some canned food. Then they left. There was no means to communicate with the outside world. We were forced to resign ourselves to the fact that we had to start a new life although we were never told when, if ever, we could return to our homeland."[33] Much like Bikini and Rongelap, Enewetak remains highly radioactive today.

Prior to this, the Marshallese had managed the small amount of land of their atolls in a way that empowered and nurtured most of the community members; travel to nearby atolls was essential for marriage and trade. Anthropologist Jack Tobin wrote an assessment for the National Research Council of the United States on how traditional land tenure had sustained Marshallese culture in a badly resourced and interdependent society for centuries: "The Marshallese system of land tenure provides for all eventualities and takes care of the needs of the members of the Marshallese society. No one need go hungry for lack of land from which to draw food. There are no poor houses or old people's homes in the Marshall Islands. The system provides for all members of the Marshallese society; it is, in effect, its social security."[34] For the people of Bikini Atoll and Enewetak Atoll, this tradition was being fatally assaulted. Even before irradiation, evacuations began to erode community cohesion: "Since the people of Bikini had been evacuated in 1946, the people of Rongelap lost their life companions," writes Satoe Nakahara. "The people of Bikini were related to the people of Rongelap by blood and marriage. Once the people of

Rongelap had no further reason to go to Bikini, they have rapidly lost their navigation skills."[35]

The Bravo test of 1954 at Bikini "completely vaporized five of the atoll's northern islands (a total of about 68 acres of land, or 4% of the pre-test lands). . . . The tests on Enewetak atoll pulverized four islands—about 10% of the land (nearly 200 acres)—and rendered about 60% of the remainder (about 1,120 acres) uninhabitable at least through the year 2026."[36] As Holly Barker has pointed out, "Land is not just a place to live and grow food: land in the Marshall Islands is the essence of life. In a country with only 70 square miles of land, land is the most valued commodity."[37] The radioactive fallout cloud that spread to the northeast from Bikini Atoll, where the weapon was detonated, proved to be catastrophic. "As radiation levels began to fall on March 1, reports of rapidly increasing readings trickled in from the atolls immediately to the east." This included the populated atolls of Rongelap, Ailinginae, and Utirik. Later that day the US Air Force sent amphibious craft to Rongerik to evacuate the 28 service personnel who manned a monitoring post there, but it took until the second day for the navy to arrive at Rongelap and Ailinginae, and the third day to Utirik to evacuate 253 Marshallese, who had spent several days living with the high levels of radiation in the fallout that had descended on their homes.[38] On Rongelap, the 82 people there were told to gather their most precious belongings and to board the USS *Philip* so that they could be temporarily evacuated to Kwajalein Atoll along with a small group of people who lived on nearby Ailinginae Atoll. A second ship, the USS *Renshaw*, evacuated the 154 people living on Utirik. The entire populations of Rongelap and Ailinginae were found to be suffering from radiation sickness.[39]

"After contamination forced the Rongelapese to leave their home islands, the community dispersed to Mejatto, Majuro, Ebeye, and other locations. Involuntary resettlement placed hundreds of people on small plots of rented land, creating extremely dense, unsanitary, and impoverished communities," describe Holly Barker and Barbara Rose Johnston. This dislocation rippled through the structural reinforcements of the community. "The Rongelap community is alienated from the land and their traditional resources. This loss of access affects diet, health, and household economy and severely inhibits the Rongelapese ability to produce or reproduce cultural knowledge about the local environment—knowledge which is

Deadly Bravo test fallout cloud drifting across the northern Marshall Islands, photo taken thirty minutes after detonation (1954), 50 nautical miles from GZ at an altitude of 10,000 ft. Thomas Kunkle and Byron Ristvet, *Castle Bravo: Fifty Years of Legend and Lore: A Guide to Off-Site Radiation Exposures* (Kirtland, NM: Defense Threat Reduction Agency, 2013): 74.

essential to the survival and long-term well-being of the community."[40] The Rongelapese were to experience both abandonment in their contaminated homes and displacement from them. Barker writes that "The testing program denied the Rongelapese use of their lands and resources."[41]

After their displacement the Rongelapese were not allowed to return to their home atoll until June 1957. They would find themselves, like the hibakusha in Hiroshima and Nagasaki, the focus of Cold War medical study. During a 1956 meeting of the Advisory Committee for Biology and Medicine of the Atomic Energy Commission, geneticist H. Bentley Glass commented to fellow committee member radiologist Charles L. Dunham on the return of the Rongelapese to their homes, "This is an ideal situation to make your genetic study. It is far more significant than anything you could ever get out of Hiroshima and Nagasaki."[42] The exposed Ronge-lapese, along with two hundred other Rongelapese not present on the atoll during the Bravo test, were returned to the atoll in June 1957. "Several radiological surveys of the Marshall Islands especially Rongelap Atoll, have been made during the past two and one-half years," wrote the secretary of the AEC in a 1957 report. "The latest survey (July 23–24, 1956) in-dicates a presence of a residual contamination on the island of Rongelap, but at a level that is acceptable from a health point of view, both for the potential external gamma radiation and the strontium-90 content in the food supply, with the possible exception of land crabs."[43] Land crabs, or coconut crabs as they are called locally, are a primary food source for the Marshallese, and especially important during wedding and first-birthday celebrations. The Rongelapese were forced to alter their culture, forgo certain foods, and avoid fishing and growing food on the majority of their landmass (the entire atoll is 21 km^2). "Even though, as pointed out, the radioactive contamination of Rongelap Island is considered perfectly safe for human habitation, the levels of activity are considered higher than in other inhabited locations in the world. The habitation of these people on the island will afford most valuable ecological radiation data on human beings," assessed Project 4.1 scientists. "Since only small amounts of iso-topes are necessary for tracer studies, the various radioisotopes present can be traced from the soil, through the food chain, and into the human being, where the tissue and organ distribution, biological half-lives and excretion rates can be studied."[44] In fact, AEC scientists found that the levels of several dangerous radionuclides that the Rongelapese had internalized, such as ^{137}Cs and ^{90}Sr, had increased after their return to the atoll in 1957, and that "Little of the body burden of the exposed group is apparently due to their initial exposure, since at present there is little difference be-tween the levels of the exposed and un-exposed populations living on Rongelap Island."[45]

Facing decades of limited use of their lands and waters, and increasing illness without adequate access to medical care, the Rongelapese sought to evacuate from the new village built for them, and were finally able to leave with the assistance of the environmental group Greenpeace, whose flagship, the *Rainbow Warrior*, evacuated them to the capital atoll of Majuro in May 1985.[46] "Leaving their home islands has had profound implications on the Rongelapese community. Fragmentation, for example, has made it difficult for the Rongelapese to remain unified as a community. Because people reside in different locations, and because the younger generation grows up in a lifestyle radically different than their parents and grandparents, community unity is a significant challenge for the local leadership."[47] During an evening commemorating the sixtieth anniversary of the Bravo test (and sixty-eight years after the initial displacement), elementary school children from the Bikini community, now also living on Majuro Atoll, sang, "The ashes burned the skin of the people living on these atolls. The bombs destroyed our coral reef, animals, soil, and plants. It poisoned our island, our home. Whenever we sit with our grandparents and they tell us stories about Bikini, we feel sad and embarrassed because we are not in the right place, and we are using somebody else's island."[48]

Over one hundred thousand people faced mandatory evacuation from their homes near the Fukushima Daiichi site after the multiple nuclear disasters following the earthquake on 11 March 2011. "Most of these people moved into emergency evacuation centers while the authorities prepared temporary housing for them. . . . About 53,000 prefabricated housing units, called *kasetsu*, were built in accordance with a national law that covers emergency disaster housing."[49] This averages to two people per *kasetsu*, which provided two small rooms, a kitchen, and a bathroom. Six years later that number remained over thirty-five thousand people, mostly over fifty years old.[50] Many younger families with children moved into apartments at a distance from their previous houses that remained in contaminated areas. Elderly family members were often living separately, many in nursing homes or other care facilities. What had been communities of people living intergenerationally, often for many centuries, had become separated generations unable to provide the kind of in-family care that had mutually benefited family members prior to the meltdowns. According to a 2014 Fukushima prefectural government survey, "Nearly 49 percent of the households that were intact before the accident are no longer under the same roof." The survey cited "housing problems, work requirements and children's educational needs" as primary causes of the generational decoupling.[51] Many *nuclear families* have also separated; concerned about children's ongoing exposures to radiation, mothers and children have fled contaminated zones while fathers stay for work. Kayo Watanabe took her children from Fukushima City and moved 45 km further away from the plants to a small apartment in a welfare building in Yamagata Prefecture. Her husband, who remained behind, visits them when he is able. "My children are getting used to the situation of not having a father around, which is not good," Watanabe commented with resignation. "There are a lot of divorces, too. When families are apart, this happens."[52]

The Shiga family was forced to evacuate its home in the small village of Oburi located about 16 km northwest of the Fukushima Daiichi plants. "The village was famous for its beautiful ceramics. Almost every single house has a pottery workshop. Mr. Shiga is descended from 16 generations of pottery masters." Oburi remains too radioactive for residents to return to their homes. "I hope my ancestors will forgive me for leaving here. It is not my fault I cannot come back. This place was stolen from us. I want to come back, I have a deep attachment to my home, but if I admit that to myself it is too painful, so I try not to think about it."[53] This becomes especially profound when one is no longer able to meet the obligations of care and honor of the graves of ancestors. " 'The remains of Fukushima's deceased evacuees are being left in limbo because radiation is preventing them from being buried. Evacuees don't want to bury the remains of family members in places with high radiation levels,' said the branch's chief priest, Shuho Yokoyama, 76," described a 2017 story in the *Japan Times*. "A 66-year-old resident of Minamisoma visited the temple branch on Aug. 12 for the Bon holidays to pray for her elder sister, who died after evacuating the area. Her remains are kept there because her family's grave is located in a no-go zone in Namie; the remains of her sister's husband, who died before the disaster, are already in the family grave. 'I am sorry that she is separated from her husband. I want their remains to be buried together,' the sister said."[54]

Part of the recovery process in Fukushima is to establish new graves for ancestors that are accessible, at least for short periods, for ritual and personal observances. Obon, the traditional Japanese Bon festival, "is celebrated in either mid-July or mid-August to honor the spirits of deceased ancestors. During this three-day period, family members return to their ancestral homes to make offerings to their ancestors (who are thought to return on those days) and to clean the family grave sites."[55] In the summer of 2019 the city of Futaba, where part of the Fukushima Daiichi plant is located, established a new cemetery to facilitate visits by evacuees to ancestral graves during the Obon holidays. The new cemetery hosts 258 plots, of which 33 were being used during Obon in 2019. "Katsuko Hayashi, 71, came to the site with her husband and daughter on Aug. 12, and conducted a ritual cleaning of a gravestone before offering flowers and prayers. She said she was relieved that her family grave, which was on a planned construction site for storing decontaminated waste temporarily, was relocated to the new cemetery. Hayashi's family home is also on the construction site, where black plastic bags filled with waste are piled high."[56] In 2019, a local Fukushima branch of the Association of Shinto Shrines proposed building a new shine to substitute for those still inaccessible because of radiological contamination.[57]

Severing Ancient Knowledge Chains

When people are dislocated from traditional lands, their losses transcend the material. Community and familial identities can be deeply tied to the historical and mythic origins of specific places. This includes connections to landscape, food sources and flavors, weather patterns, and countless other dynamics that shape local patterns of living and meaning-making. "Songs encode rich knowledge of the social and ecological worlds of Aboriginal people living in the arid interior of the Australian continent, a desert with one of the most variable rainfalls in the world," explained a 2019 study. "People have shaped the ecology of this region in continuous feedback loops over many generations such that there is nowadays a complex system of interdependence between cultural practices and the local ecosystems. Singing traditions are an integral part of the spiritual health of the ecosystem and the means by which biocultural knowledge is carried on over many generations and through shifting social and ecological contexts."[58] The Pitjantjatjara people, who live in the region of South Australia where the British tested nuclear weapons, have lived on their lands for tens of thousands of years. This part of the desert outback is a difficult environment in which to live, with limited wildlife or plants for food and very little water. However, communities have lived and thrived in this area, in part because of the high value of information that is passed from generation to generation via stories and songs:

> Returning to country with *nguraritja,* people belonging to that place, is like walking into the land as a multi-dimensional text . . . ; the marks of the ancestors' footprints are clear to see for those who have memorised the long song sagas that recount the ancestors' activities at sites along their travelling routes. A trained eye notes the subtle signs of the human hand in the clearing of vegetation around sacred sites, stone arrangements, engraved or painted marks on rocks or cave walls. The cultural landscape is not one of constructed temples and monuments, but rather the land itself is imbued with religious significance. The interconnectivity of humans and the sentient land is celebrated in song, story and dance. The land comes alive as the places, food and water sources created by the ancestors are re-energised through caring for *Tjukurpa* in place and spirit. . . . Western Desert peoples lived lightly on the land, their only possessions those that they could carry as they traversed the land seasonally. The desert environment is characterised by low rainfall with cycles of plenty followed by long droughts, cycles of boom and bust. Survival for humans depended on high mobility and knowledge of water and food sources across vast tracts of country.[59]

When the British, and Australians, removed the families of the Maralinga and Oak Valley communities from their traditional lands to use them for nuclear weapon testing, they were also fracturing bonds that extended through both space and time between the families, ancestors, and the ecosystem. These bonds can be rebuilt, but the trauma of the disruption is a violence braided into the radiological violence of the nuclear weapon tests. The ancient web of sustenance provided by the songs was itself attacked. The people who had been displaced from their land were not envisioned by the authorities as being embedded in the landscape or of thriving because of the living memory culture they inherited through mythic narration and song; they were dismissed as primitive. "In the interests of the testing program, it was decided to curtail the movements of those Aboriginal people traversing the Maralinga area. In addition, a number were taken to a reserve which had recently been established at Yalata, some distance to the south, across the transcontinental railway line," writes political scientist Peter Grabosky.[60]

In addition to being removed from their homeland, they were confronted with how the land was then treated: "Equally significant was their awareness that their land and its sacred sites and water-holes were being devastated by enormous explosions; the emotional and psychological stress that this certainly engendered has never been, and can probably never be, properly evaluated." Anthropologist Kingsley Palmer notes the deliberate nature of this neglect: "The potential effects of the bomb tests on the sacred sites and on the socio-religious life of Aborigines was ignored because it was considered irrelevant. According to [Australian anthropologist Adolphus P.] Elkin, in time Aborigines would leave their traditional ways and lifestyle behind, and all of its associations, and would develop into 'modern' men and women with singular and identifiable characteristics."[61] Witnessing the nuclear detonations being conducted on sacred land, modernity appeared barbaric.

When the Marshallese were removed from their home atolls and unable to return, both knowledge and connection to place were lost. "Knowledge of the tide, the climate, the weather, the winds, the stars have accumulated independently on each atoll. Voyages relying on specific stars differ for every atoll. And this knowledge is not shared by all residents of each atoll but basically handed down from generation to generation from one person to one person." This knowledge loss was accompanied by less use and manufacture of canoes. These simple and powerful tools of transportation and food production marked a significant disruption of cultural and interpersonal bonds, as they were the primary means of visiting family on distant atolls and arranging marriages.[62] Musicologist Jessica Schwartz analyzes Marshallese songs and points out how the relationship of the community to land and ancestral space, as well as historical violence, is embedded into the lyrical structure of many traditional and modern songs. She describes the lyrics of the song "Ailiñ eo in" (This is our atoll), which was "composed as a celebratory gesture in anticipation of the Rongelapese visit to their homeland around 2004–5." Schwartz explains that "The abstracted territory is the core of the song, and the spatial instability of an abstract territory is reaffirmed in the final verse: 'Where are you now? Where am I now? Between death and life, phase one, and phase two, they are still in front of us.' The phrase 'between death and life' (ikotaan mej im mour) references 'Ioon, ioon miadi kan,' and it refers to both the people and the land as being between death and life. The land and the people, as between death and life, are the relational space themselves."[63]

Mary X. Mitchell comments on how the people of Enewetak explained to US military commanders that seeing nuclear detonation craters on their land was equivalent to a medical patient awakening to find that doctors had amputated a limb without their knowledge or consent: "They explained, 'In Enewetakese culture man and his atoll environment—land, lagoon, sea and sky—are one integrated whole."[64] Nuclear disruptions rippled throughout Pacific society: "I think that the issue of nuclear testing here has accelerated that rejection of our traditional medicine," explained Roland Oldham, the leader of the French Polynesian hibakusha group Moruroa e Tatou, reflecting on how traditional practices and wellness were slipping away, "especially when you get told, you have to take this tablet or you're gonna die. Today, we have to go through certificate doctors. If your grandmother is a healer, no good. Even if she has a cure, it's just no good. You don't exist."[65]

Those Who Are Counted and Those Who Don't Count

When researchers and policymakers assess radiological events, they typically frame the impact on human beings by counting cancers and deaths. We may argue about those numbers, but these are the units of our assessment. How many were killed in Chernobyl? It is generally agreed that around thirty-one people died during the accident itself and from radiation received during the attempts to put out the fire in late April and early May of 1986. The 2008 report by the United Nations Scientific Committee on the Effects of Atomic Radiation (UNSCEAR) concludes that the deaths of twenty-eight plant workers can be directly attributed to radiation received from the accident in the years that followed, making roughly sixty deaths among those at the plant at the time of the accident and the "liquidators" who worked on the site in the immediate aftermath. It also explains that six thousand thyroid cancer cases can be attributed to Chernobyl, with fifteen deaths resulting from these cancers by 2005.[66] Other studies and reports present different numbers:

> In its 2005/06 assessment "Chernobyl's Legacy: Health, Environmental and Socio-Economic Impacts" the *World Health Organisation* (WHO) estimated that the total number of long-term deaths will be around 4,000. However, this figure is related only to the proximate populations of Ukraine, Russia and Belarus which were exposed to high radiation levels; if extended to estimates of those exposed to low-level radiation across the region, this number rises to 9,000. Other studies have suggested higher figures. A study in the *International Journal of Cancer* by Cardis et al. (2006) estimates a total of 16,000 deaths across Europe. . . . Radiation scientists Fairlie and Sumner provide some of highest estimates, predicting between 30,000–60,000 deaths.[67]

Each of these studies lists a different number of thyroid cancers and may or may not list numbers for other solid cancers and leukemia. Always, the impact of the Chernobyl disaster is described in a statistical presentation of mortality and cancer data sets.

There is a nod in the official studies to the psychological tolls of the disaster: "The Chernobyl accident is known to have had major effects that are not related to the radiation dose," explains the UNSCEAR report. "They include effects brought on by anxiety about the future and distress, and any resulting changes in diet, smoking habits, alcohol consumption and other lifestyle factors, and are essentially unrelated to any actual radiation exposure."[68] Additionally, the WHO report advises that "Mostly these mental health consequences in the general population were subclinical and did not reach the level of criteria for a psychiatric disorder. Nevertheless, these subclinical symptoms had important consequences for health behavior, specifically medical care utilization and adherence to safety advisories. To some extent, these symptoms were driven by the belief that their health was adversely affected by the disaster and the fact that they were diagnosed by a physician with a 'Chernobyl-related health problem.' "[69] In both cases we can see that the WHO report titled "Mental, Psychological and Central Nervous Effects" can be reduced to specific categories of self-harming behaviors attributable to an irrational response by people to the actual epidemiological risks they faced.

This footnoting of the emotional and psychological effects of the Chernobyl disaster on those living in communities blanketed with radiological fallout would become formalized in a cruel and infantilizing description of them as suffering from *radiophobia*. In her brilliant article on the politics around radiophobia, anthropologist Magdalena E. Stawkowski describes that "Radiophobia was first recognised by science experts and industry specialists after the 1986 Chernobyl disaster in Ukraine and used to describe public reaction considered out of proportion to the real risk of the accident."[70] Radiophobia—defined as an irrational fear of radiation—is widely used to dismiss and discredit the anxiety that follows exposure to radiation or proximity to a radiological event. With the caveat that their psychological suffering was "subclinical," it was nevertheless assertively diagnosed (most often by nonclinicians). Speaking at an International Symposium on Radiation Education held in Nagasaki in 2004, physical chemist Klaus Becker described radiophobia as a "mental disorder" that could be "cured."[71] Yehoshua Socol, a physicist who has written enthusiastically about radiation hormesis, declared about Chernobyl that "About 4,000,000 people living in the 'contaminated' areas were officially declared victims. . . . And after being declared thus they became very real victims. Radiophobia—irrational fear of even small radiation doses—led to extremely traumatic decisions and results."[72] Risk consultant David Ropeik advises that "our excessive fear of radiation—our radiophobia—does more harm to public health than ionising radiation itself."[73] However, psychiatrist Michael Edwards, writing about invocations of radiophobia after Fukushima, points out that "In reality, the psychological trauma actually observed is widespread, making the idea of a pathological response inappropriate."[74]

Throughout the history of the nuclear age, much derision has been placed upon those who fear radiation, a fear declared by nuclear industry promoters and technocrats as definitionally irrational.[75] Radiation risk advisor to Fukushima Prefecture Shunichi Yamashita notoriously commented that "The effects of radiation do not come to people that are happy and laughing. They come to people that are weak-spirited, that brood and fret," just ten days after the disaster began.[76] The Fukushima mothers who worked to ensure food safety both in the home and in school lunches that Aya Hirata Kimura writes about were frequently described as hysterical and referred to dismissively as "radiation brain moms."[77] Yet radiation actually is dangerous, and its health effects are stochastic. How is it irrational to fear something that can cause life-threatening diseases and is imperceptible to human senses? When people fear an oncoming hurricane, the onset of military conflict, or contracting COVID-19, some of what they fear may be clearly delineated, but much of their fear is about the potential for harm and the unpredictability of the outcomes. Anxiety about natural disasters, war, and pandemics are understood as natural responses, yet the fear of radiological contamination is pathologized. To expect a calm, measured, and calculated response to something dangerous, imperceptible, asymmetrically distributed, and long-lived is itself irrational. Derision of this anxiety is largely based on the acceptance of the risks to health from radiation articulated by the LSS. As we have seen in the previous chapters, the LSS does not account for the risks to health from internalizing radionuclides, which is precisely the danger faced by most dismissed as "suffering" from radiophobia.[78]

Ross Pastel of the Uniformed Services University of the Health Sciences explained in 2002 that "The label of 'radiophobia' has been used widely in the former Soviet Union to denote a fear of radiation, usually with political overtones and with a pejorative manner. Radiation is invisible, so in the absence of physical dosimetry or biodosimetry, exposure can be uncertain. . . . Given these uncertainties, it is impossible for someone to know if their fear is excessive or unreasonable." Pastel cautions that "Although the psychological distress was originally labeled radiophobia, the term is a misnomer—radiophobia is not a clinical phobia. Still, it is important to remember that both the reported physical symptoms and the distress are real. Although the physical symptoms may be due in part or in whole to psychological effects (e.g., depression, anxiety disorders, somatization disorders, or PTSD), the pain is real and not 'in their heads.'"[79] Even though the study cited in the WHO report that dismissed psychological legacies of Chernobyl as "subcritical," the authors take the suffering of the victims seriously: "Chernobyl is thus a stress factor oriented toward the future, and in this it is different from most other stress factors. For people who have experienced an earthquake or fire or military actions, the stress factor assumes the form of reminiscences about the past. The life of people living in a contaminated area becomes quite different: it is a 'life under stress' without limit. Even if a person leaves the danger area, the danger represented by any harm he has suffered remains with him."[80]

The term "radiophobia" was fraught from the start. Adolph Kharash, the science director of Moscow State University, was meeting with evacuees from Pripyat in the Troyeshchina neighborhood of Kyiv in November 1986, just seven months after the Chernobyl disaster. He explains that the first thing that the group told him was, "Don't promote this terrible word radiophobia!" Kharash related that "For those who were at the epicenter of the Chernobyl cataclysm this word is a grievous insult. It treats the normal impulse to self-protection, natural to everything living, your moral suffering, your anguish and your concern about the fate of your children, relatives and friends, and your own physical suffering and sickness as a result of delirium, of pathological perversion."[81]

The pejorative nature of the use of the term "radiophobia" to dismiss the anxieties and uncertainties faced by those exposed to radiation or living in contaminated zones after Chernobyl was quickly revived by Western journalists and analysts to dismiss the fears of those exposed by the Fukushima nuclear disaster. A 2013 article in *Nature* quotes neuropsychiatrist Hirooki Yabe of Fukushima Medical University saying, " 'radiophobia' remains a major problem among the Japanese refugees." The article advises that experts predict minimal if any medical consequences from the radiation contaminating Fukushima Prefecture, yet "uncertainty, isolation and fears about radioactivity's invisible threat are jeopardizing the mental health of the 210,000 residents who fled from the nuclear disaster." Part of the blame, according to Yabe, is on the community members themselves, who are slow to accept mental health support because "Northern people are a very closed people."[82]

A 2018 article in the *Japan Times* explains that *fūhyōhigai,* or "harmful rumors," remain the primary culprit in public anxieties related to the Fukushima accident: "Groundless fears about radiation may be becoming even more difficult to overcome now after seven years, especially as memories of the disaster are beginning to fade, an expert said." The expert cited,

Nobuaki Yoshizawa, is a physicist and the research director at the Mitsubi-shi Research Institute (MRI), both a think tank and also a consulting firm for the Mitsubishi Group, a global manufacturer of nuclear power plants. "Although recovery efforts are progressing, many people outside the crisis-hit zone tend to fear the risks of radiation based on information from years ago, because now fewer people follow up-to-date information about Fu-kushima's reconstruction."[83]

Looking back on the Fukushima disaster, we are already counting can-cers. There is debate about whether the abnormally high rate of childhood thyroid cancer is reflective of the radiological burden suffered by children, or a statistical outgrowth of increased testing.[84] We will add to and debate the changes in solid cancers and leukemias reported over the next decades. These assessments and their causation will be hotly debated. What are the *real* numbers? This debate will mirror the debates that still rage about the true impact of the Chernobyl disaster. Much as the millions of people whose lives and communities were disrupted by the massive and widespread distribution of radionuclides have been dismissed, or blamed for their woes, the hundreds of thousands of people who were displaced from their homes by the fallout clouds of the triple nuclear meltdown and numerous explo-sions of the Fukushima disaster will suffer below the din of the counting of cancers and will be largely out of sight of global attention. These hun-dreds of thousands of people's lives will never be the same, and their com-munities are unlikely to revive. The evacuees will be forgotten. It's part of a continuum of not seeing how survivors not yet presenting disease suffer, and how community disruption degrades well-being.

In Japan, evacuees have long been told by their government that current radiation levels in previously off-limits towns are now "acceptable" and that they should return. Reiko Hasegawa, an expert on internally displaced people and forced migrations, notes that "In the aftermath of the Fuku-shima accident, as early as 19 April 2011, the government raised the dose limit for public exposure to radiation from 1 mSv/year to 20 mSv/year. Accordingly, the authorities, including the Fukushima prefectural govern-ment and several affected municipalities, began to emphasise that it was safe to return and live in areas with an annual radiation dose of less than 20 mSv."[85] As Hasegawa points out, prior to 3/11 the dose limit for pub-lic exposure was 1 mSv/year, while the dose limit for an adult male nuclear industry worker was 20 mSv/year. Eight days after the disaster began, the dose limit for the public—including children—was raised to what had previously only been legally allowed for adults directly employed at nuclear power plants. This new measure, rapidly instituted since it was no longer possible to keep the exposures of those in the path of the fallout clouds below 1 mSv/year (and therefore liability had to be eliminated), would now be the benchmark for the "safe" return of civilians to their former homes.[86] What had previously been illegal was now a public safety benchmark: what had been unsafe was now the definition of safety.

Much like the roughly 25 percent of habitable land on Rongelap that was remediated and rebuilt for returning residents, "About 40% of Okuma, which sits immediately west of the plant, was declared safe for residents to make a permanent return after decontamination efforts significantly reduced radiation levels" in 2019. However, "most of Okuma . . . remains off-limits due to high radiation levels," and part of the city is "being used as an interim storage site for millions of cubic metres of toxic soil."[87] Returning to their homes will be considered the closing of the circle that began with their evacuation: things will have returned to normal, or at least a normal embodied by limited habitability and including the establishment of a nuclear waste site within the town limits. Those who don't return will fade from public memory. Our historical representations and political arguments will focus strictly on those who develop illnesses. After nuclear disasters, we squint our eyes so the visible victims are reduced to a small group that present disease or mortality with clear causation. We either dismiss or pathologize the emotional and community disruptions suffered by those who don't develop cancer or die. The affected become a group so small that it can make ecological disasters seem manageable. How many died at Chernobyl?

Conclusion

The millennia-long relationship of the Sámi people of Scandinavia to their food-source animal was ruptured by clouds that deposited radioactive fallout on their communities from a fire burning 1500 km away. Gerd Persson, a Sámi woman quoted earlier in this chapter, gave a vivid description of the depths and subtlety of the interpersonal human bonds that emerge out of the relationship of the community members to the deer: "There is a South Sámi practice where the deer are calmed after roundup and before the slaughter. They are brought into the separation corral and at first they run, run in a circle inside the fence. We have a name for the smell of the dust they kick up, mixed with their breath and sweat. This is called *dremschie*. The best thing I know is when my man comes home from the deer smelling of *dremschie* in the cold autumn air. This is what I call love—it is a sense and feeling of connection that moves out from the place and the deer to us and our children. What will happen if those connections are no longer real? How can my children learn what it means to be Sámi if the reindeer work is only pretending?"[88]

Reducing people and cultures to data points that can be displaced as authorities dictate, whose communities—embedded in traditional cultural and ecosystems—are separable from their health outcomes and subordinate to state liability management, is structural violence. Psychiatrist Michael Edwards writes, "the psychological trauma becomes one of *what may happen,* together with distressing background emotions such as fear in the atmosphere of heightened environmental stress. The size of the psychological response would seem to bear little relation to that of the (mathematically) predicted radiation-related physical harm."[89] Treating people as though they should adhere to clinical assessments that consider only their statistical risk for cancer rather than honoring the emotions that accompany displacement, loss of community, and toxic exposures is dehumanizing. Exposure, not sickness, is the threshold for determining effect.

4

Cloaking Contamination

Managing People and Discourse after Radiological Disasters

Even before the first nuclear test in July 1945, the control of public perceptions about radiation risks was structured into management of the weaponry. *New York Times* science reporter William Laurence was brought into the Manhattan Project secretly by Leslie Groves, the military commander of the Manhattan Project, to manage public relations about the new weapons and to set the tone for press coverage of the nuclear attacks set to be conducted in August of 1945. "One of Groves' first assignments for Laurence in mid-1945 was to prepare news releases that would disguise the detonation and resulting radiation of the first atomic bomb in history," the Trinity test on 16 July 1945. "Groves instructed Laurence to write four press releases covering a spectrum of four eventualities that ranged from the least worrisome case (a loud explosion with no property damage or loss of life) to the worst case (a mammoth explosion causing widespread destruction of property and loss of life)." Retaining a prominent *New York Times* reporter to work as a public relations professional inside of the Manhattan Project allowed Groves to control perceptions about the radiation resultant from Trinity: "Laurence's assignment at Trinity's ground zero permitted him to observe the elaborate plans Groves and his staff had made to measure, track and provide safeguards against radioactive fallout—but these observations were virtually omitted in his exclusive newspaper stories."[1] When Laurence did publish about his experiences at the Trinity test in September, he referred to the blast as a "new birth in freedom" and neglected to mention the teams dispersed around the region to measure the resultant radioactive fallout, the fact that one family received a high dose, or the cattle burned by beta particles.[2]

The dual role of prominent *New York Times* science reporter and inside public relations manager positioned Laurence to publish numerous scoops on one of the most important news stories of the twentieth century: Laurence witnessed the Trinity test, interviewed the *Enola Gay* crew immediately after their return to Tinian from the attack, and was on board *The Great Artiste,* one of the three planes that conducted the nuclear attack on Nagasaki. However, just as during preparations for the Trinity test, Laurence used his front-row seat to generate authoritative narratives about issues that were difficult to understand, framed as the US military preferred. Australian journalist Wilfred Burchett became the first Western reporter to reach Hiroshima after the attack and published a landmark article in the *London Daily Mail* on 5 September 1945, titled "The Atomic Plague." Burchett described the nature and scale of radiation sickness following the nuclear attack: "Many people had suffered only a slight cut from a falling splinter of brick or steel. They should have recovered quickly. But they did not. They developed an acute sickness. Their gums began to bleed. And then they vomited blood. And finally they died. All these phenomena, they told me, were due to the radio-activity released by the atomic bomb's explosion of the uranium atom."[3] Concerned that people would grasp the ongoing health effects of radiation, Groves took steps to counter the story. Journalists Amy Goodman and David Goodman describe how Groves organized a press tour to the Trinity site four days later for select members of the press: "Groves took the reporters to the site of the first atomic test. His intent was to demonstrate that no atomic radiation lingered at the site. Groves trusted Laurence to convey the military's line; the general was not disappointed."[4] On 12 September 1945 Laurence published a story on the front page of the *New York Times* with the title "U.S. Atom Bomb Site Belies Tokyo Tales: Tests on New Mexico Range Confirm That Blast, and Not Radiation, Took Toll." Laurence's story proclaimed that the visit to Trinity's ground zero "gave the most effective answer today to Japanese propaganda that radiations were responsible for deaths even after the day

of the explosion, Aug. 6, and that persons entering Hiroshima had con-
tracted mysterious maladies due to persistent radioactivity."[5] Laurence won
the Pulitzer Prize for Reporting in 1946 for his series of nuclear stories.
In 2005, Amy and David Goodman called for the revocation of Laurence's
Pulitzer. In an op-ed published in the *Baltimore Sun* they declared, "Mr.
Laurence won a Pulitzer Prize for his reporting on the atomic bomb, and
his faithful parroting of the government line was crucial in launching a
half-century of silence about the deadly lingering effects of the bomb. It
is time for the Pulitzer board to strip Hiroshima's apologist and his newspaper of this undeserved prize."[6]

The US government effectively limited understandings of radioactive
fallout in the press for almost ten years until the *Daigo Fukuryu Maru*
pulled into Yaizu port in March 1954 after the Bravo test. Ongoing propaganda minimizing "residual radiation" became part of the management
strategies for all nations developing nuclear technologies. Born in the
Manhattan Project and the early nuclear weapon programs of the nuclear
armed nations, secrecy and the management of public perceptions and
understandings of the health effects of radiation—specifically of long-lived
radionuclides in the environment—would later be embraced and taken as
operational gospel throughout the nuclear industry, both inside the weapon
complexes and in the companies that produced nuclear-generated electricity worldwide.

As the American public became increasingly aware of radioactive fallout
being deposited downwind after nuclear weapon tests at the Nevada Test
Site in the 1950s, the US government took specific steps to manage the
perceptions of Americans.[7] On 19 May 1953, a nuclear test named Harry
(and known within both the nuclear weapon complex and the downwind
community as "Dirty Harry") at the NTS blanketed St. George, Utah,
with radioactive fallout. This would become the only time that the AEC
advised Americans to take shelter from a fallout cloud. The event stirred
a great deal of anxiety in the downwind communities.[8] The AEC hesitated
to continue with the Upshot-Knothole testing series (of which Harry was
a part), but US President Dwight Eisenhower authorized the tests to
continue. Nine days later, on 28 May, Eisenhower told AEC chairman
Gordon Dean that in its press releases to the public the AEC should generally avoid the use of words like "thermonuclear" and keep the public

confused about the differences between "fission" and "fusion."[9] This was not to inform or protect the public, but to facilitate the unimpeded operation of nuclear weapon testing. Frank Butrico, a Public Health Service officer working as an off-site radiation monitor in the St. George area on the day of the Harry shot, later testified in a court case that after the incident, he overheard a high-ranking AEC official at the Nevada Test Site convey, "Let's try to get this thing quieted down a little bit because if we don't, then it's likely that there may be some suggestions made for curtailing the test program. And this, in the interests of national defense, we cannot do."[10] Historians of the US Atomic Energy Commission Richard Hewlett and Jack Holl describe how AEC chairman Lewis Strauss's views "reflected the Commission's consensus that fallout was not a matter of health or science but rather a public relations problem."[11]

Many people living downwind from the test site worked outdoors in agriculture and ranching. "Of the approximately 11,710 sheep that had wintered within 40 miles north and 160 miles east of the Nevada Test Site in 1953, 1420 lambing ewes and 2970 new lambs would ultimately succumb to a painful and mysterious death in the ensuing year," writes Jim Kichas, an archivist at the Utah Archives and State Records office. Kichas describes how a "team of USPHS and AEC investigators observed sick lambs in Cedar City and initially concluded that radiation and malnutrition were the most likely candidates for problems afflicting the herds." However, "These reports were immediately classified by the AEC and not provided to Cedar City sheep owners or local Iron County authorities."[12] The ranchers were not compensated for the loss of their herds from the radiation deposited by the Upshot-Knothole tests at the NTS that spring. "The losses devastated Cedar City's sheep industry. Ranchers took out loans, sold off land, and declared bankruptcy. Many eventually withdrew from the business," writes historian Sarah Fox; "at the time it proved a devastating blow—not only to the local economy but also to the spirit of a community whose identity was bound up in its ranching industry."[13]

While there was active and ongoing public relations management of populations at a domestic nuclear test site like the NTS, in colonial and postcolonial test sites there were more-blanket dismissals of local concerns over radioactive fallout. At the start of French nuclear testing in French Polynesia, officials at the Centre d'expérimentation du Pacifique, which conducted the nuclear tests, told members of the local Territorial Assembly that it would send inspectors to measure radiation levels on atolls and islands near the Moruroa test site, and would test food in local communities; however, this was impossible as "the French National Radiation Laboratory was forbidden to send any experts to French Polynesia."[14] This lack of actual research data allowed the French to hide the scale of radioactive fallout and the exposure of the people of the Southern Pacific for decades. In documents related to the years of nuclear testing, declassified by the French government in 2013, a pattern of deception in warning the public about incidents of dangerous levels of radioactive fallout can be seen; for example, a July 1974 test "exposed Tahiti to 500 times the maximum allowed level of plutonium fallout."[15] The local government, which still operates today under a semicolonial status of the French government, was complicit in this public deception. President of French Polynesia Edouard Fritch remarked to journalists in 2018, "For 30 years we lied to this people that these tests were clean. It was us who lied and I was a member of this gang!"[16] This official admission came four years after the French newspaper *Le Parisien* had unearthed a classified map showing that, counter to similar French deceptions at the time, the 1960 Gerboise Bleue test in Algeria, of a weapon that at 70 kt was more than four times larger than the weapon used in Hiroshima, "had spread fallout as far as Chad and across Southern Europe."[17]

As thermonuclear weapon testing created massive fallout clouds depositing radiation far from nuclear test sites, even globally, efforts to manage public perceptions about the dangers of fallout became a matter of global public relations on the part of the nuclear weapon states. A few hours after the Bravo thermonuclear weapon detonated on 1 March 1954, the mushroom cloud from the test was drifting to the northeast from Bikini Atoll. About 100 km away, the crew of a Japanese fishing boat, the *Daigo Fukuryu Maru,* noticed the sky above them darkening. "It was as though a high fog were forming. Then a light rain or drizzle started to fall," wrote physicist and former assistant director of the Metallurgical Laboratory in the Manhattan Project Ralph Lapp three years later in 1957. "They were puzzled at first when tiny bits of sandy ash came swirling down onto the decks. 'It looks like the beginning of a snowstorm,' one of them said. The men kept working, paying little attention to the unusual rain. But it became bothersome, and the men blinked as the irritating grains of whitish sand got in their eyes."[18] This rain contained ash from the Bravo explosion that was laden with radioactive particles. It blanketed the crew as either gritty rain or simple ash for three hours. Crewmember Matashichi Ōishi recalls, "Although our ship was all of a hundred miles from the point of the explosion, the pure-white 'death ashes' fell like snow; our feet left footprints on the deck."[19]

When the ship docked at Yaizu in Japan on 14 March, the entire crew was hospitalized and diagnosed with radiation sickness, its twenty-eight thousand pounds of tuna having already been taken to market.[20] Newspapers in Japan, and then around the world, reported the story of the contaminated and sickened crew, and the word "fallout" entered into the global vocabulary.[21] Six months after the *Daigo Fukuryu Maru* crew was hospitalized, radio operator Aikichi Kuboyama died of medical complications from his exposure, demonstrating that you can be over 100 km away from the detonation of a thermonuclear weapon and be killed by its effects. "It has been revealed recently that this incident was much larger in scope and much worse than what was publicly revealed at the time," wrote biochemist Eiichiro Ochiai in 2014 about the radiological disaster that accompanied the Bravo test. "More than 900 fishing vessels were exposed to the fallout, and they had to dispose of their catch. Two hundred and forty one fishermen were exposed to radiation, and 77 died by May 1988; 61 of those 77 died of cancers."[22]

After the world first got wind of the radiological disaster caused by the fallout from the Bravo test, the United States government tried desperately to discount the story and then to shift the blame. AEC chairman Strauss told the president's press secretary James C. Hagerty that the ship was not really a fishing boat, but a "red spy outfit." Strauss asserted that the twenty-two-year-old captain, Hisakichi Tsutsui, was too young to have sufficient experience to captain a boat and that his presence was likely part of a "Russian espionage system." He also discounted the presence of radioactive fish, saying it could be a brilliant Soviet tactic of spreading radioactive fish around the oceans of the world to sow panic.[23] Although the *Daigo Fukuryu Maru* was outside of the fifty-mile safety zone established around Bikini Atoll, Strauss claimed in a press conference on 31 March that the ship was clearly "well within the danger area."[24]

To counter continuing anxiety over thermonuclear testing, three years later, at a 26 June 1957 press conference, American president Dwight Eisenhower opined that in four or five years the United States would perfect "an absolutely clean" hydrogen bomb that would reduce radioactive fallout by 96 percent. Eisenhower's description was based primarily on the Cherokee shot of the Redwing nuclear test series conducted at Bikini Atoll a month earlier on 21 May. This first air-dropped test of a thermonuclear weapon was detonated at a sufficiently high altitude that the fireball of the detonation did not touch the ground, which reduced the scale of its fallout. Eisenhower was enamored by promises made to him by weapon scientists who wanted to stifle the growing clamor to halt weapon tests because of public anxieties over fallout. Eisenhower shared his fantasy with the press of a nuclear war in which both the United States and the Soviet Union used only "clean bombs," so that the war would be a purely military affair, and "there will be no fallout to injure any civilian or any innocent bystanders."[25] A week later, Eisenhower expanded his discourse and spoke of how thermonuclear weapons could be used for geoforming the Earth, such as "moving mountains," and repeated that weapon scientists had reduced the radioactive fallout of weapons 90 percent and were working to reduce them to 95 percent or even 96 percent.[26] Nevertheless, governments around the world would soon be issuing advisories and pamphlets to their populations instructing them how to protect themselves from fallout in the case of nuclear war. What had been called a "bomb shelter" in the first decade of Cold War America became a "fallout shelter" in the second decade.

Many of the early radiological disasters involving nuclear power plants happened within the military plutonium production facilities of nuclear weapon states and were largely hidden from the public. The first fully operational nuclear power plants were built by the Manhattan Project at Hanford, Washington. Hanford was the site of ongoing releases of radiation into the ecosystem, sometimes accidental and often intentional. There was no public awareness of the substantial radiological contamination of much of Washington and Oregon during the secret Green Run at Hanford in 1949, when radiation was deliberately released from the T Plant chemical and separation facility in what the US Department of Energy's ACHRE report termed "an experiment conducted for intelligence purposes."[27] This experiment was in response to the Soviet acquisition of nuclear weapons in the fall of 1949. Desperate to assess the scale of Soviet plutonium production, on 2 December the United States conducted the Green Run, slightly more than three months after the Soviet test, and exactly seven years to the day from the successful operation of CP-1 at the University of Chicago in 1942, the first sustained nuclear chain reaction.[28]

Plutonium is created through the maintenance of a nuclear chain reaction in uranium fuel rods in a reactor. When the rods are removed the plutonium can be chemically separated and then deployed in fission weapons. After World War II, the United States began to improve its procedures for this chemical separation, to make it radiologically safer. This meant allowing the fuel rods to "cool" for eighty-three to one hundred days so that short-lived radionuclides (like ^{131}I) could decay. The US military assumed that the Soviets would be on a rapid production schedule and unlikely to be cooling the fuel before separating the plutonium. The design of the Green Run was to process a load of fuel rods "green" (without waiting for one hundred days) and to monitor radionuclides gathered by aircraft flying downwind from the plant during the processing. They would be able to link the amounts of materials collected by the planes with the quantities of plutonium being produced, and then fly similar missions on the borders of the Soviet Union. The amount of material gathered in the air there should provide a functional sense of the scale of plutonium production inside the Soviet Union and, thus, intelligence about how quickly it could manufacture nuclear weapons.

On 2 December, the Green Run processed one ton of fuel rods that had cooled for fifteen days. Air scrubbers used to reduce radioiodine in the effluent of the plants were turned off. The experiment released an immense quantity of ^{131}I that deposited over much of Washington and Oregon. The levels of radioiodine found on garden vegetables in the region registered over a thousand times above the daily legal intake limit, yet locally produced dairy products (the primary pathway for radioiodine into the human body) were not monitored or dumped.[29] A 2002 US DOE report described it as an experiment "to measure how an airborne release of radioactive materials spread through the environment in hopes of monitoring the Soviet nuclear weapons program," while an earlier Government Accounting Office study provided the caveat that it was "not *intended* to be a radiation warfare experiment or a field test of radiobiological effects on humans."[30] The widespread contamination of the Green Run was hidden from the public until the early 1980s.[31] It is also important to note that this event has become notorious because it involved the deliberate release of radiation into the ecosystem during peacetime: the United States engaged in "green" processing all of its production of plutonium at Hanford during the Manhattan Project, when it, like the Soviets, was in a rush to manufacture nuclear weapons. The Green Run released approximately 7,800 curies of radioiodine, yet that was only 2.3 percent of the cumulative radioiodine released from Hanford into the community between 1944 and 1951. From 1944, when plutonium production began at Hanford, until 1948, when the initial nuclear weapon stockpile of the United States was established, over 440,000 curies of ^{131}I were released from the plant.[32]

The Kyshtym disaster in 1957 at the Mayak plutonium production facility in the former Soviet Union released radiation in a vast cloud along the Ural Mountains, with no public statement made to those living in the affected areas. In Ozersk, where the plant was located, "city leaders expressed more fear of rumors and panic than of radioactive contamination."[33] Serhii Plokhy writes that "the Soviet leaders refused to release information about the Ozersk explosion, thereby endangering the lives of hundreds of thousands of their own citizens who went on with their daily routines."[34]

Just eleven days after the Kyshtym disaster at Mayak, one of the plutonium production reactors at the Windscale facility in northwest England caught fire. As uranium temperatures rose inside Pile No. 1 at Windscale, "preparations were being made to discharge the fuel from the suspect channel but, when the plug was removed from the access hole, all four channels were seen to be red hot," writes historian Lorna Arnold. "More plugs were removed, revealing yet more glowing fuel elements, and they were found to be so distorted by heat that they could not be pushed out." The fire in Windscale unit 1 burned for three days: "In the course of the accident the first of the two major releases of radioactivity to the atmosphere occurred about midnight on Thursday, and shortly afterwards, when the uranium was burning."[35] Milk at downwind dairies had to be dumped because of contamination with ^{131}I.[36] The Windscale fire was not kept from the public as the Kyshtym disaster had been; however, its radiological impacts were downplayed. In the official UK government report presented later in 1957, investigators, including Sir William Penney, a board member of the United Kingdom Atomic Energy Authority, which oversaw the production of British nuclear weapons, concluded that "There has been no immediate damage to the health of any of the public or of the workers at Windscale, and it is most unlikely that any harmful effects will develop."[37] Yet a more complex computer analysis of wind and weather conditions at the time chastises the report for not taking meteorological data into consideration: "the main emphasis of the report was on local impacts. . . . This lack of data may have contributed to the optimistic assessment that contamination was restricted to the local countryside. Later assessment showed that a wider area was affected and it is generally accepted today that some deaths in the UK and elsewhere in Europe were likely attributable, in a statistical sense, to the radiological release."[38] These two disasters will be discussed further in chapter 7.

Tønnessen, Mårdberg, and Weisæth point out that "Being exposed to ionizing radiation constitutes a 'silent disaster' because ionizing radiation cannot be seen, heard, smelled, tasted, or felt. The individual is totally dependent upon others for information. Accordingly, the information becomes the stressor. Thus, it may become tempting for the authorities not to provide information about such risks."[39] Following accidents at nuclear reactors used to generate electricity, efforts to actively manage perceptions of the risks to the public from radiation are more actively engaged. In each case information about the severity of the accident and the danger to public health has been minimized, the true scale of the incidents only coming to light long after the public had opportunity to take protective measures. "Virtually all democratic nations, international organizations, and labor and consumer groups affirm the public's right to know any information that could cause serious harm. This negative right is a protection against any person or group who might attempt to withhold, misrepresent, or manipulate information that is necessary for people to protect themselves," writes philosopher Kristin Shrader-Frechette. "Yet, official government, industry, and UN International Atomic Energy Agency (IAEA) pronouncements—about the 2011 Fukushima (Japan), 1986 Chernobyl (Ukraine), and 1979 Three Mile Island (US) nuclear accidents—have misrepresented information about the severity and consequences of all these accidents." Shrader-Frechette concludes that "As a result, they have violated citizens' rights to know and therefore likely increased the death and injury rates associated with these accidents."[40] Repeatedly we see out-of-control events accompanied by intense efforts to firmly control public awareness.

"The imperceptibility of Chernobyl radiation by the human senses means that individuals' experience of it is always highly mediated," emphasizes Olga Kuchinskaya.[41] The government of the Soviet Union did not inform local governments, populations, or neighboring countries about the nuclear explosion and subsequent fire belching radiation into the atmosphere from Chernobyl unit 4 until it became impossible to deny. Even as rumors quickly spread through the town of Pripyat, where most of the workers and their families lived, 3 km away from the plant, no notice of the accident was issued to the residents. Ludmila Dyatlova, a bookkeeper for the local police station, and her husband, a railroad engineer, were hosting a birthday party for their eleven-year-old daughter, whose birthday is the day of the Chernobyl explosion. She heard from her neighbor, an engineer, that there had been an accident and the town was becoming radioactive. He advised them to keep the children indoors, to put wet towels on the windows, and to leave town as quickly as possible. They did have a car, but didn't think to leave until they noticed that all of their neighbors who had cars were already gone; the whole family packed into the vehicle and fled later that day. They were turned back at one checkpoint and then detained at another. Even though she worked at the police station, Ludmila did not recognize any of the officers suddenly staffing the checkpoints. Eventually the cars were allowed to proceed; the Dyatlova family drove all night. Early on the morning of the 27th they stopped at a petrol station. As she stood beside the gas pump, Ludmila watched as a caravan of hundreds of buses accompanied by police cars headed the opposite way—toward Pripyat. "I realized they are going to evacuate my city." In that moment, Ludmila understood that her family would never return to their home.[42]

Even as news of the crisis began to seep out, the Soviet government continued to minimize its descriptions of the disaster and labored to control the narrative. "During the second phase, which lasted from 27 April until 14 May, Soviet leaders took steps to protect their citizens, particularly through evacuations, while simultaneously attempting to keep news of the disaster as secret as possible to protect themselves from criticism," describes

RAND historian Edward Geist. Still, the central government insisted that the annual May Day parade proceed in Kyiv, even though the winds on that day were blowing radioactive smoke from the fires directly toward the city. "This strategy, which the Soviet government had utilized successfully in earlier nuclear disasters, failed spectacularly due to the discovery of radioactive contamination by western observers, whose exaggerated reports reached Soviet citizens and undermined official accounts."[43]

Historian Kate Brown, in her seminal 2019 book *Manual for Survival,* describes how the Ukrainian Ministry of Health published five thousand copies of a manual for residents living in areas contaminated by the disaster in August 1986. It told them that there was no risk to health from radiation, yet cautioned against eating numerous traditional foods. Thus, notes Brown, even while the Soviet government continued to minimize the impact of the radioactive fallout from Chernobyl and to insist to its citizens that there was nothing to worry about, "never before Chernobyl had a state been forced to admit publicly to the problem and issue a manual with instructions on how to live in a new, postnuclear reality."[44]

After the initial earthquake and tsunami at Fukushima, a week of cascading meltdowns and hydrogen explosions inside nuclear reactor buildings unfolded in front of news cameras and webcams trained on the complex. While the plants were experiencing meltdowns and other catastrophes, the government of Japan and TEPCO, the owner and operator of the plants, repeatedly assured the public that there were no releases of radiation outside the site and that they had only detected "fuel pellet melt" (their disingenuous term for melted nuclear fuel) of small percentages in the various reactors. "In an effort to avoid arousing fears, the government also deliberately withheld crucial information—a fact that confirmed the suspicions and inflamed the distrust of many when the omissions came to light," write Lochbaum, Lyman, and Stranahan. "It wasn't until June, for example, that officials publicly acknowledged meltdowns at the three reactors—information the government had possessed since March 12, based on the off-site detection of tellurium 132, a fission product that could only have come from a melted core."[45] The Japanese government continued to mislead the public, to placate anxiety, and as the disaster proved vexing to manage, focused on the performative value of competence and calm in the months and years after the disaster.

The United States Nuclear Regulatory Commission defines the "cold shutdown" of a nuclear reactor as "The term used to define a reactor coolant system at atmospheric pressure and at a temperature below 200 degrees Fahrenheit following a reactor cooldown."[46] Seeking to appear to be competently managing the ongoing disaster at the Fukushima complex, the government of Japan and TEPCO chose to employ the term to describe reactors whose fuel was in a molten mess beneath the reactor containment vessels: "Nine months after the accident began, in late 2011, TEPCO announced the three melted reactors were in 'cold shutdown.' Yet months later, high radiation levels still prevent workers from entering the entire plant. Even outside the buildings, TEPCO says radiation levels could kill someone within 15 minutes."[47]

While many in Japan, and specifically in Fukushima Prefecture, were enraged at the public relations manipulations of TEPCO and the government, others saw anxiety over radiation and distrust of authorities as a more worrisome source of danger. Ryoko Ando, who is a representative of the organization Ethos in Fukushima, sees the loss of trust in experts as the source for many of the problems for those living in contaminated communities near the Fukushima plants. "It cannot be denied that radioactive particles that did not exist before the accident are in their environment today and will continue to be there. Many residents point out that the environment has changed since the accident, and they must contend with additional risk to health from living in such an environment, however small it may be." And then he asserts that "even more serious was the disruption in their lives caused by government's approach to intervention. . . . Any major nuclear event will have similar consequences; these accidents will impair people's trust in society and the environment, and the authorities will have to impose restrictions to control the situation. Therefore, revival of trust is one of the most important goals on the road to recovery."[48] Ando took steps to legally restrain public discourse on radiation risk, filing a criminal contempt charge in 2014 against journalist and blogger Mari Takenouchi for a tweet criticizing Ethos's encouraging people to return to and live in contaminated locations. "The complaint brought against Mari Takenouchi is yet another example of the way groups linked to the nuclear energy lobby are trying to gag opposing views," remarked Benjamin Ismaïl of Reporters Without Borders. The complaint was dismissed by Japanese prosecutors.[49]

Ethos was originally organized by multiple nuclear bodies in Europe in response to the Chernobyl disaster and "became a reference for establishing post-accident management systems." Sezin Topçu, a sociologist at the French National Center for Scientific Research, describes how "The conviction of the Ethos team was that the nostalgia for the good old days before the disaster should henceforth be over. The Ethos project aimed, above all, to be a project of hope allowing the victims to live and to behave in a different manner. The victims of Chernobyl were called upon to combat their fatalism, regain confidence in the future and in themselves, and become autonomous."[50]

In Fukushima, many lost trust in official advice about radiological safety and radiation readings based on public Geiger counter "monitoring posts" being extrapolated to a large, general area. People purchased their own Geiger counters, and various "citizen science" groups, local and regional, sprang up to measure radiation throughout the region and also for specific homes, farms, and schoolyards. "Residents quickly decided to track radiation themselves as a means to keep the map of their village relevant—often finding contamination that was not evident from state mapping," reports Maxime Polleri. "In the house of one farmer, I witnessed homemade models that exhibited a 3D topography of Iitate's geographical landscape. These models had been made using 3D printers, and the level of radiation had been monitored by the citizens themselves."[51] Polleri relates how "for worried mothers, the embodiment of radiological hazards is not merely experienced through their own corporeality, but equally through other bodies (notably their children)." Thus: "Radiation completely reorganizes Fukushima residents' everyday thinking, in which the present is being torn between an uncertain future and a past where the original exposure and damage remain unknown."[52]

The certification of food as radiation-free in the vast agricultural lands in the Tōhoku region of Japan is done by spot-testing and extrapolating the results to a broader area. While this can be envisioned as a competent means of managing such an unmanageable task, for individuals it is only as effective as the ongoing health of their own family members. "Over the next few months, Nomura-san became increasingly concerned about the effects of radiation," writes Aya Kimura about a mother in Fukushima. "Her lingering fear was confirmed several months later, when her son's

Public gamma radiation monitor in Fukushima City (2015). Photo by author.

Experimental agricultural plot in contaminated soil in the Polygon, Kazakhstan (2012). Photo by author.

urine was tested and found to contain radioactive cesium, albeit a small amount. She blamed herself for believing, for even a second, the government's claims that food in general was not contaminated."[53]

For those who depict the presence of radiation ecosystems as a manageable risk, the fact that radiation exists in nature is often played as an imagined ace in a clever game. "Is radiation safe?" asks an invisible rhetorician for the World Nuclear Association, an international trade group for the nuclear industry. "Radiation is natural and found everywhere—it comes from outer space, the air we breathe, and the earth we tread. It's even in our bodies; naturally occurring radioactive elements in our bones irradiate us on average 5000 times per second. Sleeping next to someone gives us a much higher radiation dose than living close to a nuclear power station—both of which are harmless."[54] Such neutered presentations of risk are repeatedly found in the media to dismiss those expressing anxiety over radiological contamination or the presence of nuclear waste in their communities as hysterical. Another favorite rhetorical management strategy is to explain that there is more radiation in a banana than there is in . . . name your concern.[55] This discourse works to construct radiation as a singular form of energy and suggests that exposures to different types of radiation can be equated. In Fukushima, "the government distributed pamphlets that explained that radiation naturally exists in our food, such as the potassium levels present in bananas." Polleri points out the intentionally deceptive nature of such public relations: "Such information is irrelevant to the hazards of internalizing fission products from a nuclear power plant. While bananas have naturally occurring potassium, it would require eating around 20 million bananas to get radiation poisoning. On the other hand, each radionuclide released during nuclear meltdown events like Fukushima possesses specific biological signatures and presents particular risks when inhaled or ingested."[56]

Still more simplistic is the notion that if radiation is "natural," then being afraid of it is irrational. "This is of course an absurd argumentation. The fact that something occurs naturally does not prove that it does no harm to us, and neither does it prove that it is safe to increase our exposure to it," cautions Swedish philosopher Sven Ove Hansson. "We are all exposed to pathogenic bacteria, but in spite of being natural some of them do us considerable harm. The existence of this 'background' exposure to pathogens certainly does not show that we can increase our exposure to them a hundredfold or thousandfold with impunity."[57]

Weaponizing the LSS: Making Radioactive Fallout Even Less Visible

In April 2020 a wildfire raged through the heavily contaminated Red Forest in the Chernobyl Nuclear Power Plant Zone of Alienation (more commonly referred to as the Zone of Exclusion), coming perilously close to the plant site itself. The fire burned for more than a week and sent smoke across a large region of Ukraine, Belarus, and several downwind countries. Coming just a year after the widely viewed and praised HBO series *Chernobyl*, public and press attention focused intensely on the progress of the wildfires. "Chernobyl Wildfires Reignite, Stirring Up Radiation," cried a *New York Times* headline.[58] "Wildfires Draw Dangerously Close to Chernobyl Site," decried a *Guardian* headline.[59] However, amid this concern, a Reuters story reassured concerned readers that "Fires Near Chernobyl

Pose 'No Risk to Human Health,' IAEA Says." While radiation had been released by the fires, "The IAEA said it found 'the increase in levels of radiation measured in the country was very small and posed no risk to human health.'" Reporter François Murphy conveyed: "There had been 'some minor increases in radiation,' the IAEA said, adding the State Nuclear Regulatory Inspectorate of Ukraine had found 'the concentration of radioactive materials in the air remained below Ukraine's radiation safety norms.'"[60] The disparate declarations arise out of discussing two different things without distinguishing between them. When radionuclides are aerosolized in a fire and spread downwind, there is real risk to those who inhale these particles: a 1989 study "found that in small-scale fires up to 20% of the particle inventory can be resuspended and in large-scale fires up to 75%."[61] Yet when the level of ambient radioactivity being measured is low, the risk to us from external exposures may also be relatively low. Rather than informing us about those distinctions, parties can latch on to one or the other framework and present the public with either alarm or reassurance. Both can have validity, and neither negates the other.

Physicist David Brenner, who heads the Center for Radiological Research at Columbia University, told listeners to NPR one week after the triple nuclear meltdowns at Fukushima, "There are no health risks here."[62] These readings, however low, were detecting the significant quantities of radionuclides deposited downwind; if people were living in areas where these particles deposited, where they accumulated sufficiently to generate measurable levels of external gamma radiation, there were separate health risks to those who might internalize these particles. Radiogenic thyroid cancer comes from internalizing ^{131}I, not from external exposures, and has invariably followed similar fallout depositions in the past. A more realistic and useful assessment was offered by RERF radiation biophysicist Evan Douple. While explaining that the LSS predicted that there would be no health effects from external levels of radiation from the nuclear disaster at Fukushima, Douple cautioned that there were still risks from internalizing particles: "Here radioisotopes are drifting in water and air, and not necessarily producing an external whole-body exposure and are being taken up in very small doses into the body." Douple explained that while it could be instructive to extrapolate the findings of the LSS to a situation like Fukushima, there were still "big uncertainties."[63]

Wildfires burn in the Chernobyl Zone of Exclusion in April 2020. Volodymyr Shuvayev / repor.to.

As discussed in chapter 1, there were reasons the ABCC did not include internalized radionuclides in the LSS. This omission would become a tool to obscure actual health impacts from internalized radioactive particles for decades. External exposures can often be directly correlated to health outcomes, while internal exposures are seen as contributing to, not determining, outcomes. "After ingestion or inhalation, radioactive particles, depending on their size, solubility, and chemical structure, differ in their distribution through the body, their organ residence time, and the transfer, dissolution, and absorption of the radioactivity associated with the particles, and hence might be expected to vary in their effects across organ systems," caution American epidemiologists in a study of the relationship between internal exposure and subsequent cancer rates among employees at Rocketdyne/Atomics International, a primary nuclear contractor. "Because of large differences in exposure assessment and the lack of power in smaller cohorts with the most in-depth exposure characterization, comparisons of internal dose levels and of results across studies are problematic and generalization of findings may be limited." Nonetheless, the study found that, while the cohort examined typically had more robust health and better access to health care than the average American, and in spite of the limitations delineated above, "we detected increases in mortality from hemato- and lymphopoietic cancers with increasing internal radiation dose among RAI employees, a finding also reported by two previous studies of nuclear workers exposed to plutonium."[64] Whatever the reason, or cluster of reasons, for the lack of robust inquiry into the relationship between internalized particles and illness may be, this exclusion has been weaponized since the early days of nuclear weapon testing to dismiss and obscure the health impacts from living in ecosystems rich with radionuclides, and from the internalization of these particles. While it is a powerful scientific study, the LSS has been used politically to dismiss anxiety over continuous habitation in areas that have received substantial amounts of radioactive fallout, in terms of both internalized particles and continual exposures to low-dose external exposures. After Chernobyl, Magdalena Stawkowski points out that "because the A-bomb survivor studies have become an accepted 'gold standard' for understanding radiation risk, scientific studies on low-dose exposure that show negative health impacts, especially those of Soviet- trained scientists, were seen as 'ideologically tainted.' "[65]

"Veteran concern about radiogenic cancer was a major impetus for this research," explained a study by the United States National Academy of Sciences on possible increased mortality among American service personnel that participated in nuclear weapon tests. The authors assert that "Among the studies of nonmilitary cohorts, the study of the survivor experience following the atomic bomb exposures in Hiroshima and Nagasaki—the Life Span Study (LSS)—is of great importance despite the *unique circumstances surrounding these exposures;* most radiation protection recommendations are based primarily on LSS risk estimates."[66] This claim is made even though many of these veterans conducted maneuvers amid dust clouds heavy with ionized particles after the detonations of the weapons, and certainly internalized particles via respiration and possibly other routes. A 1981 study of radiation exposures of troops and observers taking part in nuclear tests at the Nevada Test Site conducted on behalf of the Defense Nuclear Agency of the United States, almost twenty years before the NAS study cited above, states clearly that "The primary example of personnel being showered by fallout was at Shot Simon, where the volunteer observers noted fallout deposition especially at 2000 to 2500 yards from GZ at about 5 to 10 minutes after burst."[67]

Invoking the LSS as a means of dismissing anxiety over living with distributed radionuclides from explosions of nuclear weapons or at nuclear power plants is ubiquitous: wherever the exposed population is politically powerful enough to merit active management of their anxieties, invocation of the LSS is inevitable. The outcomes extrapolated from the LSS are presented as definitive to silence critics: "No medical study has had such resources lavished on it, teams of scientists, state of the art equipment: this was Atomic Energy Commission (AEC) funding. Since it is assumed in epidemiology that the larger the sample, the greater the statistical accuracy, there has been a tendency to accept these data as the gold standard of radiation risk."[68] This designation—the *gold standard*—is the talisman utilized to mute all uncertainty or opposition to the dismissal of health risks from internalized radiation.

Beyond the structural problems with the LSS itself, as outlined in chapter 1, its application to all incidents of radiological exposure presents additional problems. "Studies by the Atomic Bomb Casualty Commission (ABCC) remain the gold standard for US juridical panels in determining probable causalities of illness from radiation exposure. Of course, the one-off explosions in damp and coastal Japan differed greatly from the slow drip of exposures of a different cocktail of radioactive isotopes on the volcanic soils of the arid and continental Columbia basin," where Hanford is located, points out Kate Brown. "Yet researchers made models estimating doses across landscapes and their effects on bodies that considered the contexts of Japan and the United States as interchangeable. This is remarkable considering all that had been discovered in four decades of research by hydrologists, ichthyologists, meteorologists, and soil scientists about the locally contingent pathways of radioactive isotopes. Using the ABCC studies, US government officials eventually determined that the Hanford Nuclear Reservation required a multi-billion dollar cleanup; at the same time, however, they decided that people exposed nearby were largely unaffected."[69] Similarly, scientists working for the IAEA, which functions in part as a trade promotion association for the nuclear power industry, concluded after assessments of the levels of residual radiation in the villages around the Polygon in Kazakhstan, "One village had a higher plutonium deposition level than the other settlements and has been the subject of more comprehensive soil sampling. However, estimated annual doses still remain low. Intervention to reduce the radiation exposure of people outside the Semipalatinsk test site is not considered to be justified."[70] While the external dosage received by those in the village may not be considered high enough to warrant remediation, this isotope of plutonium will remain radioactive for hundreds of thousands of years, and it has long been acknowledged that "Internal exposure to plutonium, which decays via alpha particle emission, is a recognised health hazard."[71]

Among scientists the borderline of defense around the LSS may be professional pride, but its utility to public officials and managers in the nuclear power industry is its ability to dispel critics and calm victims. In these cases, a key utility of the LSS is to remove the burden of response or intervention to deposits of radiological fallout. The LSS can be invoked to facilitate dismissal of the seriousness of depositions of radioactive particles contaminating homes, schools, farms, and woodlands. The *gold standard* tells us that nothing is wrong, no one is in danger, and thus: nothing need be done. It is wielded as a super weapon to silence opponents and to confuse and disarm claimants. Those in positions of responsibility are absolved of obligation to take actions to protect people, remediate land, or compensate those affected. The only thing that needs to be managed is the public's irrationality, which can be done cheaply and with no liability. There may not have been direct malice in the repeated, flawed public representations of the LSS, but that does not mean this has not caused harm.

Nature Blesses Contaminated Spaces

Radiologically contaminated sites must be cleaned slowly and carefully, but many, especially inside the borders of nuclear weapon states, also become sites for the remediation of perceptions or "greenwashing," a process that Shiloh Krupar describes as the "military-to-wildlife project."[72] The Fernald Feed Materials Production Center was established by the United States Atomic Energy Commission in 1951, and until 1989 it processed uranium into various components used in the production of plutonium and other products. In 1988, a year before production ceased at the site, came news not entirely surprising to the local community: "Government officials overseeing a nuclear weapon plant in Ohio knew for decades that they were releasing thousands of tons of radioactive uranium waste into the environment, exposing thousands of workers and residents in the region, a Congressional panel said today." *New York Times* reporter Keith Noble described how the government had previously decided not to spend the money to clean up multiple routes of contamination: "Runoff from the plant carried tons of the waste into drinking water wells in the area and the Great Miami River; leaky pits at the plant, storing waste water containing uranium emissions and other radioactive materials, leaked into the water supplies, and the plant emitted radioactive particles into the air."[73] The DOE's Fernald Closure Project website unexplainably denigrates a lack of awareness of industrial pollution among the public: "It was not until the 1970s and 1980s that the general populace became aware of the industrial revolution's effect on the environment, and the cost and magnitude of the resulting cleanup effort."[74] Nonetheless, it simultaneously assured the public that the site had been fully remediated twenty years later: "The 1,050-acre Fernald property has now come full circle. The property's natural features have been restored using native plants and grasses. Restoration activities at the site have created one of the largest man-made wetlands in Ohio, open water, upland forests, a lengthy riparian corridor, and 385 acres of grassland, including tallgrass prairie and savanna." This natural setting was now open to public use: "The site's varied and unique habitats are accessible on a 7-mile network of trails. More than 245 species of birds have been observed, and over 100 different species have been documented as nesting at the Fernald Preserve."[75] However, as Brown points out, visitors receive a guide that warns them against stepping off of designated gravel paths, or picking up any "masonry object."[76]

Can the people of the Cincinnati metro area relax and enjoy the natural wonders of the Fernald Preserve? In 2018 approximately fifteen thousand people came to the park, which includes "a 65-foot-high *natural-looking* mound that serves as the permanent storage site of nearly 3 million cubic yards of low-level radioactive waste—about 85 percent soil and 15 percent debris excavated during the cleanup."[77] The remediation of the Fernald Feed Materials Production Center, and the development and curation of the public image of the former Superfund site, is emblematic of the management of the perceptions around radiologically contaminated sites after they are decommissioned by major nuclear weapon–producing states. Not only are the sites sanitized, but they have also been transformed into pristine sanctuaries of the natural world. Krupar describes how "Such land conversions demonstrate that military activities have not just destroyed nature but have actively produced it, and that nature—ideologically and materially—serves to contain the risks resulting from decades of domestic mobilization for war."[78]

The former Weldon Spring Ordnance Works west of St. Louis hosted both a chemical plant and a uranium ore processing facility and was operational from World War II through most of the Cold War. After the Cold War the Superfund site was remediated and became the Weldon Spring Site, winner of the 2020 Federal Facility Excellence in Site Reuse Award of the United States Environmental Protection Agency.[79] The primary attraction of this 217-acre site is the disposal cell: "The disposal cell was constructed in the area formerly occupied by the Weldon Spring Uranium Feed Materials Plant production buildings. The 45-acre cell provides long-term isolation and management of waste materials. The high point of the cell is approximately 75 feet above the surrounding terrain."[80] The 1.48 million cubic yards of waste, including radioactive debris, lays beneath the 75-foot-tall disposal cell, which visitors can climb to enjoy "panoramic views from the highest publicly accessible point in St. Charles County."[81]

Nuclear nature reserves abound worldwide. Koeberg Nuclear Power Station in South Africa, the only commercial nuclear power plant on the African continent, is also home to the Koeberg Nature Reserve. "A nuclear power plant is the last place you'd expect to find families of eland, springbok and zebra, and a thriving bird life, but Koeberg Nature Reserve is just such a place," proclaims a South African tourism website promoting the reserve abutting the forty-five-year-old nuclear power plant.[82] The former Lop Nor Nuclear Weapons Test Base in China now hosts the Lop Nor Wild Camel National Nature Reserve.[83] The former Soviet test site at Novaya Zemlya in the Russian Arctic is now the home of the Russian Arctic National Park, a polar bear reserve.[84] In 1976 a large region that was included in the Eastern-Ural Radioactive Trace, an area heavily contaminated by the 1957 Kyshtym disaster at the Mayak plutonium production site, was designated as the East-Ural Radioactive State Reserve.[85] The

Visitors hiking up the disposal cell at the Weldon Spring Conservation Area (2019). Sydney Clark. *Weldon Spring Site*. Digital Photograph. 2019. http://sydneyclark.art.

former United States nuclear test site on Johnston Atoll in the Pacific is also now a national wildlife refuge. The Chinese and Russian sites are not open to the public, but they maintain spaces for the management of endangered species or to monitor natural settings after contamination.

The Rocky Flats Plant was a key site in the manufacturing of nuclear weapons in the United States. Beginning in 1954, Rocky Flats was where the plutonium from Hanford and the Savannah River site was fashioned into "pits," which would form the fissile core of fission weapons, as well as in the primary detonation needed to trigger the secondary reaction in fusion weapons. In the course of operations, during which the plant produced thousands of plutonium pits per year, many parts of the site became heavily contaminated with plutonium.[86] "The Rocky Flats plant experienced hundreds of fires during its production years. The first large one occurred in 1959,"[87] writes Len Ackland, former journalism professor and editor of the

Bulletin of the Atomic Scientists, when it was finally shut down, its on-site waste included "more than 3 tons of the plutonium."[88] In 1989, the US FBI raided the site based on whistleblower and surveillance information, an unprecedented action by an American law-enforcement agency against a currently operating nuclear weapon production facility. After extensive grand jury testimony, the government obtained a plea agreement with Rockwell International, the corporate operator of the plant. The plea resulted in the grand jury testimony remaining sealed, although members of the grand jury—who contested the plea agreement—considered the violations over the decades of plant operation too catastrophic to conceal and subsequently leaked the report to the press.[89]

The DOE conducted a study of remediation of the Rocky Flats site, and "In early 1995, DOE estimated that cleaning up Rocky Flats would take approximately 65 years and cost more than $37 billion, making site closure seem like a distant dream."[90] Miraculously, ten years later the department completed its accelerated cleanup plan, removing 21 tons of weapon-grade materials and 1.3 million cubic meters of waste, and treating more than 16 million gallons of contaminated water.[91] In 2007 the site shifted to the authority of the US Fish and Wildlife Service, which transformed it into the Rocky Flats National Wildlife Refuge and opened it to public usage in 2018.

Located on the edge of the Denver metropolitan area, the more than six thousand acres of the Rocky Flats National Wildlife Refuge offers about 18 km of hiking trails and presents the appearance of a natural paradise on the edge of a big city. However, things did not proceed as planned. "Almost 300,000 students from metro Denver school systems will be barred from school-sanctioned trips to Rocky Flats National Wildlife Refuge after it opens this summer, with the state's largest district enacting a ban last week on visits to the former nuclear weapons manufacturing site," declared a 2018 article in the *Denver Post*. "Denver Public Schools joined half a dozen other local school districts that say Rocky Flats, with its legacy of plutonium contamination that was often shrouded in secrecy, is too much of a risk for visiting schoolchildren."[92] Compounding matters, citizen groups who had been fastidiously conducting studies of soil in the park began turning up numerous samples whose radioactivity tested well above the "actionable level established in the closure standards."[93] Researchers from Northern

Arizona University found over forty respirable plutonium particles in a sample from land adjacent to the refuge.[94]

Geographer Shannon Cram has written that "Nuclear weapons production has created a unique geography of irradiated open space in the United States. In recent years, many of these landscapes have been re-classified as national wildlife refuges in an attempt to transform the nation's atomic sacrifice zones into spaces of environmental salvation." Why are highly contaminated sites so often made to cosplay as *nature reserves* when their utility to nuclear production ceases? Cram asserts that "nature is used to rethink and remake the post-nuclear future—transforming nuclear wastelands into manageable, aesthetically potent, scientifically productive, and economically efficient spaces."[95] As the LSS is invoked to dismiss the risks of internalized particles to fallout-exposed populations, it is similarly the standard used to determine appropriate levels of external radiation in the transmogrification of heavily contaminated former nuclear production and test sites into natural paradises.

Wildfires and Wildlife

Concerns akin to those over the forest fires burning in Chernobyl's Red Forest are a constant worry for those living nearby or downwind from radiologically contaminated spaces, since "Forest fires can cause resuspension of radionuclides in contaminated areas."[96] This resuspension can then transport the radionuclides in the smoke or the breeze to downwind ecosystems and populations.[97] "During the last 28 years dead wood and litter have dramatically accumulated in these areas, whereas climate change has increased temperature and favored drought," wrote an international team of scientists about the Zone of Exclusion in 2014. "The present situation in these forests suggests an increased risk of wildfires, especially after the pronounced forest fires of 2010, which remobilized Chernobyl-deposited radioactive materials transporting them thousand [*sic*] kilometers far."[98]

In the summer of 2016, numerous large wildfires threatened to spread across the Hanford Reservation. Most concerning was the Range 12 fire that spread from Grant and Yakima Counties into Benton County, where the sprawling nuclear site is located. The fire threatened to summit Rattlesnake Mountain and spread into the Hanford Nuclear Site itself. In its report on the management of another wildfire that threatened to cross the Hanford Site in the year 2000, the DOE discussed the importance of protecting "buildings, waste sites and storage areas containing radioactive materials." Additionally, the report praised "control of deep-rooted vegetation (e.g., tumbleweeds) on radioactive burial sites," since, had these materials been more prevalent and had they burned, there would have been a significant release of radioactive materials, which would have been aerosolized, facilitating their spread and ingestion.[99] Tumbleweeds are a particular hazard: "With roots that can grow 20 feet, tumbleweeds reach down into waste dumps and take up strontium-90, break off, and blow around the dry land."[100]

Fortunately, the fire did not cross over the heavily contaminated sections of the site. However, there have been wildfires that threatened to do exactly this in 1957, 1973, 1981, 1984, 1998, 2000, 2007, 2016, 2017, 2018, and 2019. In 2020, lightning strikes kindled a large fire on the reservation itself.[101] In this era of global warming, the wildfire threat to dry sites like the land surrounding the Hanford Reservation increases year by year. Thus far, we have avoided disaster, yet many of the radionuclides from which this danger emanates have half-lives of hundreds, thousands, or tens of thousands of years. It's delusional to imagine that before Hanford is fully remediated, and before these radionuclides have passed several half-lives, there will not be a wildfire that crosses the land—aerosolizing some measure of this waste into the wind. Being successful at managing and at avoiding this catastrophe can only be a short-term accomplishment when the dangers will last longer than spending policies, longer than governments, longer than recorded history. Catastrophe may yet loom.

Similar threats abound. Recent years have seen multiple wildfires within kilometers of the Fukushima Daiichi site, Los Alamos Laboratory, the Nevada Test Site, and countless radiologically contaminated locations.[102] Magdalena Stawkowski recounts her experiences while conducting field research living in a small village just outside of the Polygon in Kazakhstan in 2010:

> Although the fire brigade seemed effective—given that we were left to deal with the problem on our own—I was overcome by

a troubling realization, something that everyone else already knew: this fire, the first one of 2010, would pass right through the Polygon, still heavily polluted with radioactivity. The test site's boundaries, represented clearly on a map hanging in my one-room house in Koyan, meant nothing. Neither fire nor radioactive particles obey borders drawn on a map. And while fires are common in this region, especially in the fall when the steppe grasses turn brown and bone dry, I was worried. Every time we hit the flame with our sheepskins or ran through the charred earth toward the next fire line with buckets of water hoisted from a well, we crossed in and out of the Polygon. But no one really knew for sure because there were no fences or signs to warn us. I imagined radioactive particles—buried somewhere in layers of ash— suspended in the air once again, covering our clothes and drawing into our lungs.[103]

Wildfires are a perennial threat on the eastern steppe of Kazakhstan.

The return or presence of animals, especially large mammals, to radio-logically contaminated sites is often presented as an intrinsic measure of the transition of those sites to a healthy or natural state. "A barn owl bursts from the tall prairie grasses. Elk skitter among cottonwood trees near an old stagecoach halt. A shrew crosses a track and hurtles into milkweed, where monarch butterflies feed," soothes a description that could come from any wilderness locale's promotional pamphlet. "Somewhere amid the rare xeric grasses are coyotes, moose, mule deer, a handful of endangered Preble's meadow jumping mice, and more than 600 plant species. 'Welcome,' says David Lucas of the U.S. Fish and Wildlife Service, 'to Rocky Flats National Wildlife Refuge.'"[104]

I have written about images of animals being utilized to instruct people how to think about risks from living near the Nevada Test Site during the era of atmospheric nuclear testing in the United States during the early Cold War, and animals have remained our mythical teachers about the natural recovery of irradiated landscapes in the post–Cold War era as well.[105] Nowhere is this more so than in stories about the recovery of the Zone of Exclusion around the former Chernobyl Nuclear Power Plant complex in Ukraine. Human residents having been evacuated in the days and weeks after the catastrophic Chernobyl disaster in 1986, animals have naturally proliferated in this area—free from pressures by the species most aggressively limiting their habitation in the past.[106] Several studies have shown animal populations increasing in the absence of human settlements in the Exclusion Zone in the last ten years. Beginning with a study of Chernobyl wildlife published immediately after the Fukushima disaster (quickly retitled prior to publication to imply that studies of Chernobyl could comfort our fresh Fukushima anxieties) and then with a series of articles in 2015, biologists documented observations of increased numbers of various animals. They cautioned that these increases did not mean that the radiation was not dangerous or that living in the contaminated zone was safe for the animals, simply that the radiation had not made the zone unlivable.[107] However, journalists quickly declared these suppositions as truths. *Washington Post* health, science, and environment editor Laura Helmuth gushed about "Chernobyl's wildlife survivors" in 2013, declaring that "the radioactive fallout zone has turned into a refuge."[108] Dr. Barry Starr fawned to listeners of KQED radio about the "benefits of radioactive fallout." Tania Rabesandratana admonished readers in *Science* that "humans are worse than radiation for Chernobyl animals."[109]

As Chesser and Baker described in chapter 2, researchers have had difficulty determining which mammals had spent time in the zone and which had recently crossed in and become subject to observation, and also that the radionuclides were unevenly distributed: no two animals likely had analogous exposures. Biologist Timothy Mousseau, who has done fieldwork in both Chernobyl and Fukushima contaminated zones for decades, notes that very little published research on the impacts of radiation are actually biological studies. When surveying the top five hundred scientific papers on Fukushima ten years after the disaster on the Web of Science database, Mousseau found that almost every paper was an examination of the distribution of radionuclides that formed the basis of "a study of calculated doses and the possible link to health impairments rather than any sort of directly measured biological consequences."[110] A 2020 study cautions against drawing easy conclusions based strictly on external radiation levels: "simplistic measurements of exposure, such as ambient dose rates or soil concentration activities, cannot encompass all the complexity of actual exposure of wildlife. Contributions of internal and external irradiation pathways to the total dose rates have both to be considered, and this balance depends on radionuclides (type and energy of emitted radiation) and on animal species (age, diet, habitat, use of the environment)."[111]

Animals, like humans, are unable to perceive radiation with their senses, so their presence in the zone reflects the absence of perceptible threats, such as human beings, not the absence of risk from radiation. "Is it really a paradise for the wildlife? Is the wildlife in Chernobyl that is thriving under the shadow of nuclear disaster really healthy internally? How could the area where scientists were only given a limited time to stay in order to minimize radiation exposure be considered safe for wildlife?" asks political scientist Mayumi Itoh about assertions in the 2011 documentary film *Radioactive Wolves*, shown on American public television. "The scientists had to wear masks when handling wolves because, the documentary states, the wolf fur is radioactive and is toxic to humans if inhaled. The new inhabitants and migrants in the deserted zone know nothing of the lurking danger of invisible radiation. The fallout from the nuclear accident seeped into the soil and was then absorbed by plants. The herbivores eat the contaminated plants, which are consumed by the predators according to the food chain in the wild. The scientists do not yet know the effects of long-term radioactive exposure on the health of the wildlife."[112] Claims made in the film that the radiation on the wolf fur was not dangerous to be near because its levels of emitted gamma radiation were low, yet very dangerous to inhale or swallow, construct the wolf as an object when describing it as healthy: certainly, the wolf cleans itself by licking its fur. Contrary to celebrations of the presence of wildlife in the Chernobyl Exclusion Zone, a 2018 study found that gray wolves had migrated as far as 369 km from the zone, and it cautioned about the "potential spread of radiation-induced genetic mutations to populations in uncontaminated areas."[113]

Biologists Timothy Mousseau and Anders Møller have focused on smaller creatures since those living in radioactive locations would be multiple generations further descended than larger mammals, and the effects of radiation might be more easily discerned. Focusing on insects and birds, Mousseau and Møller have documented clear evidence of harm from radiation in multiple species: "Recent advances in genetic and ecological studies of wild animal populations of Chernobyl and Fukushima have demonstrated significant genetic, physiological, developmental, and fitness effects stemming from exposure to radioactive contaminants. The few genetic studies that have been conducted in Chernobyl generally show elevated rates of genetic damage and mutation rates. All major taxonomic groups investigated (i.e., birds, bees, butterflies, grasshoppers, dragonflies, spiders, mammals) displayed reduced population sizes in highly radioactive parts of the Chernobyl Exclusion Zone." However, "In Fukushima, population censuses of birds, butterflies, and cicadas suggested that abundances were negatively impacted by exposure to radioactive contaminants, while other groups (e.g., dragonflies, grasshoppers, bees, spiders) showed no significant declines, at least during the first summer following the disaster. Insufficient information exists for groups other than insects and birds to assess effects on life history at this time." They explain that "The differences observed between Fukushima and Chernobyl may reflect the different times of exposure and the significance of multigenerational mutation accumulation in Chernobyl compared to Fukushima."[114] Studies of genetic impacts on insects, or mutations in bird populations, do not create the emotional response, provide the comfort, nor generate the clicks and shares that follow images of wolves cavorting in Chernobyl's Red Forest, and they tend not to be reprinted in newspapers or featured in nature documentaries. They are not as exciting as speculating that perhaps "mankind's disaster gave nature a second chance."[115]

Community Economic Interdependence and Cultural Self-Determination

Nuclear weapon testing facilities and nuclear power plants are rarely sited in urban areas; they are placed in areas with lower populations. People living near nuclear power plants may become employed in low-wage jobs at the facilities; however, many typically depend on the nearby land and waterways to produce food for their living. Contamination of soil and ecosystems can impact their means to support themselves and erode their community's vitality. Depression or loss of the value of this land can begin a process whereby the communities near these sites, or downwind where fallout deposits, degrade and become increasingly less capable of producing either food or income from the production of food. "The land contaminated by the Chornobyl NPP accident in Ukraine includes rich forests where mushrooms and berries were harvested and agricultural lands where thousands of metric tons of hay were harvested. The loss of the ability to use farmland, water resources, and forest resources because of contamination is currently estimated to be 8.6–10.9 billion rubles," explained Volodymyr I. Kholosha of the State Agency of Ukraine for Exclusion Zone Administration in 2008. "This is more than 2% of the gross national income produced by Ukraine in 1986. These figures are for Ukraine alone from 1986–1991. . . . It will take several decades for the contamination on this land to decrease sufficiently to permit use."[116]

In Fukushima, many who live in the downwind area, especially in the inland mountains, also depend on the forests for food and income. However, the Japanese government has excluded the forest areas from both attempts at remediation and compensation for losses. Hiroyuki Kaneko describes two reasons for this exclusion: "First, the forested *yama* landscape is more spatially expansive than the residential *mura* or agricultural *nora*, and decontamination of this expansive area presents serious technical and financial difficulties. A second reason flows from the conceptualization of 'life activity' advanced in the Decontamination Guidelines." These guidelines express the need for the restoration of land and community practices that contribute directly into the broader economy, and cast aside the importance of those living off the forest commons. Kaneko argues, "Minor subsistence has not only been important as a calorific supplement to village diets but also as a space that sustains social relations and communal consciousness."[117] Rather than work to support those who derived food and community from the forest commons, the Japanese government advised them to abandon this traditional dietary source. "Residents with high levels of internal contamination consumed homegrown produce without radiation inspection, and often collected mushrooms in the wild or cultivated them on bed-logs in their homes. They were advised to consume distributed food mainly and to refrain from consuming potentially contaminated foods without radiation inspection and local produces under shipment restrictions such as mushrooms, mountain vegetables, and meat of wildlife."[118]

This cleaving of communities from both their sociomythic sense of self and their methods of sustainability in nature creates rifts not easily filled in with new truths and new relationships. "Current Scandinavian attempts to regulate Sámi consumption of contaminated reindeer meat in the interests of health and reproductive safety are premised on the assumption that physical well-being and reproduction are clearly separable from cultural identity and reproduction." Sharon Stephens further explains that when Sámi chose to eat contaminated reindeer meat because of the cultural and spiritual nourishment that could only come from this traditional food, "I was repeatedly struck by the difficulties policymakers have with seeing the cultural logic behind what they frame as Sami 'irrationality' and 'ignorance.'"[119] The Sámi consider it particularly important for pregnant women to consume meat from the family herd and to be generous in serving it to others—nurturing more than just bodies, but bonds that will sustain familial and community interdependence deep into the future. While authorities may envision their interventions as critical to community sustenance, they are often blinkered from grasping what is being stripped away. "This is not just a matter of economics," explains Sámi Gerd Persson, "but of who we are, how we live, how we are connected to our deer and each other. Now [in the winter of 1986–87] we must buy everything. Thread, material, food, shoes are now all different things when they used to be parts of one thing."[120] I am not advocating for the removal of restrictions on the consumption of contaminated meat, but arguing that the protection of health does not end when this contaminated food is removed from the diet of the local community, and that our construct of "health" must include the bonds that weave the community together as well as the nutrients that sustain individual bodies. "In Sami society, the consumption of meat from family herds pastured in family territories is central to a substantialized notion of kinship, physicalized memories of the past, and the social construction of bodies with Sami senses and capacities for understanding. . . . State-orchestrated changes in everyday herding and consumption practices are experienced as threats to Sami ways of producing and of reproducing persons and communities."[121]

Bruno Latour has written extensively about the conflicts between emotionally and spiritually held patterns of relationship with place and society and modern "factish" interpretations that displace these older forms with newer ideations of reality, which are then imposed with the force of both a new religion and state sanctioning.[122] For those whose lives and lifestyles are disrupted by radiological contamination, the insistence that communities adhere to these new models of reality extends a colonial impulse that devalues traditional and indigenous lifestyles and cosmologies. "Sami—who, like other populations in the fallout region outside Ukraine, know of the Chernobyl event only through its conflicting representations by state officials, scientists, and journalists—are now being asked to make wide-ranging changes in everyday herding and domestic practices based on cost-benefit and risk-assessment models that are increasingly central to Western state policy formation but that are in many respects at odds with the world of the Sami people," explains Stephens.[123] Sig-Britt Toven, one of Stephens's interviewees, frames it more viscerally: "It seems sometimes that things have become strange and make-believe. You see with your eyes the same mountains and lakes, the same herds, but you know there is something dangerous, something invisible, that can harm your children, that you can't see or touch or smell. Your hands keep doing the work, but your head worries about the future."[124]

Even for populations in the region that do not suffer disruptions to their lifestyles and relationship to the land, the impact of radioactive fallout can have powerful effects on the economic and social success of members of the community long after the initial deposition. "In a comprehensive data set of 562,637 Swedes born 1983–1988, we find that the cohort in utero during the Chernobyl accident had worse school outcomes than adjacent birth cohorts, and this deterioration was largest for those exposed approximately 8–25 weeks post conception," wrote Swedish and American economists in 2007. "Moreover, we find larger damage among students born in regions that received more fallout: students from the eight most affected municipalities were 3.6 percentage points less likely to qualify to high school as a result of the fallout. Our findings suggest that fetal exposure to ionizing radiation damages cognitive ability at radiation levels previously considered safe."[125] The economists pursued this inquiry based on earlier research done on reduced IQ levels in hibakusha from Hiroshima and Nagasaki exposed between two and four months in utero. This earlier research, spearheaded by William Shull and Masanori Otake, found an array of impacts on hibakusha exposed prenatally to radiation from the nuclear attacks on Hiroshima and Nagasaki, ranging from learning disabilities to cognitive impairments and even serious brain damage.[126]

Polynesians found that such decentering of culture can occur not just in the form of interventions to protect populations from radiological health risks, but also from seduction into the nuclear weapon economy. Roland Oldham explained that after corrupting the local politicians with money, the French also corrupted "the people who are going and working in Moruroa—you get this salary three times higher for whatever job. So, all of the people from the island abandon the traditional way of life. Abandoning the plantation of taro, or whatever—their own self-sustaining agriculture, they abandon all that to work in Moruroa." For Oldham, this was the start of a process that slowly hollowed out the vitality of traditional Polynesian culture in the islands: "It's just like an explosion of a bomb, you can imagine how it explodes in the heads of the Polynesian people. Blow up the whole thing, blow up this whole structure, his whole way of life, his whole philosophy, his values. How the culture, how his culture, land, ocean, the wind, cosmos—all that—is Polynesian. And suddenly all that blew up. And somewhere, he lost his soul, his deep relationship that he always had with the land the ocean, because all his life comes from there, from the cosmos. Now, there's another value, his life is turning to money and so this really completely changed our society. And that is irreversible." Oldham saw this as an active effort of cultural subversion: "If we abandon who we are . . . maybe the colonizer will tell you 'this is what you are now.' And, you start believing him. You really lose yourself. . . . If you don't know anymore, if you forget where you come from, if you don't know where you are, how can you know where you're going?"[127]

Barker and Johnston have done extensive work on militarist and colonial assaults on Pacific communities and postcolonial resilience at nuclear sites, particularly in the Marshall Islands. "Environmental contamination has robbed the Rongelapese of their customary access to natural resources used to sustain households, communities, customary exchanges for other goods, or income generation," they concluded in 2000. "As a result, the Rongelapese are unable to practice principles of responsible stewardship, and they are unable to transmit their knowledge of sustainable access and use of resources to younger generations." This situation was compounded by the ongoing radiological dangers to habitation on the atoll. "The longer the people are off their home islands, the more difficult it is for people to exercise their land rights and ensure that the land is passed on to future generations. Rongelapese who grow up away from their islands lack the knowledge about essential cultivation areas and dangers that are important to survival if they return to the land."[128]

At Enewetak Atoll, the second nuclear test site in the Marshall Islands chosen by the United States after the Crossroads tests of 1946 rendered Bikini Atoll too radioactive for immediate use, and residents were evacuated before the testing to Ujelang Atoll. They were later paid from the interest on a small trust fund of $150,000 established in 1956 that would grant the US perpetual lease rights. "Islanders' sadness upon moving was just the first of a compounding series of hardships. Ujelang was not only the westernmost atoll in all the Marshall Islands, it was also much smaller and less fertile than Enewetak. This limited possibilities for food and cash crop production leading to periodic starvations during the 1950s," explains Mary X. Mitchell. "Although islanders could purchase supplemental food, their financial resources were extremely limited. . . . The sole other source of cash income—copra produced for sale—rotted while islanders waited in vain for ships that often skipped their scheduled stops at Ujelang. . . . According to a 1969 estimate, Ujelang's copra sales and interest payments combined totaled only about $25 per person per year. By the close of the 1960s, the people of Enewetak were once again starving."[129]

The Enewetakese remained dependent on the United States, not just for processed food, but also for the tools necessary to produce traditional foods. Since they had been forcibly moved from Enewetak to Ujelang, they no longer had access to the materials required for building their own food production tools. Jane Diblin notes that "Their fishing canoes fell into disrepair when, after repeated requests for materials, fishing hooks and nets, the Trust ships brought only enough sailcloth for two canoes and enamel garden furniture paint instead of marine paint for wooden hulls."[130] An Enewetak *iroij* (chief) commented that "Canoes and fishing are the life of Ujelang. Without the canoes we cannot get to the other islands in the lagoon to harvest coconuts. Without the fishing equipment, we cannot catch fish to get enough to eat."[131]

These changes in lifestyle resulted in the Marshallese becoming increasingly dependent on processed food shipped in from the United States. Changes in diet were accompanied by new health challenges, resulting in part from traditional-work and body-usage declines. Geographer Jeffrey Sasha Davis observes that since the Marshallese who work on the US military base on Kwajalein Atoll "are blocked from accessing the resources on most of the atoll, and the remaining resources around Ebeye Island are so scant, imported foods are the main source of nourishment. Most of the foods that can survive the long voyage to Ebeye are canned: Spam and corned beef are major staples. The local water supply is so unreliable and of such low quality that soda is the preferred drink. Diabetes is rampant, as are the associated amputations caused by the lack of treatment."[132] These changes are the direct result of becoming a nuclear test site and the military base of a foreign power. "Although diabetes is not considered to be a radiogenic disease in the medical literature, the increasing prevalence of diabetes globally is certainly related to social and cultural factors," advises Seiji Yamada, a public health researcher and doctor practicing in Hawai'i. "In the Marshall Islands, the social disruption caused by nuclear testing has been a primary feature of its modern history."[133] Eventually most of those displaced from Kwajalein, Enewetak, Rongelap, and Bikini would settle on Majuro, the capital atoll of the Republic of the Marshall Islands, with a current population of more than twenty-seven thousand people on less than 10 km^2 of land.

As most nuclear test sites were built in remote locations with limited local economies, residents would find themselves reliant on sources of income located in areas that they knew to be contaminated. Kazakhs living in the villages that border the Polygon became increasingly dependent on resources from inside the test site after the Soviet economic collapse. "Residents, who possess skills that are deemed useless in the new free market economy, have come to depend on the nuclear test site for their own economic survival," writes Stawkowski. Their "survival depends on their proximity to the Polygon and what they have access to because of this—black-market diesel and coal from mines, scrap metal, pasturelands, and water."[134] In this case those utilizing the former test site were aware that the site was contaminated and that they were taking a risk they deemed acceptable for the income. Stawkowski describes how many people living in the area developed mythologies of themselves as "radioactive mutants": "By this they meant not only that they could survive it but also that it helped them stay alive; others who had left, they said, died as a result. They warned that if I stayed long, I too might come to depend on radiation." She sees this as a natural result of being abandoned by their own government and invisible to the global economy: "because they are left to deal with environmental problems on their own, their only option is to become (or believe themselves to be) enhanced human beings who can survive in toxic environments."[135] Those who leave remote contaminated areas often become relegated to the bottom rungs of a global neoliberal economy, finding their work options limited to being menial laborers. "Most people who live in the area are some of the country's poorest and most marginalized. Lacking marketable skills, they are seen by the urban elite as too 'backward' to participate in the country's new free-market economic system and enjoy its rewards," writes Stawkowski. "There is no work for them in the cities, or if there is, it is often working in the bazaar or cleaning streets or toilets. Younger residents who have moved to the cities end up living in ghettos and rely on the village network for food and financial contributions. . . . Those in the Polygon region have seen parents, children, and other family members die in dark peripheral buildings from alcoholism, drug overdoses, or other illnesses."[136]

Radiation has also caused health problems as a result of the lack of security at former test sites. "Radioactive material is seeping out from this Sahara Desert mountain where French scientists conducted nuclear tests in the 1960s, contaminating the soil and poisoning relations between France and Algeria," claimed a 2010 Reuters article. "According to Algerian data, radiation in some areas near the test sites is 20 times higher than the norm. . . . Algerian officials say France is refusing to give them access to archives about the tests, leaving them in the dark about the extent of the threat from radiation and preventing them from taking effective measures to contain it."[137] While abandoning their Algerian base in the 1960s, the French simply buried tons of radioactive material at the test site. A simple fence was placed around the facilities, and no signs warning of the radiological dangers were posted. "The fact that people were not aware of the dangers of this material for years is criminal," said French documentarian Larbi Benchiha. "From the abandoned nuclear testing bases, people have recovered plates, beams, electrical cables and equipment of all kinds, all of which is radioactive. . . . They have incorporated them into the construction of their homes."[138] Just as radionuclides transport into the surrounding ecosystem after fallout deposition, these abandoned radioactive materials did not remain where they were discarded. Their health impact was far beyond that intended, imagined, or of concern to the French nuclear authorities. They were simply another legacy of colonialism, purposefully ignored.

Learning to See the Unseeable

After radiological exposures, diseases that have been sparked by internalized particles and continual exposure to low-level radiation may begin to present as illnesses. Slowly, many notice that the disease burden of their families and neighbors is unusual. "Residents in the downwind and uranium-producing regions began to describe patterns and clusters of disease observed in their personal geographies—unique, interconnected networks of families, coworkers, friends, and acquaintances and recreational activities and neighborhoods—and they began to observe the ways these patterns and clusters could be connected to nuclear testing or uranium extraction upwind," writes historian Sarah Fox.

> Narrators began to describe patterns of disease linked by various factors of interdependence. A cluster of disease in a family might be described as crossing generations, then connected in the next sentence to a cluster of diseases in a school class, and then linked to a group of illnesses or cancers in a neighborhood or among individuals who labored at the same kind of work, like uranium mining or ranching. Narrators also referred to clusters of specific diseases, such as the leukemia cluster Kay Millet discovered in the hospital waiting room in Salt Lake and the lung cancer cluster Fannie Yazzie found in the VA hospital in Albuquerque. Occupying as they do an imagined space, one dictated by physical proximity as well as the geographies of emotional attachment and genetic ties, these clusters and the stories describing them grow more extensive with the passage of time.[139]

Fox devotes a whole chapter of her book on the downwinders of the American West to the act of neighbors "writing down names." Among the first people to begin making maps of the strange clusters of neighbors becoming sick downwind of the Nevada Test Site was Irma Thomas of St. George, Utah. "She pointed through the living-room walls toward the homes of neighbors in the residential area," remember journalists Harvey Wasserman and Norman Solomon in 1982. "She had compiled a list of thirty-one cancer victims who lived in the houses within a block radius."[140] Fox describes how after drawing up her own maps of illness in her neighborhood, Thomas visited her friend Loa Johnson in Cedar City, whose daughter had recently died of leukemia. "By the time Irma left, Loa had agreed to become a gatherer of names."[141]

After Lois Gibbs, a member of the Love Canal Homeowners Association, discovered that her Niagara Falls subdivision was built over a toxic chemical waste dump, she began to organize with her neighbors, many of whom had sick family members. "I kept thinking there must be a way to prove the illnesses were caused by the canal," remembers Gibbs. "I took out our health-survey notebook and started to put squares, triangles, and stars on a street map, with a different symbol for each disease group: central-nervous-system problems, including hyperactivity, migraines, and epilepsy; birth defects and miscarriages; and respiratory disorders. Suddenly a pattern emerged!"[142]

Various forms of learning to see the unseeable have been essential to victims of radiation exposures grasping and documenting what experts tell them cannot be true according to the LSS. Tom Bailie is a farmer whose land sits across the Columbia River from Hanford. "Bailie's white farmhouse was within what locals called 'the death mile,' where only one of ten farms had escaped cancer," writes Karen Dorn Steele. "One couple, Juanita and Leon Andrewjeski, showed me a 'death map' they'd compiled to track the fate of their neighbors. It was marked with Os for cancer deaths and Xs for early heart attacks."[143]

Community mapmaking around radiological exposures has been enhanced by digital-age tools. Jenell Wright was connecting with friends from high school through Facebook when they began to notice that many of their former classmates had died from a small number of similar diseases. She thought that there may have been some toxic agent in the school itself, until she realized that this was also true for friends from elementary school who attended different high schools. The group began to make digital maps and place different color-coded dots on a map for each case of a specific disease. They noticed that the dots clustered around Coldwater Creek, which ran through the neighborhood where they all grew up.

The Mallinckrodt Chemical Works processed uranium during the Manhattan Project and the early years of nuclear weapon production in the United States, at its St. Louis complex. When it shut down operations, it left 133,000 MT of radioactive waste behind. This material was stored at various places around the region, and ultimately 8,000 MT of waste and soil was dumped into the West Lake Landfill adjacent to Coldwater Creek. The waste leaked from the landfill, contaminated nearby soil and water, and created numerous disease clusters along the creek.[144] The community currently living near the landfill has concerns about exposures but far more dire concerns about a catastrophic possibility: a fire has been detected burning in the landfill, an "ongoing major proximal subsurface fire."[145] An earlier fire inside the landfill in 1993–1994 was smothered with a "concrete slurry cap."[146] The second fire was announced to the public in 2013, and in 2021 was still burning, approximately one thousand feet from the location of the radioactive waste. The main radionuclide in the landfill is ^{230}Th. Bob Alvarez writes, "With a half-life of 77,500 years, thorium-230 is about 60,000 times more radioactive than uranium," and if the thorium in "the landfill comes in contact with a fire and releases breathable particles into the air," there is a high risk to those living nearby to the site.[147] Dawn Chapman, a mother of three and the cofounder of Just Moms STL (an NGO that advocates for the removal of the radioactive waste and remediation of the West Lake Landfill) describes her preparations if the fire reaches the nuclear waste: "I have my escape route planned, I literally do, I have my grab bag, I have it in the car, I have a little emergency kit, and we go that way, the landfill is that way," she says pointing first in one direction for her planned escape and then in the opposite to indicate where the aerosolized radionuclides would be coming from.[148] Each day brings ongoing risk as the fire smolders.

As Fox has commented, some people's maps were of "imaginary spaces." This was specifically true for cohorts of people who were exposed to radiation as a group but who did not live in one place, such as military service personnel that took part in nuclear weapon tests. Many such veterans who had returned to civilian life in various locations in their home countries reached out and found fellow "atomic soldiers," at first through groups of friends from the service, then through traditional veterans associations, and later through associations of similar hibakusha, like the

National Association of Atomic Veterans (US) founded in 1979, the British Nuclear Test Veterans Association (UK) founded in 1983, and the Association des Vétérans des Essais Nucléaires (France) founded in 2001. Through associations, newsletters, and eventually online, radiation-affected vets have created databases of the health trajectories of those who participated in and were exposed to the same nuclear tests.

Conclusion

Novelist William Gibson asserts that the future has already arrived, it's just unevenly distributed.[149] Communities that have had generations-long encounters with radionuclides in their local ecosystems may be living in a future that is seeping—slowly migrating throughout the planet—into wider distribution. As radiologically contaminated lands become sites of historical rhinoplasty, an industrial construct of nature is superimposed upon these epicenters of nuclear mismanagement—a facsimile of remediation coated with a patina of memory loss. The front lines of the future may already be behind these communities, as a more dramatic, even biblical, cosplay is slowly being constructed around the deep geological repositories slated to hold the hundreds of thousands of metric tons of spent nuclear fuel from both the plutonium and energy production of the world's nuclear power plants. The high-level waste facilities that will receive this staggering (current) amount of spent nuclear fuel are intended to effectively contain the highly toxic and radioactive materials for millennia in sites that will sculpt surface caps depicting a natural setting marked only by effective, competent messages to future generations. This will be discussed in greater detail in chapter 7, but the *remediation* of the earlier-discussed sites of production and contamination model the discourse and curation of the more-permanent sites planned and under construction: they serve as totemic stage sets, signifying recovery and telescoping long-term health and well-being. They will hold a deep and monstrous seed that may well sprout outside of its designer containment, because, while we cycle through hundreds and thousands of generations forward in time, these fuel rods carry our waste—our sins—forward and ensure that no matter how widely the future is distributed, the past that we cloaked in these repositories will be ever present.

Invisible, imperceptible radiogenic materials have wreaked havoc throughout the world as they have migrated, been transported, been deposited, and made their way inside of our communities, our homes, and our bodies. At the first Indigenous World Uranium Summit held in 2006 at Window Rock of the Navajo Nation, a clear statement was issued about the effects and desires of the attendees regarding nuclear technologies: "Past, present and future generations of Indigenous Peoples have been disproportionately affected by the international nuclear weapons and power industry. The nuclear fuel chain poisons our people, land, air and waters and threatens our very existence and our future generations. Nuclear power is not a solution to global warming. Uranium mining, nuclear energy development and international agreements . . . that foster the nuclear fuel chain violate our basic human rights and fundamental natural laws of Mother Earth, endangering our traditional cultures and spiritual well-being."[150]

PART

III

Warlords

5

Selecting the Irradiated

Zeroing in on Bikini and Enewetak

After the nuclear attacks on Hiroshima and Nagasaki, the United States began to search for a postwar nuclear test site. "We just took out dozens of maps and started looking for remote sites," recalled Horacio Rivera, one of the members of the newly formed Naval Office of Special Weapons.[1] "Above all, it had to be away from population centers of the US . . . and yet in an area controlled by the US," testified officials to Congress.[2] In January of 1946, the US Navy announced that Bikini Atoll in the Marshall Islands, a site that "may accurately be described as one of the most remote places of the earth," was chosen.[3] Privately, however, Admiral William Blandy, who would oversee the first test series, explained that "It was important that the local population be small and co-operative so that they could be moved to a new location with a minimum of trouble."[4]

The Marshall Islands had passed from colonial power to colonial power since the late nineteenth century and were occupied by the Japanese during World War II. The United States defeated the Japanese there and took control of the chain in 1944. After the war, the islands became one of eleven trust territories under the administration of the UN Security Council. The Trust Territory of the Pacific Islands was the only one of these designated as a "strategic" trust in 1947, as it was already under military control and use by the United States.[5]

On 10 February 1946, the newly appointed American military governor of the Marshall Islands, Commodore Ben H. Wyatt, traveled to Bikini to inform the Bikinians of their fate. It was, of course, described as their "choice," but it was clear that the choice could only align with the decision that the new military occupiers had made a month earlier. The Marshallese, having been visited for centuries by streams of missionaries, were religious Christians. Wyatt waited until the 167 Bikinians were leaving church services that Sunday morning and called them to assemble so that he could address them. According to an official US Navy account, Wyatt "compared the Bikinians to the children of Israel whom the Lord saved from their enemy and led unto the Promised Land. He told them of the bomb that men in America had made and the destruction it had wrought upon the enemy."[6] Wyatt explained that the United States was now intent on testing this new weapon so that they could "put an end to war," and without explaining why, that Bikini Atoll was the very best place in the world to test this weapon. He addressed King Judah, the leader of the Bikinians, and asked him if he would agree to the people leaving their homes "temporarily" so that the United States could test this weapon "for the good of mankind and to end all world wars."[7]

King Judah consulted with his people and replied, "If the United States government and the scientists of the world want to use our island and atoll for furthering development, which with God's blessing will result in kindness and benefit to all mankind, my people will be pleased to go elsewhere."[8] Later, on 6 March, Wyatt had the Bikinians reenact this encounter eight times so that movie cameras could record it from multiple angles.[9] It was then put to effective use in US domestic news presentations of this appropriation of Bikini. According to one military official quoted in *Time* magazine, "It was one hell of a good sales job."[10] Even before the nuclear tests, MGM ran a ten-minute short in cinemas about the preparations for Operation Crossroads and the evacuation of the Bikinians. Shown deconstructing their homes, the narrator asks, "But while these 167 simple and unselfish human beings begin tearing up their modest households by the very roots let's ask why? Why these people of all people must sacrifice their ancient heritage, their proud traditions, even their sacred burial ground to the march of science? Why? Well, I'll tell you why." The reason he gives is to help the world to achieve peace, and to unlock the secrets in the atom that can make the world a paradise. The narrator explains that the test site had to be "necessarily at a spot distant from the world centers of population." As they are taken aboard the ships that would remove them from their homeland forever, we are told, "the people of Bikini join their pathetically humble belongings onboard."[11]

In March, the Bikinians were removed from their homes, never to permanently return. Even after the seizure, the sense of dislocation felt by the Bikinians was utterly dismissed by Americans, as the Marshallese were not considered sophisticated enough to discern one atoll from another. Geographer Jeffrey Sasha Davis reminds us of a contemporary 1946 *New York Times Magazine* article that asserted, "As for Juda and his people, now living on Rongerik Atoll, they probably will be repatriated if they insist on it, though United States military authorities can't see why they should want to: Bikini and Rongerik look as alike as two Idaho potatoes."[12]

The seizure and irradiation of Bikini Atoll typifies the selection of nuclear weapon test sites around the world. Aspiring and established nuclear weapon states select locations for testing their nuclear weaponry in the far reaches of their military empires or domestic landmasses, displacing and contaminating marginalized populations with little recourse, information, or compensation. Historian Gabrielle Hecht has written, "Once the weapons were built, the imperial cycle began anew, with atomic bombing—more palatably referred to as 'nuclear testing'—of the Marshall Islands, the Sahara, the Navajo Nation, Maralinga, Moruroa, and other colonized spaces."[13] These marginal populations had no say in the use or seizure of their land and were left to deal with displacement or being abandoned to live with contaminated homes and food sources for generations. They have been kept in the dark about the nature of what was being done to their land or waters and about the extent of the contamination, and often they had access to medical assistance denied or tightly controlled to impede them from understanding the nature or origin of their families' ongoing medical problems and to shield information about their experiences from the world.

"The ultimate expression of sovereignty resides, to a large degree, in the power and the capacity to dictate who may live and who must die," explains Achille Mbembe. "To exercise sovereignty is to exercise control over mortality and to define life as the deployment and manifestation of power."[14] The selection of communities to bear the brunt of nuclear weapon testing by nuclear weapon states during the Cold War was a seminal expression of Mbembe's necropolitics.

"Who are these radiogenic communities living adjacent to and downwind from the Cold War nuclear complex?" asks anthropologist Barbara Rose Johnston. "Typically, they are the marginal and powerless groups in society: indigenous people and other social or political minorities."[15] However, they were not subject to the extraction of raw materials for profit by their colonial occupiers; what they possessed was open space into which the effects of nuclear detonations could be unleashed—a space that could be contaminated without political recourse. Their colonialization was that of a *nonplace*, a dumping ground for the most long-lived toxic substances on Earth, which would plague them and will plague their descendants for untold generations. Their marginality, their powerlessness, was the resource that drew imperial occupiers to their lands and seas.

In a meeting of the Advisory Committee on Biology and Medicine of the AEC in January of 1956, Merril Eisenbud, the director of the AEC's Health and Safety Laboratory in New York, spoke about the useful information to be gleaned through the study of the Marshallese who had just been returned to their homes on Utirik Atoll after the 1954 Bravo test, a place he referred to as "by far the most contaminated place on Earth." Eisenbud laid bare his view of the dignity and humanity that he afforded to the Utirikese: "While it is true that these people do not live, I would say, the way Westerners do, civilized people, it is nevertheless also true that these people are more like us than mice."[16] Eisenbud valued the human bodies of the irradiated, as they would yield information more useful to "civilized people" than would mice. Thus, the nuclear subaltern was identified, marginalized, and brutalized.

As described in chapter 2, the underwater Baker test of Operation Crossroads conducted at Bikini Atoll in 1946 resulted in a radiological disaster. The residual radiation that would typically disperse downwind via a mushroom cloud in an atmospheric detonation instead lingered in the lagoon at Bikini and spread throughout the islands that make up the atoll. This contamination would lead the AEC to label Bikini a "radiobiological laboratory" for marine biologists; however, further testing at the atoll was problematic.[17] The AEC designated Bikini Atoll as being on "interim status" and began to look for another site to continue nuclear testing. While it did consider potential sites outside of the Marshall Islands, it selected nearby Enewetak Atoll in 1947, in part because it offered similar advantages to Bikini Atoll but also had a landmass large enough to build a runway for military aircraft.[18]

The official report of the US DOE asserted that the "scheduling of the first Enwetok [*sic*] nuclear test in the near future necessitated the immediate removal of the people" who were displaced on 21 December 1947.[19] The Nuclear Claims Tribunal, which later adjudicated claims by the Marshallese against the United States government for health and property damage, described how "In December of 1947, the people of Enewetak were removed from Enewetak Atoll and transported to Ujelang Atoll. Representatives of the U.S. government represented to the people that the relocation would be temporary, in the likely time frame of three to five years, at which time they could return to Enewetak." The US would conduct forty-three nuclear tests, including large thermonuclear tests, at the atoll. The report then clarifies that when the land was returned to the Enewetakese in 1980 (thirty-three years later), of the original 1,919.49 acres that they had lost, only 815.33 acres were returned to them. Ultimately, 949.8 acres were considered to be unusable, and 154.36 acres "had been vaporized."[20]

The "unusable" land includes a site on Runit Island where, "~545 GBq of contaminated topsoil, vegetation, and debris (concrete and metal) . . . was subsequently entombed within an unlined crater produced by an 18 kt surface test and capped with a concrete dome. The site is now known as the Runit Dome."[21] A team led by oceanographer Ken Buesseler of Woods Hole Oceanographic Institute conducted radiological assessments on Enewetak, and specifically at Runit Dome in 2018, and found that the groundwater seepage from the burial site accounted for half of the plutonium found in the sediment of the lagoon in Enewetak.[22] Rather than encapsulating the waste, it became time-released.

The United States did not test thermonuclear weapons inside the continental US, but at multiple sites in the Pacific, primarily the Marshall Islands and at Christmas Island in Kiribati. While devastating for local communities, the fallout of this testing had a global fingerprint: "During the premoratorium years of 1952–1958, large-scale testing programmes of the USA were conducted at the Pacific Proving Grounds, and these tests account for 50% of the total fallout inputs" globally of those years.[23]

Throughout the Cold War additional sites would be chosen by nuclear weapon states to test their weapons. I have discussed the history and irradiation of communities at those test sites in other parts of this book, but here I want to look specifically at the selection of the sites of atmospheric nuclear testing during the Cold War. Starting with the Soviet Polygon site in Kazakhstan (a direct response to the American acquisition of nuclear weapons) and the Nevada Test Site (a direct response to the Soviet acquisition of nuclear weapons in 1949), I will also examine the second primary Soviet site (of many) in Novaya Zemlya, and the various test sites of the United Kingdom and France, and finally the Chinese test site in Xinjiang Uyghur Autonomous Region.

Kazakhstan and Novaya Zemlya (Soviet Union)

On 20 August 1945, in the shadow of the successful American nuclear attacks on Hiroshima and Nagasaki, the Special Committee (Spetskom) of the Soviet State Defense Committee met in Moscow with the mission of accelerating the slow-moving wartime Soviet atomic bomb project.[24] The chairman of Spetskom was Lavrenty Beria, the universally feared chief of Stalin's secret police. Two years later, as the project accelerated from lab work toward an eventual test of the weapon, Beria was put in charge of selecting a site for the final assembly and detonations of the Soviet weapons. Beria chose a part of the steppe in the eastern corner of the Soviet Republic of Kazakhstan along the Irtysh River, on the edge of Siberia and near the Chinese border. Beria is said to have favored the site because it was "uninhabited," ignoring the twenty thousand people living close to the test site in villages and over one hundred thousand people living 140 km downwind in the city of Semipalatinsk.[25] The Kurchatov Institute, the headquarters of the site, first called "Test Site Number 2," then "Moscow 400," and then "Semipalatinsk-21," was renamed as the city Kurchatov by the newly independent Kazakhstan after the fall of the Soviet Union.[26] From 1949 until 1989 the former Soviet Union conducted 456 nuclear tests in the Polygon (as the large test site area was known) including 111 in the atmosphere, including its first two thermonuclear weapons.[27]

The first test in 1949 resulted in heavy contamination of communities to the east of the Polygon, as the wind had shifted and the fallout cloud was carried directly over their villages. During the earliest Soviet tests, the nearby Kazakh villagers were often not informed about nuclear tests and were both terrified and unshielded during the explosions and subsequent deposition of fallout. The government often required them to come out of their houses and stand in the street, claiming that the most significant danger was that the shock wave of the blast could damage their houses, ignoring that the lack of any shelter from passing fallout clouds would significantly increase people's exposures. No barriers or warning signs were erected to keep villagers from accidentally wandering into highly con-taminated test site areas, and to this day access to the Polygon is essentially unrestricted, and high radiation areas are unmarked.[28]

The new Kazakh government, in its application for UNESCO World Heritage status for the former test site, claimed, "All territory of former Semipalatinsk province suffered contamination by the products of nuclear blasts, and 1.2 million people got additional irradiation within various dosage range[s]." The World Heritage application describes how during the 1953 first test of a thermonuclear weapon in the Polygon: "Forty persons were left in Karaul village for [the] study of radiation effects on living organism[s]."[29] A recent epidemiological study concluded: "In terms of actual health effects, the Kazakh and Japanese scientists noted that the rate of cancer in those living in eastern Kazakhstan, the area most exposed to radiation, remains 25–30 percent higher than elsewhere in the country; they also reported a higher chance of mental deficiencies in children born to parents who were exposed to radioactive fallout from testing."[30]

To this day thousands of Kazakhs live in villages that remain heavily contaminated by radioactivity from the Soviet testing. Most people live a traditional lifestyle in which they grow most of their own food and raise most of their own livestock for meat and dairy. The horse milk that is a staple for these Kazakh villagers comes from horses that graze on con-taminated grasses, and the fish that they catch come from contaminated streams: many people live permanently with high levels of radiation and high levels of birth defects and radiation-induced illnesses.[31]

While the Soviet Union had tested its first thermonuclear weapons in the Polygon, with the subsequent large fallout clouds depositing radionu-clides on the villages to the east of the test site, it decided to establish a site specifically devoted to thermonuclear testing in the Arctic. In July 1954 the large archipelago of Novaya Zemlya in the European Soviet Arctic was decreed a nuclear test site (originally named "Object-700") by the USSR Communist Party Central Committee and the USSR Council

"Stronger Than Death," memorial to those who suffered from Soviet nuclear weapon testing in Kazakhstan, located in Semey, formerly Semipalatinsk (2010). Photo by author.

of Ministers. The decree also declared that all inhabitants of the archipelago should be deported from their homes and villages. The 536 indigenous Nenets and Russian residents were first gathered in the village of Lagernoe, about halfway up the archipelago.[32] "The evacuation of the islanders to this one settlement was by force," writes Leonid Serebryanny. "People were expelled from their homes and forced to leave boats, motors, nets and other property. Dogs were bought at low prices by the military or shot." For two years the Novaya Zemlya residents were detained in Lagernoe, in huts without windows, and were not allowed to hunt or work. In 1956, all were forcibly deported to the Russian mainland.[33]

In September 1955 the first nuclear test was conducted at Novaya Zemlya, an underwater test (the site has always been under the control of the navy), detonated on the southern island at the Chernaya Bay site, the first of three test zones. The Soviets were to conduct 224 nuclear detonations on Novaya Zemlya, including 88 atmospheric tests. The largest nuclear weapon ever detonated on the face of the Earth was the RDS-202, an atmospheric test known as the Tsar Bomba in 1961 (code-named "Big Ivan"), with a yield of 50 mt (the largest US test was the Bravo weapon at 15 mt). The Tsar Bomba was detonated close to the former Lagernoe settlement (now abandoned) at the southern tip of the northern island.[34] The monstrous bomb had an immense impact: "The zone of lethality and destruction of dwellings was out to 120 km; the zone of eye damage was out to 220 km; a shock wave in air was observed at Dickson settlement, at 700 km; windowpanes were partially broken to distances of 900 km." Radioactivity from the blast was recorded worldwide, while "two main trajectories of significant radioactive fallout were observed: (1) directly to the South as far as the Caspian Sea, and (2) towards the Sea of Okhotsk stretching several thousand km to the southeast from Novaya Zemlya."[35]

Multiple studies have found a range of radionuclides permeating the ecosystem throughout the region. In addition, the seas around Novaya Zemlya were used extensively for dumping military nuclear waste, which has further contaminated the Russian Arctic.[36] In a 2009 report, Russian scientists found that there had been "the dumping of substantial quantities of solid nuclear wastes into shallow bays located along the margin of the islands of Novaya Zemlya." These include "Nuclear reactor compartments, ships, a submarine and containers of low level radioactive waste." The waste dumping has led to even more substantial contamination of ^{137}Cs and several isotopes of plutonium throughout the region.[37] In 2018, scientists surveying the radiological impact of the dumping in the Kara Sea, into which the Novaya Zemlya archipelago stretches, also discovered significant amounts of radionuclides from the atmospheric tests conducted at the site entering into the sea, as glaciers where fallout had deposited were melting from global warming, creating a new radiological transport vector.[38]

Nevada (United States)

The United States conducted the first nuclear test in 1945 in New Mexico, several hundred thousand kilometers south of the Manhattan Project laboratory at Los Alamos, where the weapon was designed and assembled. As noted above, the United States opened its Pacific Proving Ground test site in the Marshall Islands in 1946. Once the Soviet Union acquired nuclear weapons in 1949, the United States sought a second nuclear weapon test site that could include battlefield maneuvers by military personnel in preparation for fighting a nuclear war. Additionally, weapon designers complained that the long amount of time that it took to travel back and forth to the PPG was delaying advances in weapon design in the push to beat the Soviet Union to a functional thermonuclear weapon. Toward these goals, the United States decided in early 1950 to establish a domestic nuclear weapon testing site.[39] Several sites were considered, guided by the criteria of being flat, desolate, and far from large population centers, but military planners quickly decided on a site about 100 km north of Las Vegas, in part because of its proximity to the nuclear weapons lab in Los Alamos.[40] The Nevada Test Site was established by late 1950, along with Camp Mercury, where troops who would participate in nuclear tests could cycle in and out. The first test at the NTS was in 1951, and subcritical and virtual testing continue there today.[41]

Originally described as being virtually uninhabited, the area downwind from the NTS was populated primarily by Native Americans and Mormons, both communities with little political agency in the United States in 1950. Downwinders in southern Utah and Idaho claim to have been sent an AEC document by a whistleblower that refers to the site being chosen, in part, because the residents in the downwind communities were "a low-use segment of the population."[42] The policy of the AEC, which oversaw US nuclear weapon testing, was to test weapons at the NTS only when the wind was blowing to the east or north. Typically, the wind in this area would blow either toward the east or toward the south, where Las Vegas and beyond it Los Angeles, with their sizable populations, were located.[43] To the east, however, were thousands of Native Americans and Mormons; thus, it was the explicit policy of the AEC to test nuclear weapons in Nevada when the wind was blowing directly toward the communities of eastern Nevada and southern Utah.[44] The most populated town in this region was St. George, Utah, a city of forty-five hundred, which was frequently irradiated by passing clouds of fallout. In an AEC meeting in 1955, chairman Lewis Strauss complained about the *inconvenience* of the frequent contamination by the tests of St. George, "which they apparently always plaster." AEC fallout expert John Bugher advised that the problem of St. George was primarily one of "public relations," while another AEC commissioner, Willard Libby, remarked glibly about living downwind of the NTS that "People have got to learn to live with the facts of life, and part of the facts of life are fallout."[45]

While those living to the east of the Nevada Test Site were considered more acceptable to irradiate than the people living in the urban centers to the south, they were still American citizens and, as such, were protected more than the Marshallese. The United States had an explicit policy of testing only lower-yield fission weapons at the NTS, and testing all of their much higher-yield thermonuclear weapons at the PPG in the Marshall Islands. They used this colonialist prioritization to help pacify the people living downwind from the NTS in official government documents. In a 1953 AEC publication titled *Assuring Public Safety in Continental Weapons Tests,* the AEC claimed, "Since the larger test detonations could not be held within the United States with the requisite degree of safety, construction of firing areas and supporting facilities at the Pacific Proving Ground at Eniwetok [*sic*] proceeded."[46]

Another AEC pamphlet distributed to downwinders sought to comfort the local residents by telling them that the "forces released by test detonations in Nevada are very small compared to the tremendous forces released by the large fission and hydrogen weapons tested in the Pacific. So-called 'H-bombs' are not tested in Nevada."[47] While the downwinders at the NTS may have been considered "low-use," they were primarily white

Americans. Radioactive fallout from high-yield weapons, while more than the United States was willing to inflict on its own communities, was not envisioned even as a "public relations" problem for the Marshallese.

US service personnel were routinely exposed to radiation at both test sites. Thomas Saffer was a young marine who had been sent to the NTS to take part in military maneuvers during Operation Plumbbob in 1957. He was one of 1,133 US military personnel to participate in Shot Priscilla on 24 June 1957 from a "two and a half feet wide and five and a half feet deep" row of trenches. One of the "radiation protection procedures" of the exercise stipulated "maintaining minimum safe distances from nuclear detonations," yet Saffer's trench was located just 3.2 km from the weapon, which at 37 kt was more than twice as powerful as the Hiroshima bomb. "The trench lines in front of us collapsed. They had to dig the people out. After the gyrating and the shaking stopped, we stood up and looked at the fireball escalating into the troposphere," recalls Saffer, and then, "All of a sudden, the fallout started. . . . The radiological safety person from the Army came up with a Geiger counter and put it over us and we could hear it clicking, clicking, clicking. Then one man on each side dusted us off with a broom because the prevailing thought was if you get rid of the dust you get rid of the radiation. So, we were 'decontaminated.' "[48]

At the NTS, 928 nuclear tests were conducted, including 24 tests conducted jointly by the United States and the United Kingdom.[49] Multiple scientific studies have correlated radioactive fallout from the tests with heightened cancer and disease rates and early mortality across the United States.[50] In 1990, the United States Congress passed the Radiation Exposure Compensation Act, which provides "one-time benefit payments to persons who may have developed cancer or other specified diseases after being exposed to radiation from atomic weapons testing or uranium mining, milling, or transporting" to Americans who had lived in specific, stipulated locations.[51]

Australia and Kiribati (United Kingdom)

As participants in the Manhattan Project, and possessing shared nuclear information with the Americans (from the Quebec Agreement between Roosevelt and Churchill in 1943), it was relatively easy for the

British to develop their own nuclear weaponry after World War II, even after the US cut off collaboration.[52] Determining where to test that weaponry was more complicated. The United Kingdom was considered to be too small and populated for testing nuclear weapons, even though they did briefly consider the possibility of a site in the Scottish Highlands. The British had hoped to test weapons at one of the American test sites but also began to explore sites in Canada and the uninhabited Montebello Islands off the western coast of Australia. In 1951 the British formalized an agreement with the Menzies government in Australia to begin testing at Montebello the following year.[53] The first test, code-named "Hurricane," was conducted at Montebello in October 1952, and another two tests were conducted on the islands in 1956. Two weapons were detonated in South Australia at Emu Field in 1953, resulting in significant fallout along the east coast of Australia. Finding Emu Field too remote, the British requested a permanent weapon test site that afforded easier access. In 1955 it was announced by both governments that this site had been chosen at Maralinga in South Australia, not far from Emu Field. A total of seven weapons were tested at Maralinga in 1956 and 1957.[54]

At the time of the testing, the Aboriginal people of Australia had not been granted citizenship and had virtually no political power or standing. Emu Field and Maralinga were both home to several different Aboriginal tribes, although they were considered by the Australian government and the British military to be nomadic people and the land was deemed "virtually uninhabited." Walter MacDougall was a native patrol officer with the Commonwealth Department of Supply in Australia, whose job was to ensure the British testing effort (and nearby rocket testing range) would not infringe on the Aboriginal communities that lived in the test site area. MacDougall, who was genuinely sympathetic to risks to Aboriginal communities, with one assistant attempted to survey thousands of square kilometers of desert, an impossible task. Anthropologist Kingsley Palmer commented of MacDougall that he "was required to determine the numbers of Aborigines in the desert regions which later became the test sites. One man was scarcely able to give an accurate account of the Aboriginal population over 100,000 square kilometres."[55] Any recommendations that MacDougall made or any cautions that he raised were crisply dismissed. In the 1985 *Report of the Royal Commission* that investigated exposures of

military personnel and civilians from British testing in Australia, it was stated
that Richard Penney, the chief British scientist on the mission, had complained that MacDougall placed the "affairs of a handful of natives above
those of the British Commonwealth."[56] Nonetheless, MacDougall necessarily declared the desert test site to be abandoned, and testing proceeded
while many people were still living in the prohibited areas.[57] Today, at the
site of the original detonation of the Totem test in 1953, a stone marker
stands at Emu Field to warn of the dangerous radiation present. It proclaims,
"Radiation Hazard / Radiation levels / for a few hundred metres / around
this point may / be above those / considered safe for / permanent occupation." Although those who traditionally occupy or live in this area are
Aboriginal peoples, the marker is written only in English.[58]

Historian Roger Cross recounts the specific journey of Eddie Milpuddie
of the Maralinga Tjarutja people and her family into the test site area: "In
1955, one year before the Maralinga tests began, the Milpuddies set off
from the Everard Ranges for Ooldea, not knowing that the Aboriginal
people had been removed to Yalata. They made use of a new system of
dusty tracks that passed for roads, as well as using local knowledge gained
over thousands of years. They followed the old rock waterhole route en-
suring access to water and crossing apparently waterless terrain, eventually
they came to Maralinga. 'WARNING. You Are Approaching A RADIO-
ACTIVE AREA. Read All Notices,' proclaimed the signs placed around
the site—signs that Eddie and her family could not read." The Milpuddie
family, invisible in multiple frameworks to the British and Australian forces,
would intimately encounter this radiation and the dehumanizing manner
in which Aboriginal people were treated by these military forces and their
scattered radiological particles:

> The second 1956 bomb was a ground burst of 1.5 kilotons on 4
> October at the Marcoo site. It left a large crater. On 14 May 1957,
> Eddie, her husband and children were camping in the crater itself.
> She tells how soldiers came and took her and her family to the
> Maralinga village. It was a shocking trip; they had never ridden in a
> vehicle before and they vomited everywhere. At Maralinga they
> were given the first showers of their life, with soap that got into
> their eyes and blinded them. . . . The family were showered four or

five times before the Geiger counter showed an acceptable level of radioactivity and they were given clothes. The soldiers shot their four hunting dogs. No medical follow-up occurred. Eddie was pregnant at the time of the incident and she subsequently buried a stillborn baby, which she believed was because of poison in the ground. Her next baby died, aged two, of a brain tumour in 1963 and the next was born very premature.[59]

This series of incidents was described in the later inquiry into the history of testing in Australia conducted by a Royal Commission, and has become known as the "Pom incident."

The British and Australian governments had an agreement that the UK would not test hydrogen bombs in the nation, and so when the British began to prepare for its first thermonuclear weapon test, the Grapple test, an alternative test site had to be established.[60] Following the lead of the Americans, their attention turned to small islands in the Pacific Ocean deemed far enough away from "populated" areas to allow detonation of immense weapons and their massive fallout clouds.[61] The site they chose for the first thermonuclear test was Malden Island in the central Pacific. The test was staged from Christmas Island; both Malden and Christmas Islands were part of the Line Islands, which had been incorporated into the colony of the Gilbert and Ellice Islands in 1919. The Gilbert Islands became independent of the UK in 1979 and became the Republic of Kiribati. The Line Islands, which were being administered by the US, became independent in 1983 and joined the Republic of Kiribati.[62] Speaking on a BBC documentary in 1983, Royal Air Force Bomber Command Squadron leader Roland Duck recalls, "When we were first given the task of finding a site, we decided the nicest way and the easiest way was to find a vast expanse of water. The largest amount of water with the least land is the Pacific. So, we took a large map of the Pacific and we really settled our finger and put it down rather like picking a winner and we got Christmas Island and Malden Island."[63] Journalist Nic Maclellan quotes from a British military report assessing the possibilities of radiological contamination to populated areas from thermonuclear testing in the Gilberts. Imagined radiological doses were described as being culturally distinct: "For civilized populations, assumed to wear boots and clothing and to wash, the amount of exposure necessary to produce this dosage is more than is necessary to give an equivalent dosage to primitive peoples who are not assumed to possess these habits. . . . It is assumed that in the possible regions of fallout at Grapple there may be scantily clad people in boats to whom this criteria of primitive people should apply." At a subsequent meeting in which the findings of the military report were presented to the UK defense minister, this became reduced to the statement: "Only very slight health hazards to people would arise, and that only to primitive peoples."[64]

The United Kingdom conducted nine nuclear tests on Malden and Christmas Islands during 1957–1958. They had varying amounts of success in achieving thermonuclear detonations with a few of the tests, but the third nuclear weapon state did fully succeed in becoming the third thermonuclear weapon state. Grapple X was tested on 8 November 1957, detonated 1 km above the southeastern portion of Christmas Island and yielding 1.8 mt. The Grapple Y test, on 28 April 1958, yielded 3 mt and was the largest British nuclear weapon ever detonated. Grapple Y was air-dropped about 2.5 km above the sea and 1.6 km off the southeastern tip of Christmas Island. Yet "the explosion was closer to sea level than expected. The detonation sucked up quantities of water and debris into the mushroom cloud, irradiating them in the process," subsequently spreading fallout across the naval flotilla and the main military base on the northern, inhabited part of the island. In 1998, journalist Nic Maclellan interviewed Sui Kiritome, a young woman living on Christmas Island at the time of the test, who was briefly removed, like most residents, onto a British boat. Kiritome recalls that "Some time after the test, something happened to my head and face. Every time when I combed my hair, I was losing strands of my hair and something like burns developed on my face, scalp and parts of my shoulder. My face was the worst affected because I was looking up at the black cloud from the blast which was directly above us when the light shower fell on my face."[65] Paul Ah Poy, a member of the Fiji Royal Naval Volunteer Reserve, was assigned to assist the British military during seven of its nuclear tests at Malden and Christmas Islands in 1957. Ah Poy was told that if he opened his eyes during the test his "eyeballs would melt out of his head," and so he held his hands firmly over his eyes during the detonation, during which he could see the bones of his fingers through his closed eyes.[66]

Equipment abandoned by the British at their nuclear testing base on Christmas Island, Republic of Kiribati (2017). Photo by author.

Teeua Tetoa, the leader of the association of hibakusha on Kiritimati (Christmas Island) in Kiribati, recalls how in 2003 British nuclear veterans and BBC reporters who had come to Christmas Island told the locals that they and their children had suffered from cancer as a result of the exposure to radiation during nuclear tests on the atoll, and that the Kiritimati people were likely to be similarly suffering. The Kiritimati did not immediately believe them, but "In 2009 and 2010 some of our kids died of cancer—many, many in our association. And then we remembered the time when they told us about the problem."[67] The Kiritimati hibakusha have unsuccessfully tried to seek compensation, medical care, and simply an answer to why the British came to their island on the other side of the world from the UK to test thermonuclear weapons.

In 1958 the United Kingdom and the United States signed the UK-US Mutual Defense Agreement, which "enables the two countries to exchange classified information to enhance each party's 'atomic weapon design, development and fabrication capability.' "[68] The US, USSR, and UK observed a nuclear test moratorium from late 1958 to the summer of 1961. After the abandonment of the moratorium, this agreement facilitated the UK's continuing nuclear testing at the Nevada Test Site, where the United States also resumed the testing of fission weapons. As per its policy, all American thermonuclear testing was kept in the Pacific. Since the United States was not inclined to resume testing at the heavily contaminated sites on Bikini and Enewetak Atolls in the Marshall Islands, they moved their thermonuclear testing to the UK site on Christmas Island and Johnston Atoll to the northwest.[69] In 1962 the United States conducted twenty-two nuclear tests on or near the island, including five in the megaton range.

Algeria and French Polynesia (France)

France has tested 210 nuclear weapons and, like the UK, has never tested a nuclear weapon within its own borders. Disarmament expert Tariq Rauf has pointed out that while "France claims that all its tests have been conducted on 'French' territory," that is just an imperialist claim that colonial spaces are actually part of France.[70] As the French government was deciding to pursue nuclear weapons in the late 1950s, a key problem in determining a nuclear test site was that the French empire was declining and offered limited and often combative options for potential colonial locations. The preferred choice of the French military was in the Sahara Desert in the then French colony of Algeria. However, in 1954 the Algerians began a war of independence that made the investment in a large military infrastructure in the desert to stage and support nuclear weapon testing a questionable investment. Another option was French colonial holdings in the South Pacific, an area that the French had largely ignored and that lacked the infrastructure, such as airfields, power generation, and intact structures to house the materials and personnel to stage the tests. Additionally, it was on the other side of the world from France, and it would be impossible for large cargo transport planes to reach without stopping at foreign airfields en route, a scenario the French wished to avoid since the planes would be carrying militarily sensitive and occasionally radioactive cargo.[71] The French conducted studies of sites in the Alps and Pyrenees, but only for underground testing. Ultimately it was determined that there was no way to ensure radioactivity from the tests would not enter the groundwater sources for French cities.[72] No such extensive studies of ecological impacts were made of the proposed colonial sites.

In 1956, the government of France decided to proceed with testing in the southern Sahara region of Algeria, even though the war of independence made long-term use of the site doubtful. This was a provisional step: at roughly this same time it was decided to begin developing the infrastructure and, more importantly, the political receptivity to French nuclear testing in French Polynesia.[73] This dual plan was due to two key considerations: first, the likelihood that the French would lose access to its military bases in Algeria; and second, they did not believe that they could atmospherically test thermonuclear weapons in Algeria without broad opposition. Thus, a site in the South Pacific was pursued even as preparations for the start of French testing in Algeria began.[74]

The first French nuclear test was conducted in 1960, an atmospheric test detonated near Reganne in southwestern Algeria. This was a dramatic political move internationally, as the test came after a full year of the nuclear test moratorium, agreed between the three existing nuclear weapon states, had halted testing elsewhere.[75] The French conducted four atmospheric tests followed by thirteen underground tests at two additional sites farther to the southeast.[76] This period of testing was noted for the absolute lack of warning or safety precautions for the indigenous Tuareg people, many of whom received significant exposures. "Estimates of the number of Algerians affected by testing range from 27,000—cited by the French Ministry of Defense—to 60,000, the figure given by Abdul Kadhim al-Aboudi, an Algerian professor of nuclear physics."[77]

Lax procedures also exposed military personnel and test site workers to dangerous levels of radiation, especially during the botched underground Beryl test in the spring of 1962.[78] Jean-Claude Hervieux, a former electrician who worked at the various French nuclear test sites, recalls that the Beryl test blew off the side of a mountain and exposed soldiers, workers, and visiting French ministers to a radioactive dust cloud. "Everyone ran, including two French ministers. At military barracks, the group showered and had their radiation levels checked as a crude means of decontamination. 'You don't see nude ministers very often,' Hervieux chuckled. . . . 'The showers cleaned our bodies and clothes . . . but not what we breathed in or swallowed.'"[79]

The testing regimen was complicated by the victory of Algeria in its war of independence in the summer of 1962.[80] In the Treaty of Evian, in which France recognized Algerian independence, France was able to negotiate a secret deal that allowed it to continue with certain military activities at its bases in Algeria for five years after independence. It was this agreement that allowed the French to continue with underground testing until 1966.[81] When the French finally withdrew from their Sahara test site facilities, they did little to clean up the contamination, and even less to mark or warn the local population of the presence of dangerously radioactive areas near their villages. The French "dug big holes in which they put all the material," recalls M'hamed Zengui, an ex–base worker, "and then they covered it with sand."[82]

In 1958 the French military began to pave the way for later nuclear testing in French colonial South Pacific territories. Except for the island of Tahiti, these holdings were of rather small islands and atolls scattered across the Pacific. None were ideal for a nuclear test site without significant development. As French military and political leaders surveyed possible locations in their colonial holdings, one area, New Caledonia, was quickly removed from consideration as its proximity to Australia and New Zealand presented the likelihood of political opposition that could not be contained through colonial mechanisms. Historian Jean-Marc Regnault cites a letter sent from General Jean Thiry, head of the research commission on underground sites, to Jean Robert, the director of Military Applications, in 1961, stating that in selecting a test site in the Pacific it was necessary to "make a choice based primarily on political considerations."[83]

That political choice became Moruroa and Fangataufa Atolls in the Gambier Islands of the Tuamotu Archipelago at the far southeast of French Polynesia. Here the French conducted a total of 193 nuclear tests between 1966 and 1996. To secure the sites for nuclear testing, the French had to control local Polynesian independence movements to eliminate the possibility that they would lose the site due to anticolonial agitation, as they had lost their Saharan test site. The strongest opposition leader in French Polynesia was Pouvanaa a Oopa, whom the French promptly threw in jail on trumped-up charges of inciting a riot to burn down Papeete (the capital city of Tahiti). With Pouvanaa in captivity in France, local Polynesian opposition was leaderless and uncoordinated.[84] Pouvanaa's party split into two, with John Teariki leading the Rassemblement démocratique des populations tahitiennes, an independence party that held a majority of seats in the local Territorial Assembly. Although it was the seat of indigenous power, the assembly itself held little real power. Teariki's party drafted a bill requiring a referendum on nuclear testing in French Polynesia.[85] But the French-appointed governor rejected that move, which it dismissed as "meddling in defense problems."[86]

"The only portion of Moruroa available for underground testing was a 23-kilometer strip of the southern half of the reef ring, since the rest of the island was covered with laboratories, warehouses, airstrips, and living quarters," describes anthropologist Bengt Danielsson. "Over the next five years, according to official statements, 46 shafts were drilled, 800–1,200 meters deep, depending on the size of the bomb to be tested."[87] Soon after the commencement of French testing in 1966, President Charles de Gaulle visited French Polynesia intent on observing a nuclear test.[88] The Bételgeuse test was set for 10 September but was delayed because the wind was blowing in the wrong direction—toward the more heavily inhabited islands of French Polynesia rather than toward the east, where the islands and atolls had smaller and more indigenous populations. Tired of waiting and eager to return to Paris, de Gaulle ordered the test to proceed regardless of weather conditions, on 11 September, which resulted in a radiological disaster. "Monitoring stations set up by the New Zealand National Radiation Laboratory in the Cook Islands, Niue, Samoa, Tonga, Fiji, and Tuvalu—to the west of French Polynesia—immediately registered heavy radioactive fall-out."[89]

Although government officials and test site scientists claimed that "not a single particle of radioactive fallout will ever reach an inhabited island," before the Bételgeuse test, the French continued testing, fully aware that they were irradiating the populations of nearby and downwind islands.[90] If precautions were taken, they were enacted in ways that also communicated the subaltern nature of the population. Dannielsson has written, "As for the fifty inhabited atolls in the Tuamotus, shelters had to be built in a great hurry on three of the easternmost ones. Each time a test was made in 1968, the hapless islanders were locked up in these shelters for a day or two and each time their homes also had to be 'decontaminated' by spraying them with sea water."[91]

Radiological protection for soldiers, contractors, and those living in the region of the test sites was also exceedingly lax. On 12 June 1971, the French conducted the Encelade test on Moruroa Atoll with a yield of 440 kt. Yann Cambon was a photographer working for the radiological security service on Tureia Atoll, about 110 km to the south. In the afternoon, several hours after the test, heavy fallout came down with rainfall, and radiation levels on the atoll spiked. Cambon turned his Geiger counter from millisieverts to sieverts, and yet it was "still ringing." Cambon remembers: "I saw children taking things from the ground and eating them, soldiers cleaning their teeth with rainwater" from catchments—water that was also used for all showering and washing.[92]

Tanemaruatoa Michel Arakino was born on Reao, an island 450 km from Moruroa Atoll, and worked as a diver in a French military unit tasked with gathering samples from the underwater craters of nuclear tests. On the sea floor at levels of 60 m, a 1 km shaft would be drilled down in which the weapons would be detonated. Arakino and other divers were sent down to the sea floor to gather biological samples to test for radiological effects. The divers would go down immediately after the detonations, returning for additional dives "every six hours." The dives during the first nuclear test were done without diving suits. Arakino, who suffers from radiation-related illnesses (officially recognized by France), worries that his children will inherit his radiogenic diseases.[93]

As detailed in chapter 4, recent admissions by the French government reveal that the radiological contamination of the South Pacific, and of Tahiti specifically, was far more dangerous than had been previously admitted.[94] Revelations released just as this book is going into production provide devastating evidence of the widespread contamination of the whole of French Polynesia, as well as a systemic effort on the part of French political and military institutions to hide those facts.[95] A 2020 study found that New Zealand sailors that had been on ships sent in protest of the testing by the New Zealand government showed significant increases in a range of cancers. The ships had stayed upwind and beyond a twelve-mile exclusion zone for the detonations of relatively small nuclear tests. The study of crew members of HMNZS *Otago* and HMNZS *Canterbury* also found fertility problems in the children of the sailors.[96] Recently, French Polynesian politician Moetai Brotherson testified before the UN Special

"Place du 2 julliet 1966" marae (sacred place) memorial honoring those who suffered from French nuclear testing in French Polynesia, located in Papeete, Tahiti (2017). Photo by author.

Stones from Semipalatinsk, Hiroshima, and Nagasaki, part of the "Place du 2 julliet 1966" marae (2017). This shows the global hibakusha spirit of the memorial. Photo by author.

Committee on Decolonization (officially called the Special Committee on the Situation with regard to the implementation of the Declaration on the Granting of Independence of Colonial Countries and Peoples) in 2019 that radioactive waste dumped into shafts on Moruroa Atoll was in danger of leaking into the Pacific because "cracks suggested it was only a question of time before the atoll collapsed."[97] This waste management practice is at odds with practices for the disposal of radioactive waste enforced inside France: "solid radioactive wastes (sRAW), classified both as low level and high level, have been stocked in the higher part of shafts and then sealed with cement. This practice is not in accordance with French standards for nuclear waste disposal, which normally require such waste to be vitrified and buried at much greater depths within a stable geological structure."[98] The cleanup demonstrates a final imperial disregard.

Lop Nor (China)

The Chinese nuclear program began with the establishment of the Committee of Atomic Energy in the Chinese Academy of Science in 1953, and the signing of an agreement to cooperate on scientific and technological matters between the Chinese government and the Soviet Union on 12 March 1954.[99] In 1958 the first experimental nuclear reactor was brought online near Beijing.[100] Later that year, China began to build its first plutonium production reactor and chemical separation facility in Jiuquan in Gansu Province.[101] After the Sino-Soviet split in 1960, the Soviet Union withdrew further assistance for the Chinese nuclear program.[102] When the Chinese government decided to construct its primary nuclear laboratory and production site later in 1958, the Northwest Nuclear Weapon Research and Development Academy, known as "Plant 221" located in Jinyintan in Qinghai Province on the Tibetan Plateau, it forcibly removed thousands of Mongolian and Tibetan herders living in the area: "herders and farmers who were moved for the project endured starvation, executions and brutal expulsions." Yin Shusheng, a police officer who participated in the forced migration, wrote in a 2012 memoir that "officials imprisoned about 700 herders around Jinyintan, accusing them of joining counterrevolutionary gangs. Seventeen died under brutal interrogation. Up to 9,000 herders were expelled in forced marches, given only a day or so to prepare and allowed to take just a few yaks per family. Hundreds died on the journey, beaten and abused by guards."[103]

Soviet advisors accompanied Chinese military officials as they searched for a test site. The Soviets suggested the desert area in Xinjiang region, and in 1959, China selected a nuclear weapon test site at Lop Nor, a dry salt lake bed in the Xinjiang Uyghur Autonomous Region, located in China's westernmost region and bearing a population of over twenty million people. The Chinese test site would be established approximately 1,000 km away from the Soviet test site outside of Semipalatinsk.[104] At over 100,000 km² the Lop Nor site was the largest nuclear test site in the world, more than twice the size of the NTS.[105]

The first successful test of a Chinese nuclear weapon was on 16 October 1964 at Lop Nor. "The Chinese code-named the development of its first atomic bomb the '596' project as 'a reminder of the 'shameful' date' in June 1959 when the Soviet Union decided to withdraw support of China's nuclear program."[106] All of China's forty-five nuclear weapon tests took place at the Lop Nor site, including twenty-two atmospheric tests, and eight in the megaton range. China tested its first thermonuclear design on 9 May 1966 and its first full thermonuclear weapon on 17 June 1967.[107] This progression from an initial fission test to a fusion weapon in three years is the fastest trajectory of any nuclear weapon state.

China did not have a military empire projected beyond its national borders, and so, like the Soviet Union, and as in the choice of a domestic testing site in the United States, internal politics determined where the nuclear test site would be placed. The Xinjiang region is the Chinese region farthest from the main centers of Chinese population located along the eastern coast of the country. Additionally, it is not the traditional home of the Han Chinese, the dominant ethnic group in China. The majority of the population of Xinjiang is Muslim and made up of many ethnicities, predominantly the Uyghur people of Turkic origin. The population is a diverse mix that reflects the history of the Silk Road trading period and various conquerors and occupiers, such as the Mongolians and Turkmen. There has been a substantial history of discrimination in China against the people of Xinjiang, and especially of the Uyghur population, which has led to frequent violent protests and clashes in Xinjiang and other parts of China. Some of the protests have included accusations of the deliberate radioactive contamination of the local population during the period of nuclear testing.[108] Lawrence Wittner, the preeminent historian of antinuclear movements, writes that "in March 1993, when Chinese troops opened fire on a crowd of a thousand demonstrators outside the test site at Lop Nor, the enraged protesters stormed the complex—damaging equipment, setting fire to military vehicles and airplanes, and tearing down miles of electronic fencing."[109] In recent years, reports of over a million Uyghurs being held in concentration camps and prisons styled as "re-education" have led to an international human rights crisis.[110]

In 2008, the Chinese state news service Xinhau announced that undisclosed payments were being made to military personnel affected by radiation from nuclear weapon testing, but there was no mention of civilian victims or payments to civilians.[111] Chinese medical doctor and Xinjiang native Enver Tohti and Japanese physicist Jun Takada have recently calculated that atmospheric nuclear testing at Lop Nor may have resulted in hundreds of thousands of deaths in China and 1.2 million people receiving doses high enough to cause leukemia or other cancers.[112] Chinese researchers conducting soil analysis near Jiuquan discovered alarmingly high levels of plutonium in 2015.[113] The last atmospheric nuclear weapon test on Earth was detonated at Lop Nor on 16 October 1980.[114]

Conclusion

The French did not test nuclear weapons upwind of Paris; the Soviets did not test between Stalingrad and Moscow; and the British did not test in the Midlands. Testing was carried out at the extremes of empire or the remote corners of national boundaries. The populations subjected to exposure to radioactive fallout, subjected to the contamination of their land—subjected to having their bodies being primary sites where the Cold War turned hot—were selected because of their subaltern status. This was determined by their race or ethnicity, their socioeconomic status, and their location at the peripheries of what was defined as civilization. While the Cold War was narrated as a war between economic systems and political ideologies, it was practiced as an asymmetric war between the technologically enabled and the technologically uninitiated, between the planet's wealthy and their colonized or "low-use" populations. Their use, their utility, was in bearing what fell out of the air onto their lands and homes, and living with what migrated into their water systems and food production, without political repercussions. Henry Kissinger made this clear in the early 1970s when speaking about the Marshall Islands: "There are only 90,000 people out there. Who gives a damn?"[115]

Within the military planning divisions of each of the nuclear weapon states, criteria used for the selection of test sites included scientific and strategic measures—small populations, dry soil suited to testing with little burnable foliage, access to infrastructure, materials, and troops. However, these criteria were not applied without limitation; in each case those nearby populations were of a different race, ethnicity, or religion than the majority populations of the nuclear power. "It was clear that colonialism remained central to the nuclear order's technological and geopolitical success," writes Hecht. "Even a short list of atomic test sites makes the point: Bikini Atoll, Semipalatinsk, Australian Aboriginal lands, the Sahara, French Polynesia."[116] These choices were not made for scientific reasons—these were political choices expressing dominance and subjugation.[117]

There were other communities specifically irradiated during the Cold War by nuclear weapon development. Sites of production were sometimes predetermined by the deposition of necessary materials, such as uranium mines. Refining and enriching of uranium, the production of plutonium, and the assembly of weapons were shaped by practical needs, such as the availability of water or electricity or integration to transportation grids. We can still find traditions of marginalization and social hierarchy in those sites, such as the different living and working conditions of white compared to black and Hispanic workers at Hanford.[118] But the necessities of ongoing production required that all of these workers be maintained in relative health and at least minimal remuneration. Test sites were locations where indigenous populations served no purpose to the machinery of nuclear war besides endurance. The degradations of their cultures, their communities, and their bodies were immaterial to the structures of nuclear weapon production.[119]

During the Cold War, the world was politically divided into bipolar blocs, and the ultimate targets of nuclear weaponry were enemy nations of the nuclear weapon state. However, the nature of developing and testing nuclear weapons necessitated some divisions within domestic populations.

When distant imperial property was secured, the inhabitants of those colonies often became the population targeted with the residual radiation and destined to endure the contaminated lands and seas resulting from weapon testing. Within large nations, minority ethnicities or religious communities were targeted. The people who live in these nuclear test site locations globally make up a virtual nation: victims of nuclear war rehearsals conducted with live weapons. They live in the contaminated margins of the imperial Cold War standoff. Their value to their colonial occupiers or national governments is that they could be ignored.

6

The Cold War Was a Limited Nuclear War

Nuclear Weapon Testing Seen Holistically

The cover of *Life* magazine on 15 September 1961 presented a man in a "civilian fallout suit" holding his hand up for protection, just below a headline that proclaimed, "How you can survive fallout." Inside was a letter to "all Americans" from President John Kennedy. "The government is moving to improve the protection afforded you in your communities through civil defense," comforted the president; "we have begun, and will continue throughout the next year and a half, a survey of all public buildings with fallout shelter potential."[1] The cover story, which followed the president's letter, warned that after a nuclear detonation, "Hundreds of miles from the target, people would come into contact with destructive fallout, which they could not necessarily see, touch or smell. They could get enough on their skin to cause burns and sickness. Fallout might also contaminate their food and water and damage their vital organs."[2]

Kennedy's letter was dated 7 September 1961. One month later, on 6 October, another letter came from the president, this time to state that officials working on civil defense warned that "Radioactive fallout, extending down-wind for as much as several hundred miles, could account for the major part of the casualties which might result from a thermonuclear attack on an unprotected population."[3] Yet six months later Kennedy would approve Operation Dominic, which would authorize the detonation of thirty-six nuclear weapons (nine of them thermonuclear) at Johnston Atoll and Christmas Island.[4] Kennedy would approve the "testing" of ninety-six nuclear weapons in total in 1962. Each of these weapons, especially the H-bombs, would send fallout clouds into downwind communities. The president warned some people of the dire risks of fallout while simultaneously inflicting it on untold numbers of other human beings.

Historians Matthew Wald and Benjamin Zeimann explain that the Cold War, as constructed in Western minds, was primarily an *imaginary* war. Protagonists and the public envisioned a direct nuclear conflict between the United States and the Soviet Union in which "the 'bomb' itself became the *central metaphor.*"[5] This imaginary war might involve an exchange of weapons as part of a land war in Europe; or it might become apocalyptic, with the full weaponry of both sides, utilizing multiple delivery systems, directed against each other's arsenals, infrastructure, and populations. The many rungs of this imaginary war were fastidiously mapped out by nuclear strategists on both sides. The final rung was imagined to leave the planet "a republic of insects and grasses," unburdened by human politics, society, or beings.[6] The World War III fever dream never came true; still, millions of people suffered harm from the real-world operations of the superpowers in feverish Cold War preparation. The mechanics of production and testing of the weapons created regions in which human beings, indeed, whole communities, directly experienced the radiological effects of nuclear weaponry. After the Cold War, we can see the scars of this warfare without functional narratives to grasp what happened.

In the previous chapters of this book, I tracked some of this devastation in multiple nuclear test site communities and ecosystems around the world. People in these communities have grappled with the assaults and shown resilience and dynamic resourcefulness.[7] However, what has been arrayed against them has been the relentless degradations of weaponized radiation, which can remain invisible, stochastic, and enduring. Scholars have most often considered these devastations on a case-by-case basis; they have examined, for instance, American nuclear weapon testing, or French nuclear weapon testing. Such work forms essential parts of national, colonial, and postcolonial histories. But denationalized and considered globally, it's clear that this was a limited nuclear war. Let's look at the nuclear testing conducted during 1962, from a single-nation perspective.

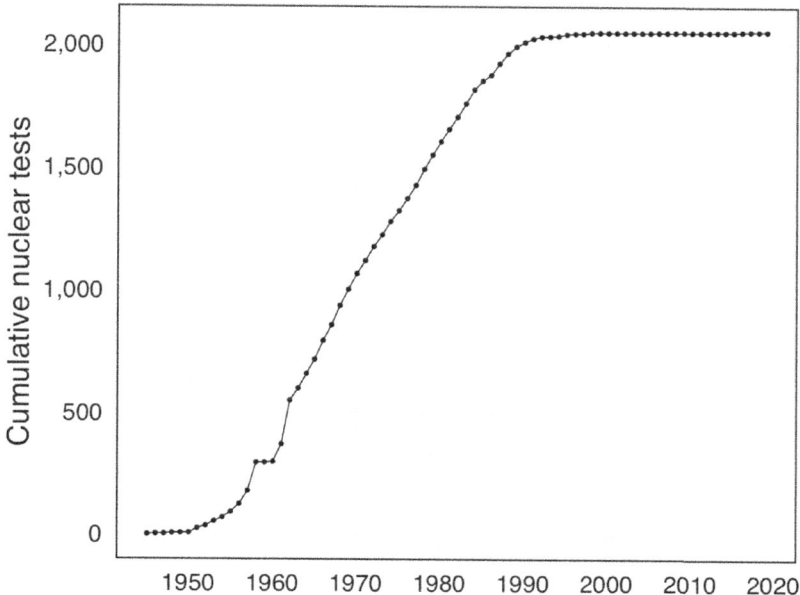

Total cumulative nuclear weapon tests shown globally (denationalized). David Montgomery, "A History of Nuclear Testing in Nine Charts" (2021): http:// dhmontgomery.com/2018/02/nuclear-tests/.

In 1962 there were 177 nuclear weapons tested on the planet: 96 (as mentioned) were tested by the United States, 78 by the Soviet Union, 2 by the United Kingdom, and 1 by France. Weapons were tested on four continents and multiple Oceanian territories; 26 tests were of thermonuclear weapons. The 177 nuclear weapon tests roughly average out to a nuclear detonation every other day in 1962.[8] The inability to keep the effects of these detonations confined to the actual test sites resulted in substantial distributions of fallout radionuclides in multiple locations globally, and into the stratosphere. Scholars of US testing would focus on the specific test series in either Nevada or the Pacific territories, while historians of the Soviet Union would parse the differences between tests in Kazakhstan and Novaya Zemlya. There were significant radioactive fallout distributions in Europe, Asia, North America, and the Pacific. These tests can be analyzed as local or individual events; seen holistically, they reveal a global military endeavor and environmental disaster. For the long-lived radionuclides distributed into the ecosystem, such as ^{239}Pu or ^{235}U, the life

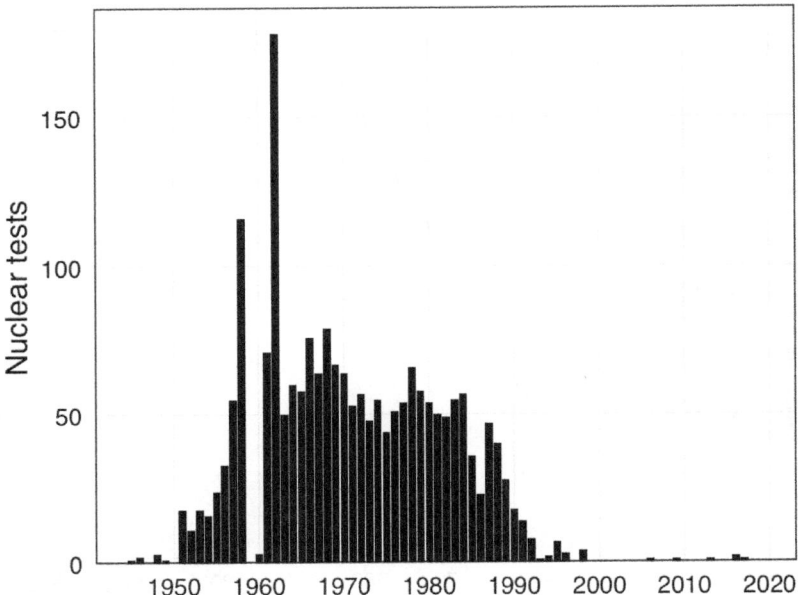

Nuclear weapon tests by year shown globally (denationalized). David Montgomery, "A History of Nuclear Testing in Nine Charts" (2021): http://dhmontgomery. com/2018/02/nuclear-tests/.

cycle of these particles (in the hundreds of thousands or millions of years) makes the specific locations of their initial deposit, or the political motivations of the testing nations, less important than their quantities. From any framework, 1962 represented a period of massive distribution of long-lived radionuclides at multiple locations and high into the atmosphere, resulting in global transport. Given the pernicious and enduring nature of the toxic materials released, even without the weapons being detonated in a direct nuclear attack, their long-term impacts can be seen as analogous to the ecological legacies of a nuclear conflict.

Whether we call this a group of nuclear tests or a limited nuclear war depends on our vantage point. From a political perspective, this can be called a sequence of test series by separate nations, as there was no direct exchange of weapons from opposing powers or direct use against people. From an environmental perspective, this could be called a limited nuclear war because it resulted in the large-scale distribution of a wide range of radionuclides into the ecosystem at multiple locations. We know from the discussion in the previous sections that the real-world impacts on people's health and on the integrity of their lands and seas included widespread suffering and social dislocation in multiple communities from these military actions.

World War III: The Global Thermonuclear War

During the Cold War there was widespread anxiety over the possibility of a global thermonuclear war (GTW). This would have involved a substantial portion of the arsenals of both superpowers being detonated in the course of a few days or weeks. During the last decades of the Cold War this may have involved thousands of weapons, many of them thermonuclear weapons in the megaton range. Herman Kahn referred to this imagined war as a "spasm or insensate war" in which both sides reflexively launched their weapons in an attempt to completely destroy their opponent regardless of the cost—a desperate attempt to slaughter a whole society and somehow survive.[9]

The GTW was popularly called "World War III," but what was imagined was far beyond the scale of earlier world wars. While World Wars I and II involved vast carnage and destruction, whole sections of the world, away from the conflict zones, or *military theaters,* escaped harm: the whole of North and South America in both wars, for example. Casualties in World War I, the Great War, were mind-boggling at the time. Millions of soldiers died in the *meat grinder* of industrialized trench warfare. In World War II, warfare expanded out of the trenches and into urban centers and led to the aerial bombardment of cities by most protagonists. This would expand to the use of incendiary weapons, nuclear weapons, and rockets. The wholesale slaughter of human beings in the Holocaust by the Nazis added twelve million victims, almost all noncombatants, to the dead.

The practice of "total war" redefined warfare as a conflict of whole societies as "legitimate" targets: combatants, noncombatants, workers, and their families. Military historians cite the centrality of automatic weapons in the wars of the nineteenth century as the birthplace of this theory of conflict. American Civil War general Philip Sheridan, who was observing the conduct of the Franco-Prussian War from the vantage point of German headquarters, remarked to Otto von Bismarck that "The proper strategy . . . consists in inflicting as telling blows as possible on the enemy's army, and then in causing the inhabitants so much suffering that they must long for peace, and force the government to demand it. The people must be left nothing but their eyes to weep with."[10] The concept of total war has been central to historical interpretations of warfare in the twentieth century. Political scientist Mark Duffield describes how such thinking evolved and became envisioned more holistically: "It is a modus operandi of violence that, in the broadest sense of the term, demands the destruction of an enemy's environmental lifeworld. The targets of this war include the climate regimes, vital urban infrastructures, ecological systems, and social networks, together with the neurological and cellular processes that collectively support life and make it possible."[11] The much-planned GTW of the Cold War was just such a total war, however, completely without boundaries. World War III was envisioned as obtaining the "end of the world" for all human beings, regardless of nationality, military status, or even cognizance of the fact that a war was being fought.

The catalyst of this civilization-concluding destruction was, again, to be advances in the technologies of warfare: specifically, thermonuclear weaponry, ballistic missiles, and MIRV configurations on those ballistic missiles. This apocalyptic imaginary tracked closely with the actual nuclear warfighting plans of the Cold War protagonists. From the mid-1950s on, each side had hundreds, then thousands, then tens of thousands of nuclear weapons, many of which were on alert status, allowing their use within minutes of a command launch order. Both had developed targeting lists that would have slaughtered the population of their enemy nation, with profound, civilization-disrupting radiological spillovers.[12]

The concept of a "nuclear winter" being triggered by the GTW was first postulated in the 1980s, theorizing that a full thermonuclear exchange would have consequences beyond the lethality directly resultant from the weapons' effects. The entry of masses of particles into the atmosphere from the burning of cities would trigger a nuclear winter, as the particles shielded the surface of the Earth from sunlight and heat, provoking worldwide crop failures and mass starvation lasting for decades—devastating global populations and ecosystems.[13] Recent climate science has modeled the reality behind such theories far more accurately and revealed that even a "small" nuclear exchange of ten weapons, if targeted at cities that would then burn, would be likely to trigger nuclear winter.[14] Whether by design or by unintended consequences, the GTW would likely have led to a massive collapse of human structures of civilization and the deaths of the majority of human beings alive at the time.[15]

The Cold War as a War by the Soviet Union and the United States against Their Own Citizens

Counterintuitively during the Cold War, the apocalyptic risks of directly harming "enemy" soldiers and citizens created a "Bizarro World" dynamic in which it made more sense to nuclear weapon states to harm populations under their own care.[16] This helped to cloak the violence, as the aggressors left the enemy untouched and instead self-harmed with nuclear weaponry.

The United States maintained numerous communities of workers who made homes in toxic locations, harming the workers, their families, and the broader populations in the region. The Green Run at Hanford in 1949 released radioactive ^{131}I across most of the US Northwest. Nuclear weapon tests in Nevada peppered fallout clouds across most of the states in the union. After the radiological disaster that was the Operation Crossroads Baker shot in 1946, over forty thousand US military personnel had to be evacuated from the Marshall Islands and Central Pacific. A large segment of the population of the northern Marshall Islands, people who were in trust care of the United States at the time, was exposed to radiation, many at dangerous levels. While the fallout clouds from tests like the Bravo test in 1954 harmed people outside of American responsibility—for example, killing a Japanese fisherman—the majority of people that the United States harmed in the Cold War were either American citizens, or trust populations directly under US care.

The Soviet Union conducted all of its nuclear production and testing within its own borders. The Kyshtym disaster at Ozyorsk in 1957 contaminated a vast area of the Chelyabinsk Oblast to the northeast of the plutonium production facility and exposed several hundred thousand people to radiation.[17] When space in underground waste tanks for high-level liquid waste was unavailable at Mayak, the waste was dumped into the Techa River for several years in the early 1950s, leaving a deadly legacy.[18] The Totskoye test in 1954 exposed forty-five thousand Soviet troops to dangerous levels of radiation.[19] Massive tests of thermonuclear weapons were conducted at the Polygon in Kazakhstan, where continuously populated villages lay only tens of kilometers away. The Tsar Bomba in 1961 spread a fallout cloud across the Soviet Arctic and south to the Black Sea. Just as did the United States, the Soviet Union self-harmed with nuclear weapons.

After the explosion and during the subsequent fire at the Chernobyl Nuclear Power Plant in Ukraine, officials in the former Soviet Union realized that a fallout cloud laden with radioactive particles was moving northward from the plant site and could reach large cities in the republic of Russia. Officials in the Soviet State Committee of Hydrometeorology decided to take emergency action and sent planes to seed the clouds so that rain brought the fallout down before the clouds reached Moscow and nearby cities. The cloud seeding succeeded in bringing a substantial amount of the fallout particles down in the Belarus region of Mogilev, which remains significantly contaminated today. There may be an overarching logic to exposing smaller populations to radiation to forestall the exposure of larger populations threatened by exposure to the same particles, but it is a clear example of one of the superpower governments choosing to sacrifice the health of a segment of their own population. "In the path of the artificially induced rain there lived several hundred thousand Belarusians," writes Kate Brown. "No one told the Belarusians that the southern half of the republic had been sacrificed to protect Ukrainian and Russian cities."[20] In the heat of the moment, these Belarusians were sacrificed to protect the health and lives of their fellow citizens, living in places that the national leaders considered more important.

Beyond the physical harm that both superpowers inflicted on their own populations, each depleted significant amounts of public funds to bankroll nuclear weapon programs, money that could have been dedicated to the social welfare of their citizens. American Cold War triumphalists proudly cite the strategy of funding the Strategic Defense Initiative (popularly known as "Star Wars") during the Reagan administration to destabilize the Soviet Union by forcing them into economically precarious defense spending.[21] While there is debate about whether this strategy was the "cause" of the fall of the Soviet Union, the SDI, but more directly, nuclear weapon spending, did take a large toll on American society as well.

A Brookings Institution study of American nuclear weapon spending during the Cold War reveals that "The amount spent through 1996—$5.5 trillion—is 29 percent of all military spending from 1940 through 1996 ($18.7 trillion). This figure is significantly larger than any previous official or unofficial estimate of nuclear weapons expenditures, exceeding all other categories of government spending except nonnuclear national defense ($13.2 trillion) and social security ($7.9 trillion). This amounts to almost 11 percent of *all* government expenditures from 1940 to 1996 ($51.6 trillion). During this period, the United States spent on average nearly $98 billion a year developing and maintaining its nuclear arsenal."[22] It is debatable exactly what security was purchased with these expenditures that could not have been secured more affordably, but there can be no argument that Americans would have experienced vastly more secure lives had these public moneys been invested into policies like providing universal access to health care, debt-free higher education, or secure retirements.

While the United States and the Soviet Union had tens of thousands of nuclear weapons minutes away from killing tens of millions of each other's people—and threatening the whole planet—the two countries primarily irradiated their own citizens and trust populations. It is not unreasonable to understand the Cold War as a war by the United States government against the American people, and by the government of the Soviet Union against the Soviet people. Both sides irradiated their own populations and eroded their own social welfare in an effort to "win" the nuclear arms race. As both the United States and the former Soviet Union spent vast sums of their publicly derived budgets on nuclear weaponry and military spending, while at the same time subjecting their citizens to exposures to radiation from the production and testing of their nuclear weapons, it can be said that while the US and the USSR did not fight each other directly in the Cold War, they did subject segments of their own populations to degradations not unlike those of wartime.

Thinking about Nuclear Weapons as Radiation Weapons

In the immediate aftermath of the nuclear attacks on Hiroshima and Nagasaki there was an intense debate within the military and political leadership of the United States as to whether its new weapon was "revolutionary" or just a bigger bomb.[23] Much of this debate was instrumentalized around postwar planning and funding allocation, but the core question about whether the weapons forced a fundamental reassessment of the nature of warfare and conflict endured. As the Cold War unfolded, a key site where this question became consequential was in the testing of nuclear weapons. Was a nuclear weapon test just a bigger weapon test? Or was there something so *revolutionary* about these weapons that each detonation was of profound consequence? We can see both views comingling simultaneously in the histories of nuclear weapon testing that followed. In terms of the preparations for nuclear tests, the remediation of the test site locations, and the responsibilities of the nuclear testing nations to the populations local to the test sites, the testing nations acted as if these were simply bigger bombs, requiring no extraordinary care to community or ecosystems. Yet they took critical steps that reveal their understanding of the revolutionary nature of nuclear weapons' effects. France and the United Kingdom, for example, never tested nuclear weapons in their own countries. This was to avoid exposing their native populations to radioactive fallout. The United States primarily tested nuclear weapons in two sites, in the Marshall Islands and in Nevada. However, the US never tested thermonuclear weapons at its Nevada Test Site, so as not to subject its own citizens to the vast fallout clouds resultant from thermonuclear tests. It is easy to see the logic and the consequences of this policy: although the US conducted only 14 percent of its nuclear tests in the Marshall Islands, that 14 percent constituted approximately 80 percent of the yield of all of its nuclear testing.[24] In selecting which weapons to test in Nevada and which weapons to test in the Marshall Islands, the United States clearly considered thermonuclear weapon testing to justify a separate set of operational protocols. While it is possible to depict this policy as the United States government seeking to keep from exposing US citizens to immense fallout clouds, it can also be understood as specifically selecting the Marshallese to be deliberately exposed to those same immense fallout clouds. It was their nationality, their ethnicity that led to their being selected for this irradiation of their bodies and of their homes.

It was radiation, and specifically fallout radiation, that made nuclear weapon testing revolutionary. As described in the first chapter, a nuclear weapon delivers prompt radiation in a burst from the reaction at the core of the weapon at the moment of detonation, while residual radiation lingers for as long as the specific chemical isotope is active. Some are only dangerous for a few minutes, while others, like ^{239}Pu, one of the key fissile materials, are dangerous for hundreds of thousands of years. Blast and heat are not new to nuclear weapons: all explosives deliver blast and heat.[25] While nuclear weapons have not been used in direct warfare since the attacks of 1945, the use of explosives has never stopped. Even this year, even today, bombs are being dropped and people are suffering from the blast and heat of explosives. So when we say that nuclear weapons have not been used in war, we mean that people have not been subjected to blast and heat of the magnitude of nuclear weapons since 1945. This is a matter of scale, and in regard to these two effects, nuclear weapons are actually just bigger bombs. Radiation, however, was a revolutionary and, at the time, new weapon effect.

The specific effects of internalized radioactive fallout have been integrated into nuclear war–fighting strategies since the 1940s. As part of its analysis of the first postwar nuclear weapon tests in the Marshall Islands in 1946 during Operation Crossroads, a military assessment of the radiological disaster that followed the second test, the Baker test, written in 1947, articulated a clear understanding of the military utility of radioactive fallout as a primary weapon effect:

> 3. Test Baker gave evidence that the detonation of a bomb in a body of water contiguous to a city would vastly enhance its radiation effects by the creation of a base surge whose mist, contaminated with fission products, and dispersed by wind over great areas, would have not only an immediately lethal effect, but would establish a long-term hazard through the contamination of structures by the deposition of radiological particles.
>
> 4. We can form no adequate mental picture of the multiple disasters which would befall a modern city, blasted by one or more atomic bombs and enveloped with radioactive mists. Of the survivors in contaminated areas, some would be doomed to die of radiation sickness in hours, some in days, and others in years. But, these areas, irregular in size and shape, as wind and topography might form them, would have no visible boundaries. No survivor could be

certain he was not among the doomed and so, added to every terror of the moment, thousands would be stricken with a fear of death and the uncertainty of the time of its arrival.[26]

While these military planners described the manufacturing of terror in a targeted population, their assessment applies equally to untargeted people who, by the quirks of "wind and topography" following a nuclear test or accident, find themselves enveloped in a radioactive-particle-rich environment. It succinctly describes the actual situation faced by nuclear test site populations around the world in the days, months, and years following the deposition of radionuclides into their ecosystems from the fallout clouds sweeping across their horizons.

It was clear to military planners that distributing radionuclides into an ecosystem could be particularly effective as a psychological terror weapon as well as a physiological weapon. At a 1949 conference on the "psychological effects of unconventional weapons" held by the RAND Corporation, the confidential report would detail that "If the objective is to create fear, confusion and panic among the civilian population, it would be best to withhold information about the presence of radioactive substances until severe symptoms of exposure begin to appear."[27] This was precisely what was done to human beings in every community downwind from a nuclear test site.

In the Soviet Union, Ukrainian physicists Viktor Maslov and Vladimir Shpinel proposed the development of nuclear weapons and the use of uranium as a radiological weapon against foreign troops in 1940. "As a result of a uranium bomb explosion, radioactive substances are created. It has a toxic capacity which is thousands of times greater than the strongest poison (and thus ordinary toxic agents)," they described in their letter to the Bureau of Inventions of the People's Commissariat of the USSR. "Therefore, taking into account that after an explosion radioactive debris will exist during a certain period of time in gaseous form, and that it will spread across a vast area while keeping its [toxic] capabilities for a relatively long time (for hours and for some [fission products] even days and weeks), it is difficult to say what feature (high destructive capacity or toxicological characteristics) is most attractive in this regard militarily."[28]

Residual radiation effects would be designed as a primary weapon effect during the Cold War years. David Alan Rosenberg describes a US Strategic

Air Command (SAC) briefing held at Offutt Air Force Base in Nebraska on 15 March 1954 (before the integration into nuclear strategy of the findings of the Bravo test conducted two weeks earlier), in which SAC commanders described the nuclear battle plan of the command as hitting the Soviet Union with its full arsenal in a "drop as you go" system, leaving the entire nation "a smoking, radiating ruin at the end of two hours."[29] While the blast and heat of the weapons were intended to neutralize the Soviet Union's nuclear weapons, delivery systems, infrastructure, and command-and-control facilities, and the burst of gamma radiation meant to eliminate any military personnel not killed by the blast and heat, the resulting fallout radiation would prove lethal to the entire population. And beyond.

Within a few months of the Bravo test, senior members of the American nuclear weapon establishment showed how quickly an understanding of the military utility of radioactive fallout and the potential for using thermonuclear weapons to spread it was realized. The United States military mapped the fallout cloud from the Bravo test that had irradiated downwind Marshallese and the crew of the *Daigo Fukuryu Maru* and killed Aikichi Kuboyama. In an AEC meeting held on 24 May 1954, Kenneth Fields, the general manager of the AEC, showed a map that superimposed the Bravo fallout cloud over the eastern US seaboard and described the consequences had a similar weapon been used in an attack on Washington, DC: "the lifetime dose in the Washington-Baltimore area would have been 5,000 roentgens; in Philadelphia, more than 1,000 roentgens; in New York City more than 500, or enough to result in death for half the population if fully exposed to the radiation delivered. Fallout in the 100-roentgen area, which might have been roughly comparable to the *Lucky Dragon* exposures, stretched northward in a wide band through New England towards the Canadian border."[30]

In a letter to John Bricker, the chairman of the Military Applications Subcommittee of the US Senate, sent in July 1954, United States Atomic Energy chairman Lewis Strauss opined about the dynamics of using fallout as a weapon:

If radiological contamination by high-yield weapons were used for military purposes, its distribution would be difficult to control for individual detonations. The mechanism of distribution is sensitive to wind directions and velocities up to the top of the mushroom-shaped cloud. Accurate prediction of the location of the hazardous area with respect to the detonation point would require extensive wind data at altitudes up to about 100,000 feet, and such data are difficult to obtain, especially for points in enemy territory. However, if a number of high-yield weapons were detonated within a region, the region could be generally contaminated without accurate knowledge of wind conditions for each detonation. The situation in regard to such indiscriminate dissemination of radioactive materials would be different for targets deep in enemy territory than for targets in enemy-occupied territory or near the front lines where there would be hazards to friendly personnel. In summary, the detonation of high-yield weapons of present design at or near the surface will, by radiological contamination, extend the anti-personnel effects of such weapons over considerable areas that would otherwise not be affected.[31]

Strauss makes clear that there was an advantageous military outcome to the strategic use of fallout radiation as a primary nuclear weapon effect rather than a secondary aftereffect. The Bravo test had made the utility of such "salted" thermonuclear weapons crystal clear, as summed up in a letter by Robert LeBaron, chairman of the Military Liaison Committee of the AEC, to the wider commission: "Before last year it was thought that ionizing radiation was of little importance relatively; the first shot of Operation CASTLE [the Bravo test] however proved that radiation can become a major killing force."[32] Chairman Strauss later stated that residual radiation delivered from thermonuclear fallout was now to be understood as fundamental in nuclear war planning: "Consideration of the effects of nuclear weapons, particularly those with high yields, indicates that radioactive fall-out and contamination are among their most important characteristics, and indeed may be one of the factors governing their use in some situations."[33]

We can see that such understandings facilitated the views on the use of fallout radiation as a weapon demonstrated in the letters sent by AEC officials, but we do not see any change in protocol in the testing of such large thermonuclear weapons upwind from human populations.[34] Even after the completion of the Castle test series, of which the Bravo shot was the first of six high-yield tests, the United States conducted twenty-three more atmospheric tests of thermonuclear weapons upwind of inhabited islands and atolls.[35] These tests were conducted with the explicit understanding that the subsequent radioactive fallout clouds were a direct threat to those populations, the long-term habitability of their communities, and the safety of the ecosystems from which they derived their food and water.

Blast, heat, and prompt radiation are all effects that any nation can contain to a test site during a nuclear weapon test, and thus keep their civilian population protected, but fallout is beyond effective control. Fallout is the reason to site your nuclear testing facility in the middle of the ocean on the other side of the world from your home population. Communities exposed to fallout clouds from high-yield nuclear weapon tests were being directly exposed to a military application of those weapons. This was not done to directly harm them: it was done because their welfare could be ignored. There were few serious consequences for the United States in subjecting the Marshallese or the Kiribati to such radioactive fallout. In making this determination, the United States was explicitly subjecting these populations to weapon effects that it also understood to be conditions created in active nuclear warfare. A 1964 RAND report on limited nuclear war offered a joke made by game theorists: "In tactical nuclear conflict it is much easier to annihilate our friends than our enemies."[36] Implicitly or explicitly, the United States was carrying out a limited nuclear war against segments of the Marshallese and the Kiribati populations. This was true of all of the nuclear weapon states that conducted thermonuclear weapon tests in the atmosphere: the United States, the Soviet Union, the United Kingdom, France, and China.[37] All these states were aware of the scale of radioactive fallout from testing thermonuclear weapons in the atmosphere, and yet all proceeded to carry out such tests upwind from populations under their political control.

The Cold War as the Long Peace

In 1986, a few years before the end of the Cold War, Yale historian John Lewis Gaddis counterintuitively described the Cold War decades as the "long peace."[38] While Gaddis acknowledged that the period was filled with conflict, he asserted that it was remarkable that there was no direct war between the two "great powers," the United States and the Soviet Union. Gaddis found solace in the "system" of the Cold War world that imposed a structure, much as that articulated through game theory, in which the participants found themselves both bound to, and ultimately liberated by, unwritten but tacitly agreed upon rules of the *game*. Writing twenty years later, Nina Tannenwald would find that these unwritten rules had effectively imposed a "nuclear taboo" on the Cold War protagonists: "The non-use of nuclear weapons since then [1945] remains the single most important phenomenon of the nuclear age. . . . Many reasonable observers expected that nuclear weapons would be used at some point during the Cold War. It was thus by no means inevitable that, after their use against Japan in August 1945, a 'tradition' of non-use would arise."[39] Since the turn of the millennium, numerous scholars have explored what was downplayed by Gaddis and other earlier Cold War scholars. Odd Arne Wested's 2007 tour de force *The Global Cold War* detailed the depth and ferocity of conflicts in the Third World that both were fueled by and also politically deconstructed the binary geopolitics of the Cold War superpowers.[40]

In the arguments of these scholars, nuclear weapons were used once as weapons of war in the nuclear attacks on Hiroshima and Nagasaki, after which the dangerous Cold War period was profoundly characterized by the *non-use* of nuclear weapons. With this assumption as the backdrop, an academic argument evolved in which the position that this non-use of nuclear weapons in direct conflict between the United States and Soviet Union is the central historical point—the long peace—contrasts with the view that this non-use provided cover for the fact that the two superpowers did square off on multiple battlefields, across numerous continents, often utilizing the bodies of local fighters and being subsumed under local, anticolonial movements—and therefore there was no long peace. What is not debated is the non-use of nuclear weapons.

Both schools employ traditional definitions of warfare: weapons that are not used directly against an enemy are unused weapons. The line of disagreement comes from the definition of the word "peace." For traditional Cold War scholars, "peace" is used to describe the absence of conflicts on the scale of the two world wars of the twentieth century: the imagined

"World War III." Gaddis chides the focus of scholars of peace studies who have "given far more attention to the question of what we must do to avoid the apocalypse than it has to the equally interesting question of why, given all the opportunities, it has not happened so far."[41] For Gaddis, the "apocalypse" has not happened unless it has happened to the populations of the Cold War protagonists in a direct confrontation. While acknowledging the global scale of the weaponry, this vision still restricts itself to a specific form of usage, warfare, and a specific period of duration, days, weeks, or months. There is no room in this framing for the fact that munitions on the scale of nuclear weapons, and specifically of thermonuclear weapons, can exact catastrophic tolls on populations from the simple act of their detonation at a distance, regardless of the purpose of that detonation. Likewise, while the critics of the long peace rightly assert that conflicts in the Third World or in other locations diminish any claim of the Cold War period as being peaceful, even without direct conflict between the two superpowers, the ecological devastation of the test site locations and ongoing contamination of the living spaces of the populations of those sites, especially in remote Pacific communities, extend the violence of the Cold War far beyond the locations of active warfare, the classic duration of "battles," and even the limitations of the Cold War time frame. Jacques Derrida famously proclaimed about nuclear war in 1984 that "It has never occurred, itself; it is a non-event."[42] In reducing nuclear war to a techno-scientific construct and "non-event," Derrida was dismissing it to a generation of humanities and cultural-theory scholars, alongside the historians and political scientists mentioned earlier, and the reality of those who had actually experienced the effects of nuclear weapons after 1945. All of these intellectuals could not see what strategists from the nuclear weapon states had structurally cloaked: the subaltern status of the actual nuclear victims.[43]

We count the number of nuclear weapon tests: we count the date, the time, the yield, the longitude and latitude. Can anyone say how many tests of tank shells have occurred in history? Where? What time? How many conventional bombs have been tested by any one nation? We count the number of nuclear weapons tested because they have a substantial impact on the Earth, even though they are tests and not combat usage, and also because of the political implications of nuclear weapon possession. It is largely impossible for any country to track how many test firings there are of almost any other weapon system. The test of even one small nuclear weapon anywhere in the world can be detected by seismic instruments, as well as by the radiological traces they leave in the atmosphere, even if those weapons are tested underground. A nuclear weapon test has always been a global ecological phenomenon.

Limited Nuclear War

Gaddis's threshold for describing the Cold War as the "long peace" is that there was no apocalyptic nuclear exchange "ending the survival of everyone on earth."[44] However, nuclear strategists and military planners prepared for a wide range of possible nuclear wars before spiraling up the "escalation ladder" to Kahn's spasm war, or Gaddis's World War III. There was extensive planning on both sides of the Cold War binary toward conducting a limited nuclear war, potentially taking many forms and facilitating many different outcomes. Initially, limited nuclear wars were envisioned as direct nuclear confrontations in which the use of weapons and the selection of targets was constrained, while the possibility of "total" nuclear war loomed if the initial attacks did not yield the desired result. Sometimes these wars were envisioned as the "battlefield" use of smaller, tactical nuclear weapons. Such scenarios typically involved a traditional battlefield engagement between nuclear protagonists in which low-yield nuclear weapons were used to accomplish battlefield objectives, such as breaching an enemy line to allow penetration by infantry columns.[45] It was hoped that such limited use of low-yield nuclear weapons would avoid escalation to a more robust use of large-yield nuclear weapons directed against cities and whole societies. In the US, military strategists and private think tanks devoted a great deal of time and resources to planning, or *gaming*, these scenarios, seeking to liberate the "taboo" weapons for practical military use.[46] Typically, limited nuclear wars were not envisioned as the use of nuclear weapons against nonnuclear adversaries, but rather as the calculated use of nuclear weapons by nuclear protagonists who could fight each other while effectively managing the inclination to lose control, because of their internalization of the rules of the game.

Theorist Robert Osgood described some of the strategies and dynamics of a limited war in his 1957 treatise on the topic:

A limited war is one in which the belligerents restrict the purposes for which they fight to concrete, well-defined objectives that do not demand the utmost military effort of which the belligerents are capable and that can be accommodated in a negotiated settlement. Generally speaking, a limited war actively involves only two (or very few) major belligerents in the fighting. The battle is confined to a local geographical area and directed against selected targets— primarily those of direct military importance. It demands of the belligerents only a fractional commitment of their human and physical resources. It permits their economic, social, and political patterns of existence to continue without serious disruption.[47]

Osgood later expanded on this definition, explaining that limited war could be defined as "a general 'strategy of conflict' in which adversaries would bargain with each other through the medium of graduated military responses, within the boundaries of contrived mutual restraints, in order to achieve a negotiated settlement short of mutual destruction."[48]

This is an indirect, yet accurate description of the utility of continual nuclear testing; the history of nuclear weapon testing by the US and USSR can be seen as a means of signaling nuclear capacity to each other, of demonstrating the size, scale, and ferocity of one's arsenal and means of delivery, without invoking a militarized response and while operating "within the boundaries of contrived mutual restraints." The nuclear superpowers were demonstrating their capacity to wage nuclear war, but not directly against each other—which would necessarily provoke a response—but toward politically powerless people under their sovereign control.

As detailed earlier, there have been over two thousand nuclear weapon tests since the two direct nuclear attacks on Hiroshima and Nagasaki. Anthropocene scholars have calculated that since the Trinity Test in New Mexico on 16 July 1945, a nuclear weapon was detonated, on average, every 9.6 days until 1988.[49] These detonations were bunched up more in some years than others, and the ecological impact of underground tests differed from that of atmospheric or underwater tests, but it is accurate to say that the Cold War was a period in which nuclear weapons were going off continually. Even when the United States, the Soviet Union, and the

United Kingdom halted testing from late 1958 until September 1961, the French were beginning to test nuclear weapons in Algeria.

Much as in the numbers of weapons in their arsenals, the scale of the testing programs of the United States and the former Soviet Union stands out from other nuclear weapon states. While the other states were able to effectively design and deploy their nuclear arsenals with a limited number of tests (UK—45, France—210, China—45), the Soviet Union conducted 715 nuclear weapon tests, and the United States conducted 1,054 nuclear tests. Both the former USSR and the US have each tested more weapons than the other members of the P5 (the five permanent members of the United Nations Security Council) have collectively built. Why? How were these other countries able to move from design to deployment with so many fewer tests? One answer is the size of the nuclear arsenals these two superpowers built. The US and the USSR developed many more weapon models for many more intended uses and delivery systems than the other nuclear weapon states. But the number of tests still remains disproportionate to the simple requirements of design, versatility, and reliability. The United States and the Soviet Union combined conducted 84.87 percent of all nuclear weapon tests, and produced 91.2 percent of all of the force yielded by nuclear weapons tested globally.[50] These numbers track closely with the scale of the nuclear arsenals of the two nations, who still hold (in 2020) between them 90.07 percent of all nuclear weapons even decades after the end of the Cold War.[51] The 1,769 tests conducted by the United States and the Soviet Union, during which they manufactured over one hundred thousand nuclear weapons to signal their power to each other, endangered and continue to endanger the planet as a whole.[52]

I would argue that the scale of nuclear weapon testing by the United States and Soviet Union had additional functionalities not specifically related to technological demands. Detonating weapons safely within one's own territory or the territories under one's sovereign control, demonstrating one's capacity to inflict harm effectively upon an adversary, is not dissimilar to parades of military hardware in national capitals or, for that matter, the chest beating of primates—which is effected to allow the adversary to hear the resonate echoing of a large chest cavity—indicating size and power.[53] Similarly, US and Soviet nuclear weapon testing was partly performative. However, because of the scale of the effects of nuclear detonations, such demonstrations did not occur without substantial and far-reaching physical, social, and ecological impacts in the subaltern communities where this macabre thermonuclear theater was performed.

British military strategist Liddell Hart long advocated for the use of limitations in accomplishing military goals. Writing in response to the militarily ineffective horrors of World War I by articulating the theory of an "indirect approach" in military strategy, Hart argued that direct military conflict often led to catastrophic and counterproductive outcomes. He advocated for the utility of an indirect military approach that could compel victory through a minimal allocation of resources and loss of human life. This approach relied in some measure on the psychological unsettling of one's adversary. Hart wrote that "the perfection of strategy would, therefore, be to produce a decision without any serious fighting."[54] This strategy seemed even more essential to Hart after nuclear weaponry had been added to the arsenals of modern superpowers. Military historian Brian Holden Reid notes that "the advent of nuclear and thermonuclear weapons rendered *prudence* a high virtue. These weapons, Liddell Hart affirmed, rendered any idea of 'victory' or 'total war' absurd. The nuclear age gave a new impetus to the cautious ripples of the indirect approach—and its undertow."[55] Nuclear testing allowed both the United States and the Soviet Union to make bold and threatening displays of their military capacities. The ceaseless testing of nuclear weapons demonstrated the scale of their respective arsenals. Testing of delivery systems impressed with the complex and multifaceted means of achieving the implicit threats of total nuclear annihilation and the ability for a retaliatory force to survive a direct attack.

All of this could be achieved in live and vivid demonstrations without directly provoking an actual nuclear conflict in which the two protagonists would obliterate each other's population and territory and invite retaliation. It was a "safe" means of signaling strength, capacity, and will. It was "indirect," as Hart would describe it, meant to psychologically disrupt and disorganize the enemy. And it was also successfully limited: it safely forestalled an apocalyptic encounter. Rear Admiral Sir Anthony Buzzard of the British Navy intoned, "Morally, we should not cause or threaten to cause more destruction than is necessary. All our fighting should therefore be limited (in weapons, targets, area, and time) to the minimum force necessary to deter and repel aggression."[56] Limiting the use of the weapons to test sites fit this model of morality. In Gaddis's formulation: it was peace.

Amy Fass Emery describes popular conceptions of effigies such as "voodoo dolls" as utilizing indirect, sympathetic magic on their intended targets: "An ouanga (or charm), like the infamous voodoo doll, works by affecting the well-being of its victim indirectly, through sympathetic magic."[57] One can envision a similar intention for the scale of nuclear testing by the Cold War superpowers at their respective test sites. Each detonation was like a pin inserted in an effigy to destabilize and psychologically terrify the rival, indirectly manifesting a sense of dread, powerlessness, and looming death. As the United States and the Soviet Union tested again and again—far more than was necessary to verify their weapons as functional—each subsequent test was meant to exert a "sympathetic" terror response in their adversary. The sheer quantity of pins being inserted into the nuclear test site effigies attempted to overcome "powerful adversaries while avoiding direct confrontation." Each pin coldly displayed the genocidal resolve of the testing nation. Similarly, political scientist Anne Harrington de Santana has described the role that nuclear weapons play as fetish items for their possessors: "By predicating nuclear strategy on what could be accomplished through the threat of pain and destruction (the threat-value of nuclear weapons), as opposed to what could be accomplished through the efficient application of military strength (the use-value of nuclear weapons), the association between military force and control was reasserted at the level of perception."[58] Behind this theater, behind the curtain, unfolded devastating consequences to real people.

Conclusion

The Cold War was a limited nuclear war. It resulted in casualties, some directly from the weapons usage, more from the toxic aftermath, and additional numbers from the trauma of community dissolution and the loss of home, property, and food sources. Countless more suffered deep emotional distress because of these disruptions and uncertainties about health risks to oneself and family. Hundreds of thousands of soldiers were exposed to residual radiation as they carried out maneuvers beneath the mushroom clouds of nuclear tests. Forty-five thousand Soviet soldiers were exposed to radiation during the Totskoye nuclear test in 1954. Forty-two thousand US military personnel had to be evacuated from Bikini Atoll following the radiological disaster of the Baker test of Operation Crossroads in 1946. The entire populations of several Marshallese atolls were exposed to significant radioactive fallout following the Bravo test. AEC scientists noted about the Bravo evacuees, "According to laboratory hematological data on these natives, the whole body gamma dose approached lethal levels, yet these natives who for the most part, remained in the open during active fall-out, received burns limited to unclothed parts of the body or in areas where perspiration carried contamination into clothed areas."[59] There were over five hundred atmospheric nuclear tests during the Cold War, after which populations directly grappled with the lethal military effect of fallout radiation. Hundreds of thousands more were left to live in radiologically contaminated communities downwind from nuclear test sites. The residents of the villages outside of the Polygon in Kazakhstan, the ranchers living downwind from the Nevada Test Site, the residents of Oak Valley in South Australia, the Tuareg communities living near Reggane in Algeria—all have spent generations living in communities that had experienced significant radioactive fallout following nearby atmospheric nuclear weapon tests and whose homes, schools, and communities received no remediation by the nuclear weapon states that irradiated their communities. Given the long-lived half-lives of many of the fallout products of nuclear weapon testing, subsequent generations experienced—and will continue to experience—risk from living adjacent to these Cold War *military theaters* of operation.

These victims did not suffer direct attack, but their sicknesses, suffering, pain, and anxiety came directly from military activity. The long peace of the Cold War was purchased through the deterrence that vast nuclear arsenals and copious nuclear weapon testing purchased. With a more containable technology such deterrence may actually have been reasonable, but with nuclear weapons, the effects of developing, manufacturing, and testing the weapons had consequences experienced far outside of the factories and military reservations in which these activities were conducted. Nuclear weapon effects are too powerful to contain to military sites or targets: that is precisely how their use is designed. Predicating peace on the development and deployment of nuclear arsenals is grounding it on the sacrifice of the health and welfare of test site populations. The thousands of nuclear weapon tests of the Cold War had powerful and irreversible global impacts ecologically, but the impacts that they had on local populations were immediate and devastating. This was not the global thermonuclear war that the Cold War superpowers devoted their economies to and bet their survival on being able to win: it was not World War III. This was a limited nuclear war: it killed merely on the scale of thousands; it just displaced hundreds of thousands; it spread radiological poisons across communities where merely millions lived. The Cold War was not a long peace; it was a limited nuclear war.

Strategic definitions of the *non-use* of nuclear weapons need to be revisited. If the effects of a weapon are sufficiently revolutionary that one chooses radical steps simply to test them, such as conducting the tests outside of your own national borders—this constitutes an actual *use* of the weapons. Many will argue that the health impacts on populations of being exposed to radioactive fallout is far less significant than the effects of the direct military use of the weapons, and they are right. My question is: Does that still somehow equal non-use? "War affects environment and society simultaneously because humans are shaped by and in turn shape the environments they inhabit," writes historian Emmanuel Kreike, "Environcide consists of intentionally or unintentionally damaging, destroying, or rendering inaccessible environmental infrastructure through violence that may be episodic and spectacular . . . or continuous and cumulative."[60] Kreike calls the degradation of a society's capacity to thrive in an ecosystem because of the military activities of an outside power a crime against humanity and nature. The atmospheric testing of nuclear weapons, especially large thermonuclear weapons that deposit immense loads of long-lived radionuclides into nearby ecosystems and the global atmosphere and oceans, are war crimes against humanity and nature.

IV

Heirs

7

The Slow-Motion Nuclear War

The Competent Management of Nuclear Technologies

Nuclear technologies are operated within a culture of competence. That is not to say that they are competently operated, but, rather, that in the human, professional culture in which they are embedded, a sense of competence pervades. This is true for both nuclear weapons and nuclear power plants. The people who build and manage nuclear systems have faith that the weapons will only be used by choice in ways political and military leaders intend and that nuclear power plants will function as designed.

While a significant portion of the nuclear arsenals of the United States and Russia remain on alert status at all times, the nuclear and political establishments of both countries believe that the weapons will only be used by intention: they are the masters and the weapons are the obedient servants. While nuclear weapons are military weapons, they are primarily envisioned as political tools that leverage desired political outcomes via the threat of devastating military consequences. According to the doctrine of nuclear deterrence, the purpose of nuclear weapons is to deter nuclear warfare. American diplomat George Kennan wrote in 1954, "The sooner we can learn to cultivate the weapons of mass destruction solely for their deterrent value, the sooner we can get away from what is called the principle of 'first use.'"[1] Upon the failure of that objective, nuclear weapons, and their delivery systems, are relied upon to work as designed and to inflict catastrophic damage on the military, infrastructure, and population of the enemy. However, because of the likely survivability of some enemy nuclear forces in a "first strike" scenario (based on a nuclear triad deployment),[2] their use is understood to be a failure as well. If nuclear deterrence fails, the policy description of what follows is "mutually assured destruction."[3] Although military strategist Carl von Clausewitz famously said that "war is a mere continuation of policy by other means,"[4] in the case of nuclear weaponry, war may actually be the end of policy and policymakers.

Competent management of nuclear weapons means not using them. Even their large-scale production and deployment by the Cold War superpowers was, theoretically, designed to deter their use. Usage constituted failure. American game theorist Barry Nalebuff wrote in the late Cold War, "Nuclear war is irrational; it would destroy that which we are trying to save. To threaten nuclear war, therefore, is equally irrational. Nuclear deterrence becomes credible only when there exists the possibility for any conventional conflict to escalate out of control. The threat is not a certainty but rather a probability of mutual destruction."[5] The goal, in fact the imperative, therefore, was to politically manage nuclear deterrence with the Soviet Union and thus manage the massive nuclear weapon arsenal of the United States (or vice versa).[6] This would avoid mutual destruction—a destruction that would not only be *mutual,* but holistic, global.

Toward this end, the United States and the Soviet Union established rigorous protocols for the command-and-control structures over their deployed arsenals. There were multiple occasions in which this command-and-control infrastructure nearly failed, and the world was brought to the brink of a global nuclear war by human or technological error.[7] However, both the United States and Russia, inheritor of the Soviet nuclear arsenal, believe that they have competently managed and continue to manage their nuclear weapons. I am focused here specifically on the United States and the Russian Federation because they possess over 90 percent of nuclear weapons and maintain a larger number of "on alert" weapons than the entire arsenals of the UK, China, and France combined.

Nuclear weapons were first introduced to the world through their direct use against human beings in Hiroshima and Nagasaki near the end of World War II. This has been thought of as their single wartime use. Since this usage of the second and third nuclear weapons ever manufactured, none of the more than one hundred thousand that have been manufactured have been used directly against human beings in war. This is imagined as the competent management of nuclear weapons.

On the other side of the Cold War construct of peaceful and nonpeaceful uses of nuclear technology, those who design nuclear power plants build them with backup safety systems to insure they can be controlled throughout the period of their operation and during any accidents or equipment failures. Nuclear power plant operators work each day with a sense of competence over the dynamics of the technology and the performance of their jobs. "The function of plant maintenance is to preserve and restore the inherent safety, reliability and availability of plant structures, systems and components for reliable and safe operation," informs a 2005 IAEA safety report.[8] The plant can be brought online and shut down according to the intentions of the operators: the plant is under control.

According to the International Atomic Energy Association, as of 2019 there were 451 nuclear power plants operating in the world, with another 54 under construction. Its Power Reactor Information System database tabulates that human civilization has experienced "18,014 reactor-years of operation."[9] While the IAEA's PRIS database does not include statistics on nuclear accidents, and we do know that there have been accidents at nuclear power plants, including a number classified as catastrophic, this would be a very small number beside the theoretical millennia of nuclear power reactor-years. There have surely been less than one hundred hours in which nuclear power plants have failed. This is a statistical presentation of overwhelmingly competent management.

Competent management of nuclear power plants means the controlled generation of electricity without accidents, and especially without the melting of nuclear fuel. Modern nuclear power plants are built with extensive and redundant safety systems to ensure the plants will operate without meltdowns or the release of significant amounts of radiation into the ecosystem. This is one reason that they take years, or more recently, decades, to build and bring online. This safety culture has been used to manufacture illusions of absolute control.[10] John Downer and M. V. Ramana assert, "the contention that reactors are knowably designed to levels of reliability that make catastrophic accidents too unlikely to warrant public consideration has become a foundational premise of the practices, rationalities, and logics through which all states manage atomic energy: the 'governmentality' of reactors worldwide."[11]

While there have been some incidents of nuclear fuel melting and some of complete meltdown, the vast majority of nuclear power plants have operated without any melting of nuclear fuel. Additionally, while there have been routine releases of radiation from almost all nuclear power plants, particularly in the form of tritium, there have been few catastrophic releases of radiation from nuclear power plants into nearby communities.[12] Although electrification of homes and cities, from any power source, has existed for less than one hundred and fifty years, the more than eighteen thousand reactor-years of nuclear power plant operation are a means of visualizing the competent operation of nuclear power plants for the less than eighty years since their invention.

Most nuclear power plants have not melted down, and since 1945 no nuclear weapon has been used directly against a human population.

Seeing Incompetence in Nuclear Technologies Management

As noted, there have been over two thousand nuclear weapon detonations since 1945, and because of the scale of the weapons' effects, these detonations have had a profound impact on numerous societies and millions of people. From the perspective of someone living downwind from the Polygon in Kazakhstan, or in many Pacific island nations, the management of nuclear weapons is unlikely to be envisioned as having been executed competently. As discussed in chapter 5, managers at the Nevada Test Site had instructions to postpone tests when the winds were blowing toward Las Vegas to the south: "Detonations proceeded when the prevailing winds were moving east across southern Nevada and Utah to preclude fallout over Los Angeles to the southwest, Las Vegas to the southeast, and Salt Lake City to the northeast."[13] This was a formal procedure for the competent management of the nuclear tests. Yet for those living in St. George, Utah, where, as Willard Libby said in 1955, fallout clouds "always plaster," this was a disaster.[14] Incompetent management was further demonstrated when the *anticipated* health impacts on those communities were assessed strictly in terms of their exposure to external radiation rather than the particles that would be internalized by many people, based on an overreliance on the findings of the LSS. As with the government of the Soviet Union seeding fallout clouds from Chernobyl to deposit their radioactive load in Belarus rather than over Russian cities mentioned in the previous chapter, one person's competence becomes another person's life-altering tragedy.[15]

The previous chapter discussed this history of nuclear weapon testing as constituting a limited nuclear war carried out by the nuclear weapon states against the populations living downwind from their test sites. Those who lived farther away from the nuclear test sites and who feel that they and their families benefited from the deterrent value of nuclear weapons during the Cold War may well feel that the two thousand nuclear weapon tests were conducted safely, but that is simply the virtue of perspective. To those for whom the nuclear tests were intimate, for whom the effects spilled out of the test sites and into their communities, onto their gardens, and into their homes and bodies, carelessness may be the descriptor that comes to mind, if not cruelty. When asked on CBS Radio in 1954 about the ir-radiation of the *Daigo Fukuryo Maru* crew following the Bravo test, US president Eisenhower declared, "this time something must have happened that we have never experienced before, and must have surprised and as-tonished the scientists."[16] This particular, unforeseen incompetence resulted in the entire population of multiple Marshallese atolls suffering radiation sickness and losing their homes. Despite nuclear weapons not being used in direct combat against human beings after Hiroshima and Nagasaki, they have not been competently managed during the intervening seven decades. The testing of nuclear weapons directly affected the health and well-being of millions of downwinders. Entire communities have been forcibly evacuated, and multiple villages and atolls are no longer safe for human habitation because of radiological contamination.

The deposition of transuranic radionuclides from nuclear weapon test-ing is considered a fundamental marker that geologically determines the start of the Anthropocene.[17] This deposition did not come from accidents, but from the intentional management of nuclear weapons. Recent research has shown that fallout radionuclides deposited from this testing in Arctic, Antarctic, and subarctic regions have begun to migrate as global warming continues to thaw glaciers: "As FRNs [fallout radionuclides] are released into the proglacial environment through glacier melting and retreat they may act as a secondary source of environmental contamination many years after the nuclear event of their origin."[18]

Accidents have also plagued the management of nuclear weapons. In 1966, an American B-52 bomber exploded during midair refueling over Europe, dropping its four hydrogen bombs around the small village of Palomares, Spain. One weapon landed in the Mediterranean and three on the village.[19] None of the weapons experienced nuclear detonation, but two of the weapons that landed on the village cracked open from detonation of high explosives in the weapons' casings and leaked plutonium into the local ecosystem.[20] This is only one of dozens of accidents with nuclear weapons. Multiple nuclear weapons remain "lost," as do several whole nuclear armed submarines. The fact that human beings have not initiated a nuclear war is not the sole criterion to ascribe the effective management of nuclear weapons.

As has been amply documented, the United States and the former Soviet Union came close to actual global thermonuclear war on multiple occasions during the Cold War, through misreading the intentions and actions of their opponent and then being lured to the edge of weapon launch by the strategies baked into Cold War preparedness and paranoia.[21] NATO nuclear war games held in Europe in 1983, under the title of Able Archer, created suspicion in the Soviet command that the exercises were actually preparations for imminent nuclear attack.[22] The Soviet weapon complex moved quickly into battle stance and prepared for direct nuclear conflict. Both sides had weapons hot, and were deeply misreading the intentions of their opponent. This incident brought the world to the threshold of nuclear war with the advanced arsenals and hair-trigger delivery systems that typified the late Cold War. Numerous other accidents and false readings of radar data brought the superpower protagonists close to errantly launching nuclear attacks against their adversary.[23] Soviet officer Stanislaw Petrov decided, against protocol, that instrument readings indicating American missiles had been launched at the Soviet Union were faulty, and he did not pass the alert up the chain. Petrov was widely hailed after the Cold War as the "man who saved the world."[24] One could say that the system worked in these cases, but it is far easier to see from these cases that the system was structurally flawed.

All of these weapons emerged out of military nuclear complexes in which the construction and operation of plutonium production plants were elemental. Plutonium production plants are simply nuclear reactors accompanied by plutonium separation facilities, such as at Hanford in the United States, Mayak in the former Soviet Union, Windscale/Sellafield in the United Kingdom, and numerous sites that followed as the weapons proliferated. Nuclear reactors were invented in the Manhattan Project for the singular purpose of manufacturing plutonium for nuclear weapon production. These plants operated on the Earth for ten years before they were ever used by any nation to produce electricity for civilian use. The first controlled and sustained nuclear chain reaction was in the Chicago Pile-1 reactor (CP-1) in 1942. The first industrial-scale nuclear reactors were built by the Manhattan Project at Hanford, Washington, during World War II, and the plutonium produced there was used to kill almost one hundred thousand people, including thousands of children, in Nagasaki. I have written elsewhere that nuclear power was born violent.[25] The first thirteen nuclear reactors to go online in the United States were all strictly plutonium production reactors. The US had thirteen functional military nuclear plants before it had one nuclear power plant built to generate electricity for civilian usage.[26] The main difference between a nuclear reactor used to produce plutonium and a nuclear power plant is to connect the reactor to a turbine and use the heat generated to generate electricity; the core reaction is the same.

The first truly catastrophic nuclear power disasters happened eleven days apart in 1957 in two of the plutonium production complexes of nuclear weapon states. On 29 September an explosion took place in a radioactive waste tank at the Mayak facility in Chelyabinsk in the former Soviet Union. The Kyshtym disaster happened when a cooling system for the waste tank failed, and a significant amount of radiation was dispersed over a wide area, creating a contaminated region known as the East Ural Radioactive Trace.[27] Eleven days later, on 10 October, a fire broke out inside one of the reactors of the Windscale site in Cumbria, United Kingdom, burned for three days, and dispersed radiation across the region. The fact that these two disasters took place in plutonium production reactors operated by the Soviet and British governments to manufacture nuclear weapons allowed the governments to largely control information about the disasters. Nuclear meltdowns in the reactors of Soviet nuclear submarines in 1968, 1979, and 1985 also went unreported at the time they occurred.[28]

There were major accidents at multiple research reactors, including an explosion and partial meltdown of fuel at the experimental reactor at Chalk River in Ontario, Canada, in 1952, a partial fuel meltdown suffered at the Experimental Breeder Reactor-1 at the National Reactor Testing Station (now Idaho National Laboratory) in 1955, a core meltdown at the Santa Susana Field Laboratory in Los Angeles in 1959, and an explosion and core meltdown of the SL-1 reactor, also at the National Reactor Testing Station in Idaho in 1961.[29]

There have been numerous accidents at commercial nuclear reactors, although partial nuclear fuel melts are not widely known as catastrophic events. The Lucens reactor in Switzerland suffered a partial fuel meltdown in 1968; a year later the Saint-Laurent Nuclear Power Plant in France suffered a partial fuel meltdown; there was a partial fuel meltdown at the A-1 reactor of the Bohunice Nuclear Power Plant in Czechoslovakia in 1977; and a partial nuclear core meltdown occurred at the Three-Mile Island Nuclear Power Plant in Pennsylvania in 1979.[30]

Much better known are the two catastrophic nuclear power plant accidents at Chernobyl in 1986 and Fukushima in 2011. After an explosion in reactor 4 at the Chernobyl Nuclear Power Plant on 26 April 1986, the full fuel load was either dispersed in the fallout cloud of the explosion or infused into the ruins of the reactor after melting. At Fukushima there were three full meltdowns and multiple explosions, including from a fire in the spent nuclear fuel pool of a fourth reactor that was not online at the time of the disaster. Some nuclear fuel was dispersed in the clouds of the explosions, while the remaining melted fuel (amalgamated with materials it absorbed while melted, known as corium) is believed to be below reactors 1, 2, and 3.[31] In almost every one of these incidents there was some combination of technical and human error that led to the disasters.

Nuclear risk assessment experts frequently use the probabilistic risk assessment method and predict nuclear accidents in relationship to reactor-years rather than by empirical years, and construct elaborate probability charts to predict risk. "The basic idea used by such studies is simple to describe: one enumerates the possible fault trees that could lead to an accident. For each individual component in the reactor, one can estimate a

Sculpture linking Chernobyl to Hiroshima, featuring Hiroshima peace cranes and nuclear fuel rods, located in Chernobyl village in the Exclusion Zone (2019). Photo by author.

frequency of failure. For a serious accident, some combination of these components has to fail simultaneously," explains theoretical physicist Suvrat Raju. "The theoretical problem with such estimates is obvious. Consider the Fukushima nuclear complex, which had 13 backup diesel generators. Assigning a probability of 10^{-1} for the failure of each generator per year and assuming that they are independent would lead to the naive conclusion that the probability that 12 generators would fail together in any given year is about $13 \times 10^{-12} \times 0.9 \approx 10^{-11}$. However, the tsunami did precisely this by disabling all but one of the generators at once. The point is that once the obvious fault trees have been eliminated and corrected, the dominant contributions to accident-probabilities come from unlikely sequences of events that conspire to cause failure."[32] Charles Perrow has described how simple and anticipated failures in systems built with "interactive complexity and tight coupling" lead to what he calls "normal accidents."[33]

Studies conducted using a PRA methodology by the United States Nuclear Regulatory Commission yielded probabilities of one accident per 100,000 reactor years, or in a more conservative estimate of one accident per 10,000–20,000 reactor-years. The partial list just mentioned of accidents all happened in the currently tabulated 18,000 reactor-years total for worldwide nuclear plant operation. A 1986 post-Chernobyl study conducted by Islam and Lindgren used existing data on nuclear accidents rather than probabilities to come up with a more realistic figure of one accident per 3,000 reactor-years.[34] Put in more experiential terms, there have been more core-damage accidents than there have been decades since the advent of the technology. While this history is full of instances of the competent management of nuclear power, it clearly includes dramatic and consequential failures.

One can envision incompetence in the management of nuclear power plants that extends beyond the success or failure of operating the plants to produce electricity or plutonium. The fact that many nuclear power plants around the world operate as for-profit businesses creates a disincentive to pursue safety as the highest priority. Businesses generate profits by increasing income and decreasing expenses. Decreasing expenses at nuclear power plants can often mean curtailing safety protocols and planning. The Fukushima Daiichi Nuclear Power Plant was not the closest nuclear plant to the epicenter of the earthquake on 11 March 2011; the Onagawa Nuclear Power Station was 60 km closer. The earthquake was more violent at Onagawa and the tsunami was higher, yet the three reactors there did not suffer meltdowns. The two complexes have different corporate owners, and many ascribe their destinies on that fateful day as being linked to the "safety culture" of their respective owners and managers.[35]

Many governments have provided legal limitations to the liability of nuclear power plant operators in the case of catastrophic disasters. Even before the first commercial nuclear power plants went online in the United States, policymakers realized that no private utility company could possibly afford sufficient insurance to cover the anticipated damages incurred by a "run-away reactor." Testifying before the Joint Committee on Atomic Energy in 1954, Francis K. McCune, the general manager of the Atomic Products Division of the General Electric Company, informed lawmakers that limitation of their liability was a bedrock of the industry's business plan. "Without some government-sponsored program of insurance or indemnity over and above the conventional limits of 'third-party liability' coverage, McCune predicted that power companies and their suppliers would never build or operate nuclear power plants in the United States."[36] The US Congress passed the Price-Anderson Act in 1957, which did limit corporate liability in the case of an accident. It has been revised numerous times; its current iteration extends until 2026.[37]

Even through the normal operation of nuclear power plants without disaster, there is structural incompetence. The plants have generated large amounts of extremely dangerous and long-lived waste, with structures for the permanent management of that waste. Peter Custers observes that "nuclear waste is not churned out at only one link in the chain of nuclear production but in fact at *each and every* step in the chain."[38] The largest quantity of this waste by volume is the tailings at uranium mines, while the most concerning is the high-level waste that is produced by burning nuclear fuel in reactors. This high-level waste has been piling up around the world for over seventy-five years, most often in forms of temporary storage on-site. The IAEA estimated that that the total amount of commercial spent nuclear fuel in 2013 was 180,800 MTHM.[39] This is strictly from the operation of nuclear power plants to produce electricity and does not include the vast amounts of spent nuclear fuel left over from plutonium production by the world's nuclear weapon states. For example, the United States has the world's largest fleet of commercial nuclear power plants, with 104 having been operated or currently operating. This commercial operation has produced 67,600 MTHM of high-level nuclear waste. The military production reactors of the United States have produced another 25,000 MTHM of high-level waste, or a bit more than a third of the inventory created by the commercial reactors, for a total of 92,600 MTHM from the US when the commercial and military high-level fuel is combined.[40] This US inventory amounts to half of the total produced globally by all commercial power generation alone.

High-level nuclear waste must be competently managed to protect the health and safety of living creatures around the world, today and deeply into the future. This requires that the waste be securely stored for over one hundred thousand years. The challenges to successfully and securely store all of this spent nuclear fuel for such a vast amount of time have been daunting since the technology was developed less than a century ago. In 1972, Alvin Weinberg, the former director of Oak Ridge National Laboratory, cautioned that "We nuclear people have made a Faustian bargain with society. . . . The price that we demand of society for this magical energy source is both a vigilance and longevity of our social institutions that we are quite unaccustomed to."[41] After decades of inaction, there are now strategies developed in the last few decades to address our debts in this bargain.

The Competent Storage of Spent Nuclear Fuel

The management of spent nuclear fuel has been a challenge since the first prototype reactors of the early Manhattan Project. Spent fuel contains many long-lived and extremely radioactive elements, some of which also will generate heat for thousands of years.[42] The isotope ^{239}Pu will remain dangerous to living creatures for over one hundred thousand years, and ^{235}U will remain dangerous for over a million years. During that time the radiation will slowly decline; some radionuclides will decay in the first few thousands of years, but many will decline in periods of tens of thousands of years. These dynamics make the safe and secure storage of spent nuclear fuel an imperative for public health and also an obligation for those of us that produced and will pass down this waste legacy to countless generations.

This was understood by early managers of the nuclear complex, in both governments and industry. A 1957 US National Academy of Sciences study on the problem of storing high-level radioactive waste stated, "Unlike the disposal of any other type of waste, the hazard related to radioactive waste is so great that no element of doubt should be allowed to exist regarding safety. . . . Safe disposal means that the waste shall not come in contact with any living thing."[43] As early as 1949, the AEC booklet *Handling Radioactive Wastes in the Atomic Energy Program* emphasized guarding against waste radionuclide migration through the ecosystem and bioaccumulating in the food chain: "liquid wastes which find their way into streams are also potential hazards for they might be swallowed by animals or absorbed by algae or other micro-organisms and in turn be consumed by fish."[44]

There are a number of steps to grapple with the dangers specific to spent nuclear fuel before the waste is ready for long-term storage. As nuclear fuel is "burned" in a reactor, creating heat and therefore electricity, some of the fuel is converted into "daughter particles."[45] Plutonium is one of those particles, and it was to manufacture plutonium as a "progeny" that nuclear power plants were first designed and constructed. However, other progeny, or fission products, are also created, and the accumulation of these impedes the ability of the fuel to burn efficiently. Because of this inefficiency, nuclear fuel rods are typically kept in reactors for about three years and then replaced by new fuel rods: the plant is *refueled*. The removed, or *spent*, nuclear fuel then begins a journey of decay that will take millennia.

Some amount of commercial fuel is used for reprocessing. This means that the plutonium is separated out, as is done for weapons, and then used to make mixed uranium-plutonium, or MOX, fuel. While some amount of fissionable uranium can also be separated out, it has not been economical to use MOX for fuel because newly mined uranium has typically been less expensive. Fuel is commonly reprocessed in France, the United Kingdom, Russia, and several other nuclear states, although there is also debate about whether the procedure is actually more economical than direct storage of the spent fuel. The World Nuclear Association estimated that MOX fuel accounted for only 5 percent of new fuel serving the world's fleet of nuclear power plants in 2017.[46] Although some nuclear power promoters assert that all nuclear fuel is "recyclable," one can think of this as analogous to plastics. Plastics are recyclable, but it is much cheaper to make new plastics. Hence, in spite of its recyclability, the world is awash with plastic waste.

Most spent nuclear fuel is not reprocessed and is held in various types of storage, awaiting "permanent" disposal. This sequence of storage systems begins at the moment of refueling: when spent fuel is removed from a nuclear reactor, it is immediately placed into a spent fuel pool. This is similar to a large swimming pool, and the water is continually cycled in and out to remove heat to keep the fuel from melting. The water also shields workers from the intense radiation emitted by the spent fuel. Therefore, every nuclear reactor has an SFP located in the reactor building so that the fuel can be placed into water immediately upon removal. Once the fuel has cooled for a period of time, it may be removed to a secondary SFP located outside of the reactor building or in a collective larger pool, to make room in the pools located in each reactor building to accept new spent fuel. These spent fuel pools must actively cool the spent rods for years, minimally ten. Cold water must be continually circulated through these pools and, after being heated by the spent fuel, must be dumped so that more cool water can continue the process.[47] At Fukushima, the loss of power at the complex resulted in a catastrophic explosion in a spent fuel pool that had boiled away all of its cooling water. Power must be maintained continually to keep the cooling system operating. Even during this process particles can transport into the ecosystem. Over the course of several decades, authorities have repeatedly had to cull gulls that visit storage pools at the Sellafield plant in Cumbria, United Kingdom. "Bosses say some birds may have got into open spent fuel storage ponds and become contaminated by low-level waste," described a BBC article in 2010.[48] While some were culled, many likely transported particles far from the site and deposited them through feces or, later, their carcasses.

Once the fuel has sufficiently cooled, it can be transferred to dry cask storage. These are casks that can shield workers from the radiation of the spent fuel once it has cooled sufficiently so the heat will not damage the casks. All spent nuclear fuel in the world that is not actively being reprocessed is sitting in either a spent fuel pool or a dry cask right now. A 2014

US government study describes the current and immediate future of the US spent fuel management program:

> In the future, more spent nuclear fuel is expected to be put into dry storage for two reasons. First, since most spent nuclear fuel pools have reached their maximum capacities, reactor operators must transfer fuel from the pools to dry storage to make room for newer spent nuclear fuel, a time-consuming and costly process. Second, the amount of spent nuclear fuel transferred to dry storage is expected to increase as reactors shut down and their pools are closed. . . . By 2067—after the last of the currently operating reactors have shut down—nearly all the 139,000 metric tons of spent fuel expected to be generated by currently operating reactors is expected to be in dry storage.[49]

As of 2013, 30 percent of US commercial spent nuclear fuel (~22,000 MT) was being stored in dry casks, while 70 percent (~50,000 MT) had remained in spent fuel pools since the day it was removed from a reactor.[50] As for the military spent nuclear fuel, "nearly 80 percent of the Department of Energy's (DOE) inventory of spent nuclear fuel is stored underwater at Hanford in two 4.94-million-liter (1.3-million-gallon) pools. The pools—known as the K Basins—are less than a quarter-mile from the Columbia River and close to now-defunct production reactors," explains a 1999 fiscal-year report on Hanford, adding that "The K Basins are aging and the fuel is corroding."[51] All of this waste is awaiting the Department of Energy of the United States to designate a permanent storage site and begin to accept shipments.

The permanent storage of spent nuclear fuel is a challenge that, to quote William Burroughs, "buggers description."[52] The waste must be completely and securely contained, and the site must be geologically and hydrologically stable for tens of thousands of years.[53] Additionally, there must be no breaching of the site by living creatures, particularly by future human beings. Not fully achieving each of these challenges constitutes failure—a failure with dire and irreversible consequences for untold generations. Work has also been done designing retrievable storage sites, which would have to be actively maintained for a period of several hundred years. This would allow for the monitoring of the reduction of radiation in the waste and the integrity of the containers.[54] However, this approach has been largely abandoned, primarily because of the high cost and political uncertainty of the required ongoing management and monitoring.

Most nuclear power–producing states have focused on storage in a deep geological repository as their preferred method to contain spent nuclear fuel for the required millennia. A DGR is a storage facility carved into deep and geologically "permanent" rock about half a kilometer underground. In a DGR, the waste is carefully packaged and placed in a strategically located cavern; after the packaged waste is deposited, the cavern is filled and sealed. The design intends to assure that the cavern will be geologically stable for longer than the waste remains dangerous, and that the packaging will retard or resist encounters with subterranean water that may decay the container and bring water into contact with the spent fuel rods, potentially transporting radionuclides from the site into the surrounding ecosystem. The design of the site is intended to separate the waste from the surrounding ecosystem for tens of thousands of years.

While several designs for DGRs are in development, the SKB system is actually under construction both in Sweden and Finland.[55] This process creates a multibarrier system to contain the spent nuclear fuel. First, the spent fuel is placed in copper canisters, with a cast-iron insert, that are designed to be water resistant. These canisters are then placed into slots dug into bedrock half a kilometer underground. Once the slots in the storage cavity are filled with canisters, the entire site is backfilled with bentonite clay, which expands when wet (it is commonly used in cat litter). Thus, the spent nuclear fuel is contained in a copper canister, this canister is surrounded by bentonite clay, which fills the repository, and the entire facility is shielded by half a kilometer of bedrock from the surface.[56]

Beginning in 1990, Sweden built a fully functional underground laboratory to develop the technology of the SKB method at Äspö Hard Rock Laboratory near Oskarshamn, an existing nuclear power plant site.[57] Upon refinement of the technology, both Sweden and Finland committed to the construction of DGRs based on the SKB method. Finland began to construct at the Onkalo site adjacent to its nuclear power plants at Olkiluoto, a site that is slated to begin to accept spent fuel rods later this decade.[58] Sweden has chosen its site at Forsmark, although it has still not completed the approval process.[59] Each nation that is preparing strategies to manage its high-level nuclear waste, especially the spent nuclear fuel rods, has invested large sums of money into researching the methods it will choose to isolate this waste. Each country has engaged teams of scientists, researchers, and scholars to determine the best possible plan to successfully achieve competent nuclear waste storage during the period of time in which the waste may harm human beings or other living creatures.[60]

Electricity generated from nuclear power plants, and the waste from plutonium production in nuclear power plants, begins a process that will remain physically present and impactful for thousands of generations of human beings. In the case of commercial nuclear power, only two of those thousands of generations will receive a benefit from this technology: electricity. The primary experience of more than 99 percent of the generations that will have a relationship to our nuclear power plants will be one of risk. This is an astonishing legacy to leave for a relatively small benefit. Philosopher Kristin Shrader-Frechette points out that "Allegedly permanent storage of nuclear waste is not merely problematic on grounds of temporal distributive justice. It also is questionable on grounds of participative justice, because future persons would be unlikely to consent to it."[61] It may be that we can successfully protect them from this risk with our deep geological repositories. For the benefit of the thousands of generations of people who will bear the risks, it is essential that we work to achieve as competent an outcome as humanly possible.

Deep Geological Storage as Incompetence

In 1958, physicist and engineer Bernhard Philberth presented a paper at the International Association of Scientific Hydrology conference at Chamonix, France, outlining a strategy for the long-term storage of high-level nuclear waste.[62] Philberth proposed storing all of the world's high-level waste in containers set into the permanent ice sheets of Greenland or Antarctica. Further elaborating on this original plan almost twenty years later, Bernhard's brother, Karl Philberth, also a physicist and engineer, asserted in the *Journal of Glaciology* in 1977 that the "high-level waste of the whole world for the next 30 years could be put into 3×10^7 spherical containers with 0.2 m radius and disposed of in an area with 15 km radius and a depth range of 20–100 m under the surface of either the Antarctic or the Greenland ice sheet. The deposit does not affect the stability of the sheet. Even the most upsetting natural ice-sheet instabilities and/or climatic changes could not cause radioactive contamination."[63] The Philberth brothers imagined that their nuclear storage depository in the ice sheets would place the waste in retrievable containers for the first several hundred years, so that problems with the site could be addressed as they arose. From our 2021 vantage point, it would be clear that had the proposal been acted upon in 1977, we would already be retrieving the containers from the rapidly melting ice sheets of Greenland or Antarctica.[64] While the proposal may not have led to a radiological release, what appeared reasonable in one era had not anticipated changing conditions that lay merely a few decades ahead.

The Philberths' plan was based on sound engineering and detailed study of the dynamics of ice sheets. It was presented as an alternative to deep geological storage, a strategy that was also being studied at the time. It may be that DGRs are the best waste disposal method that we can design and execute. However, this strategy is also burdened with illogic. No science from a past century would seem sufficient or complete to its contemporary practitioners, yet designers of DGRs assume that they can fully anticipate and exert competent control over physical processes for millennia based on the cutting-edge technologies of their present-day materials science, geology, and numerous other disciplines. There is no doubt they can operate at the top of their fields within the models functional in their own times, but exerting that expertise over materials stretching farther into deep time than any human construct or concept has traveled is, at minimum, presumptuous.

First, let's consider our determination of the best sites for DGRs. For countries that have a designated or suggested DGR site, almost every one happens to be on land already owned either by electrical utilities that are nuclear plant operators or by the governments tasked with disposing the waste. After having conducted extensive surveying, detailed analysis of geological and other scientific and technological characteristics, many nations have determined that the best site is one that also happens to be politically convenient: where some portion of the country's waste is currently located, and often surrounded by a community that is largely employed by the entities siting the waste. This follows an existing pattern whereby nuclear weapon sites are often built where ordnance sites were already being operated by national militaries. The United States established the Nevada Test Site for nuclear weapon testing at its existing Las Vegas Bombing and Gunnery Range in 1940. This site was chosen because it "offered excellent year-round flying weather, a strategic inland location, nearby mountains that could provide natural backdrops for cannon and machine gun practice, dry lake beds for emergency landing and an existing airfield."[65] These are not necessarily the requirements for an optimum nuclear weapon test range—it was just easy. The US has long been focused on the Yucca Mountain area as its DGR, located within the former Nevada Test Site (now known as the Nevada National Security Site), another example of a potential DGR being conveniently located on an existing nuclear complex site. The former Windscale site (since renamed Sellafield), location of the plutonium production reactors of the United Kingdom, was built at the site of a prior ordnance factory, the ROF Sellafield, which manufactured propellants for explosives during World War II.[66] Sellafield is also being considered as the primary DGR by the UK.[67]

Finland, the nation furthest along in actually building its DGR (Onkalo), has selected a site on Olkiluoto Island, adjacent to the site where two of its four existing nuclear power plants are located, and where a third nuclear plant (currently behind schedule and over budget) is being constructed. Half of Finnish spent nuclear fuel is currently located on Olkiluoto Island and will not need to be transported to the Onkalo site, significantly reducing the anticipated transportation cost and risks. The local community is well disposed to the Finnish nuclear power industry, having hosted two reactors for forty years with a local economy that has been largely dependent on the nuclear industry for a generation. Selecting a site where there was no existing nuclear power plant would likely have resulted in a more challenging political path to local community buy-in. These difficulties were sidestepped by the seemingly coincidental best location for Finland's DGR being on-site at the Olkiluoto complex.

The Finnish company that is constructing the Onkalo repository is Posiva Oy.[68] In a presentation at an International Press Day event at Onkalo in 2016, a Posiva Oy public relations officer stated that all of Finland was perfectly suited to use as a deep geological storage site, and that you could "throw a dart" at a map of Finland and wherever it landed would be as good as anywhere else.[69] He pointed out that siting in Helsinki would obviously be more difficult than in a rural area, because of land ownership and population density issues. While it may be substantially accurate, it cannot be true that all of Finland would equally meet the safety and stability requirements of a DGR without variance. However, there is no doubt that all locations in Finland are not equal in terms of economic cost and political difficulty. Onkalo may meet all of the geological and physical requirements demanded by the site selection process in Finland, but it is clear that the "dart" did not just hit anywhere; it hit the very place that is most economically and politically advantageous to the needs of the corporation constructing the site and the government that has to generate political support for the decision. Onkalo has been chosen as much for its appropriateness to twenty-first-century needs as millennia-long requirements. Speaking to a reporter in 2017, Tiina Jalonen, the senior vice president of development for Posiva Oy, emphasized the short-term benefits of the project: "here in Eurajoki and Loviisa everyone knows someone who is working there, so they know how things are handled and they have trust. They also see the benefits of hosting the nuclear facility. There is a high rate of employment and the community itself is quite wealthy."[70]

Across the Baltic Sea in Sweden, after the pioneering DGR test site at Äspö Hard Rock Laboratory worked out the dynamics of the SKB method of disposal, the Swedes have been determining the site of their actual DGR. The two sites under consideration were at Oskarshamn and Forsmark. Both sites have had three nuclear power plants operating since the 1960s (Oskarshamn) and the 1970s (Forsmark). Forsmark is the location of the SFR, the underground repository for short- and medium-level nuclear waste, and Oskarshamn is the location of the CLAB facility, which is the temporary storage site for the nation's spent nuclear fuel. Here, as in Finland, both ideal locations for the DGR are sites with currently operating nuclear power plants, where waste is already being stored and where the local population is economically dependent on the plants.

It is hard not to conclude that the essential decisions compelling the siting of the DGRs in Finland and Sweden were the original decisions to locate nuclear power plants on those sites back in the 1960s and 1970s. Once those plants were built there would be an inevitable momentum to permanently storing the spent nuclear fuel at one of the preexisting nuclear

Tunnels in the Onkalo Spent Nuclear Fuel Repository, half a kilometer underground, Eurajoki, Finland (2016). Photo by author.

sites. Or it could be a coincidence . . . two coincidences. Clearly, decisions that will have a profound impact for millennia are driven as much by the short-term requirements of twenty-first-century national and municipal politics as they are by the geological and safety requirements of millennia-long storage of highly toxic and long-lived materials.[71] While this may be unavoidable, it indicates that other aspects of our decision making may also serve our own needs above those of the people we profess to consider.

Drawing Lines That Will Become Invisible

A similar argument can be made for the constituent bodies with which we make these waste-siting decisions. Our structure for determining and executing the most well-designed deep geological storage facilities for our spent nuclear fuel is based explicitly on the current delineation of our political boundaries. In other words, Finland must dispose of Finnish nuclear waste in Finland; America must dispose of American nuclear waste in America; Japan must dispose of Japanese nuclear waste in Japan. This seems natural since the waste was generated within the boundaries of these political systems. The electricity or weaponry was manufactured for the benefit of the citizens of those states, and the responsibility for the disposal of that waste is the burden of the benefited citizens of those nations. While that makes sense to us, it makes less sense in terms of disposing the waste. When we look at Japan, there is no suitable place to establish a deep geological storage site inside of Japan.[72] The entire country exists on fault lines and in volcanic zones. There is no geologically stable location inside of Japan. However, since the Japanese waste was generated by Japan, it will be disposed of in Japan. One may say that Japan should have considered such a dilemma before it chose to generate nuclear waste, much as one may ask that question about the endeavor globally, but such questions are beside the point: the waste is here, the waste will also be there—in the future. What is the best strategy to mitigate risk to ourselves and future generations?

The Japanese government has determined the "best" locations for its proposed deep geological storage site.[73] The designated sites are considered "better" than other sites. That does not make them good, it just makes them (by some measure) the best of a bad lot of alternatives. Still, the then

Japanese vice minister of METI (Ministry of Economy, Trade and Industry), Yōsuke Takagi, exalted Japanese competence in 2019, declaring that "We were told that we could not do such a project in Japan [because of the seismic activity], but we have the scientific expertise to do so!'"[74] Ultimately, one of these sites will be chosen, and the 110 tons of spent fuel from the operation of fifty-two reactors will be deposited in the site.[75] And then time will simply unfold. For people in the future forced to grapple with this large quantity of waste having been deposited deep in an earthquake fault zone full of volcanos, the politically construed entity of "Japan" in the twentieth and twenty-first centuries may be a ghost that is long forgotten. While the concept of "Japan" may be meaningless to them, the dangers to them of "Japanese" spent nuclear fuel may become incredibly meaningful.

We are making millennial-long decisions based on temporary political boundaries. I am not suggesting that there is a more pragmatic way to proceed and site this waste underground. However, this should confront us with the fact that we are not making these decisions as competently as we imagine. We are working within temporary boundaries of political expediency. We are not envisioning, and being compelled strictly by, the constraints of geology and waste management, the scientific and technological requirements, or our obligations to our descendants.[76] We are doing what is expedient within our current social structures, regardless of their technological wisdom. Our social constructs and boundaries compel us to make technologically incompetent decisions. We assure ourselves that we are making competent decisions and choices, while wearing blinders made of national flags. As anthropologist Joseph Masco cautions, nuclear materials defy concepts of "nation-time" and "nation-space."[77]

It may be a better plan to consolidate the waste from all nations and site it in the single best geological location determined globally, keeping the threat in one location rather than distributing it in dozens of sites. It may be more pragmatic to design one form of marker for all of the waste rather than dozens of different, inconsistent markers. This clearly makes more sense from the thirty-thousand-foot perspective—but we can never do that because we are social animals living in a politically constructed world. So we will do a less competent thing, a number of less competent things. And we will tell ourselves that we have chosen the best plan(s). We will feel assured that we are being competent. And we will leave this planet believing so because the flaws in our plans will probably not become visible before we ourselves die. In every way, it will look like a highly skilled, well-planned project. We will know nothing about whether what we did was successful, but we will tell ourselves it was a total success and that we did a good job.

What we are doing here is simply being human. Human beings are not linear and rational creatures, selecting the most efficient strategies and executing them on their own merits. Human beings are social animals operating with limited data about the world around us. We construct models of reality through scientific inquiry, but that "reality" is always filtered through biological and social mechanisms. Our models reflect the medium through which that external reality is filtered and from which the capacity to build models emerges: the human brain. And as Kuhn has shown, social groups of scientists debate and agree on a paradigm that will for a period of time pass as "truth."[78] These are human dynamics, and it is impossible to strip them out of any process designed and built by human beings.

Can we accurately predict the dynamics of geological change over millennia? Maybe. Not definitely, but maybe. We can probably predict some things accurately, while not having adequately anticipated other things. That's natural. However, we pretend that we do have the capacity to calculate geological changes for tens of thousands of years, and that we can construct materials whose behavior over those periods is also predictable. Already research points out our incomplete knowledge. A study published in early 2020 argues that the materials that are proposed to be used to encase some high-level waste (for example, the fluid waste currently in the "tank farms" at Hanford) in glass or ceramics are likely to suffer "severe" and "localized" corrosion. The lead author of the study, Xiaolei Guo of Ohio State, explained that "In the real-life scenario, the glass or ceramic waste forms would be in close contact with stainless steel canisters. Under specific conditions, the corrosion of stainless steel will go crazy. It creates a super-aggressive environment that can corrode surrounding materials."[79] Also in 2020, researchers led by a team from the University of Manchester found that a previously unknown form of uranium was produced under conditions designed to mimic those of a DGR. The interaction of the waste with various microbes produced uranium in an oxidized form that had not been previously seen, some of which were extremely environmentally mobile and transported via water in a short period of time.[80] As participating mineralogist Samuel Shaw put it, "you can't sterilise the earth."[81] These studies reveal that compelling uncertainties remain around the safe storage of these materials for millennia, and also that it is unlikely for humans to perfectly predict the actual conditions these materials will experience for one hundred thousand years. SKB's own imagination of a "beyond-worst-case scenario" is limited to being the failure of the barriers in their design, the canisters and the bentonite clay layers.[82]

Brains and the human beings in which they reside are good at believing themselves to be competent; our survival has always depended on this framework. Although, if we turn our gaze back in time, it is hard to find any example in which humans had perfectly predicted what was to come, or how systems would operate over time, or for that matter, have built anything that would last. There are always uncertainties and unforeseen dynamics that enter into system operations. We plan best when we anticipate that uncertainties will be part of any dynamic system, such as in weather prediction. The risks of harm from leaking high-level nuclear waste are so catastrophic that the idea of uncertainties thwarting our plans is acceptable, so we tell ourselves that we have anticipated and planned for all potential dynamics.[83]

For the deep geological disposal of spent nuclear fuel to go well we have to get a lot of different things right. But it is not enough for us to get all these different things right once; we have to bury spent nuclear fuel in dozens of locations: more than a dozen different agencies/governments have to get all of those things absolutely right several dozen times at multiple sites for twenty-first-century societies to have been successful in managing our collective nuclear waste. In any group of dozens of government agencies and private companies, some will be more efficient and some will be less so. Getting it completely right once would be an amazing accomplishment; getting it right dozens of different times is statistically far less likely. This is human; we live in politically defined states, and so we endeavor to solve the problems those states generate within this socially constructed boundary. We will bury the waste underground, fill it in, and take pride in our accomplishment. It's who we are.

Strategies for Communication into Deep Time

Assuming that we successfully and competently deposit the spent nuclear fuel from all of our military and commercial operation of nuclear power plants, a primary challenge remains: How do we keep people in the future from breaching our DGRs either intentionally or accidentally? How do we mark these sites? While in the short term this problem does not seem so vexing, when we begin to imagine the journey of this waste in the far future, the challenges begin to seem daunting, or even impossible. The physical placement and the isolation of the waste presents a set of challenges; figuring out how to communicate with people over a period of millennia presents a separate set of challenges. The study of this dilemma is called "nuclear semiotics," and the specific means of communicating are called "nuclear markers." Peter van Wyck has referred to this second challenge as one of maintaining "temporal security."[84]

If humans breach the containment of DGRs, the containment is by definition a failure. The facilities are designed to isolate the waste, and that means from both natural forces and biota. Humans that dig into a waste site and violate the containment system will facilitate the entry into the local ecosystem of radionuclides, and the goal of shielding the waste from living creatures will not have succeeded. This can happen for a variety of reasons: construction, fracking, archaeological inquiry, or even treasure hunting. Thus, there are two prongs to the deep geological storage strategy: isolating the waste physically, and avoiding any violation of the structure from living creatures. For this reason, we are investing significant amounts of labor and planning into determining how to mark the locations and how to communicate the risk to people over millennia.

The main and obvious reason that the effort is so vexing is that we assume from the start that the simple use of language may not be fully adequate to the task. Written information, or any means of transmitting information in existing languages, may not endure or effectively communicate to human beings thousands of years from now. If people from one thousand or two thousand years ago walked up to us today and started speaking, it is doubtful that we would understand what they were saying. Languages evolve. Even though we do have texts written from thousands of years ago, we can only read some of them; this must be done by experts with a lot of time and resources. We need any written marker that stands at a nuclear waste site to be quickly understandable to whomever approaches it.

We can imagine a scenario in which future human beings are implored to convey this knowledge to those who come after them, to update the data into newly developing or changing languages, but that means we have to rely on them to willingly follow our instructions for thousands of years or longer. Even if we imagine that we could convince people to do this, like in the commonly played childhood game of "telephone," in which a series of children whisper a statement from one to another, the message rarely travels through ten iterations before it is far removed from the initial content. We would be compelling our descendants to play telephone with a loaded, radioactive gun.

The DOE conducted a decades-long study in preparation for markers that it intends to build at the Waste Isolation Pilot Project (WIPP) in New Mexico, which is already being filled with high-level waste from America's nuclear weapon production program (excluding the spent nuclear fuel). An expert panel was established at Sandia National Laboratory in New Mexico, and it identified the basic principles in its study of effective nuclear marking: "(1) the site must be marked, (2) message(s) must be truthful and informative, (3) multiple components within a marker system, (4) multiple means of communication (e.g., language, pictographs, scientific diagrams), (5) multiple levels of complexity within individual messages on individual marker system elements, (6) use of materials with little recycle value, and (7) international effort to maintain knowledge of the locations and contents of nuclear waste repositories."[85]

The DOE Office of Nuclear Waste Isolation convened the experts as part of the "Human Interference Task Force," and created two teams tasked with developing separate strategies to design temporal security measures. Team A advocated for an emotional communication. Their proposal was to build a terrifying environmental sculpture that would be monumental in scale; the sculptural elements were to be "non-natural, ominous, and repulsive."[86] The goal of this large installation would be to kindle fear in the emotions of anyone that comes upon it. This fear is intended to repel humans from the exterior of the site. The sculptural installation would be immense, extending beyond the footprint of the underground repository. Team A participant American linguist Frederick Newmeyer cautioned that the installation itself may not accomplish the goal of repelling visitors: "If the collective proposals of Team A are carried out, the WIPP site will quickly become known as one of the major architectural and artistic marvels of the modern world. Quite simply, there will be no keeping people away. We owe it to these people to explain to them why WIPP was built and its overall significance. To do so adequately would require a dedicated information center; the structures themselves are not designed for this purpose."[87] So: a massive, repellent sculptural installation, but since that may simply attract people, also a text-based information center after they arrive.

Team B advocated directly for a more information-based marker. An exterior fence would channel visitors into a central information room: "An assortment of symbolic, pictographic, linguistic, narrative, diagrammatic, scientific and astronomic messages should be used to ensure that people from any conceivable culture or future society would be able to understand that hazardous materials are buried in the immediate area and that they should not intrude."[88] Eventually the DOE largely sided with Team B and backed an informational marking. The current design of the marker prominently features a central information room with text in seven languages: English, Spanish, Russian, French, Chinese, Arabic, and Navajo (the six languages of the current United Nations and the local, indigenous language), again showing current political constructs intruding into a millennia-long design.[89]

These design competitions and judgments perform competence and capability, the belief that we can engineer future understandings and influence the behavior of people living in very different cultural constructs. As the intended outcome of nuclear marking is what Peter van Wyck describes as "a hybrid device—part time capsule, part memory theater," ongoing speculation has led to increasingly fantastical visions.[90] Two members of the task force, French author Françoise Bastide and Italian semiotician Paolo Fabbri, designed an alternative marker, which they called the "ray cat solution."[91] They proposed to genetically engineer cats so that they changed color when exposed to radiation, releasing these cats into human communities, and developing an accompanying mythology, almost a religion, that would pass down the fundamental truth that whenever cats change colors that means danger—people should leave places where this happens. Needless to say, it is easy to pick such visions apart. How do you maintain a color dynamic in a species over thousands of generations of genetic inheritance, occasionally with "non–ray cat" cats? But more to the point, how do you develop a mythology, a religious narrative, that will effectively be passed on from generation to generation not because it has some demonstrable value to those passing it on, but because it solves a problem we want solved in our time? For that matter, how do you create myths that will be faithfully carried across thousands of years of time for any reason? Many people have tried to start religions, few successfully. Religions that have been successful have succeeded for a complexity of reasons and circumstances far beyond the maintenance of a mythological story that someone thought would be useful thousands of years prior: successful religions are not just effective stories but multifaceted social institutions. Also, what would cats be doing out in the desert of southern New Mexico near the WIPP site, anyway?

Another idea included the establishment of a nuclear waste theme park modeled after Disneyland, with a villainous mascot named Nickey Nuke. The Nickey Nuke scenario was modeled on the capacity for cultural communication packets to outlive physical monuments: "Something as seemingly frail and unsubstantial as a story or poem, it turned out, was more durable than the most established social institution or the toughest metal, plastic, or stone." As the writers of the Boston Team Report appendix of the WIPP study theorized: "Long after metal had disintegrated and granite worn smooth of markings, the legends of Nickey Nuke remained in people's minds everywhere on Earth (much as Robinson Crusoe and his story were known by all peoples centuries after his creation in 1719, or as Alice in Wonderland or Mickey Mouse were universally recognized across cultures, space, and time, or even, if you please, as the story of the Garden of Eden had lasted thousands of years)."[92] It is easy, and fun, to deride the many suggestions of the task force members. American semiotician Thomas Sebeok's call for an "atomic priesthood,"[93] which would guard information about the repository as though it were items in a religious reliquary, again relies on the establishment of a functional religious institution that would endure with its secrets intact for millennia. All of these imagined means of keeping information intact over time would apparently work because they have some utility to us in the twenty-first century, not because they have any understandable relevance or value to the people who would need to faithfully maintain them and pass them on.

It is worth noting that the WIPP criteria for success for each of the two challenges of physical containment and risk communication have different timelines. The waste must be contained and isolated from hydrological degradation for one hundred thousand years, while the imperative to communicate the risk to future generations must be successful for only ten thousand years: the temporal security need only succeed for one-tenth of the amount of time that the geophysical security must endure. The early benchmark report of the Human Interference Task Force offered this explanation for the shortened time frame for message permanence: "The emphasis for transmitting information will focus on the first 10,000 years after repository closure. This period of time considers both the decreasing degree of risk of radioactive exposure over time and uncertainties due to natural phenomena. First, the radioactivity hazard associated with the nuclear waste diminishes over time. Relatively rapid decay of fission products occurs during the first 1,000 years after closure. Slower decaying transuranic elements would reach levels that approximate background radiation after ten to thirty thousand years."[94] The assertion here seems to be that our obligation to communicate the danger from the radioactive waste being contained in the repository is inconsequential after ten thousand years, yet the project retains the burden of the physical containment of the waste for ten times longer.

It may be useful to look at the tasks being considered. Those who are tasked with determining a location that is geologically predictable for hundreds of thousands of years have the skill to determine past geological activity for millions or even billions of years. They feel competent to describe the future in these timescales. On the other hand, those tasked with trying to establish means of cultural communication do not have immense expanses of time to examine for their models. The longest intact human communications have remained coherent for mere thousands of years. Even the oldest organized religions in the world have not endured for longer than five thousand years. The early years of those religions are only understood in terms of the traditions that can first be materially interpreted from thousands of years after their origin points. Those tasked with designing message durability think in terms related to the longest enduring cultural messages they can examine.

Andra, the French nuclear waste agency charged with designing and constructing the French DGR in Bure, has sponsored an art competition to develop ideas for nuclear markers for the site. It has awarded prizes each year to the top ideas, which have included a repellent earthen sculpture, a genetically modified (to look blue) forest, a children's song, and a sound installation.[95] It is hard to view such an effort as more than a process of generating support in the current population of France. Although no final decision has been made, at the Onkalo site there is a strong view that leaving the site unmarked may be the best strategy to insure there will be no later human intrusion.[96] The environmental impact statement for the Onkalo site prepared by Posiva Oy explains that the prevention of exposures resultant from future human intrusion is "achieved by placing the final disposal repository sufficiently deep in the earth so as to be outside the reach of people's normal lives. In any case, it is likely that if future generations are able to penetrate into the facility, they will also have the knowledge and skills to avoid the radiation emitted by the waste."[97]

Once a system of markers has been constructed, the burden is understood by the nations creating markers to fall on those who encounter it: "The U.S. Department of Energy has taken the position that: 'although this generation bears the responsibility for protecting future societies from the waste that it creates, future societies must assume the responsibility for any risks which arise from deliberate and informed acts which they choose to perform.'" Creating the markers is an end point of the responsibility for the society engaged in passing down a toxic legacy whose end point will endure beyond tens of thousands of years. "This society's obligation should be discharged by providing a secure isolation system that would continue to function if left undisturbed, by avoiding probable causes of disturbance, and by transmitting knowledge of the repository to future generations, thus allowing them to plan their activities accordingly."[98] As van Wyck points out, "Should future persons elect to breach the repository, they and not the present would be responsible. Therefore, the marker's ethical function works in two directions. It will allow those who should know better to avert the danger. . . . And for those who either cannot figure it out or do not care, the present cannot be held ethically negligent."[99] While there is intent to transmit a message to people in the future, there is a clear path for those who work on the issue, and the politicians who tasked them with this work, to feel that they have successfully met their own obligations. They can shuffle off this mortal coil fully without a worry about the radiological landfills left behind.[100] This concept—that those yet born bear responsibility to manage our toxic legacy, and so its impacts on living creatures in the future is their failure rather than ours—can also simply be seen as a continuation of the short-sighted selfishness that originally led us to manufacture hundreds of thousands of tons of toxic and radioactive material.

The Rational and the Irrational Humans

We envision ourselves as the communicators, as those leading people in the future toward understanding, as the masters—this, even as the critically important information that we need to communicate to them is about our catastrophic failure: our generation of waste that is lethally dangerous to them in quantities that require massive mining projects all across the globe. We will be the clever teachers and they will be the obedient students—running the gauntlet of our temporal radioactive obstacle courses.

There is ample historical evidence of societies explicitly ignoring warnings left by ancestral generations instructing them on how to protect themselves from dangers. "Hundreds of so-called tsunami stones, some more than six centuries old, dot the coast of Japan, silent testimony to the past destruction that these lethal waves have frequented upon this earthquake-prone nation," writes Martin Fackler. "While some are so old that the characters are worn away, most were erected about a century ago after two deadly tsunamis here, including one in 1896 that killed 22,000 people. Many carry simple warnings to drop everything and seek higher ground after a strong earthquake. Others provide grim reminders of the waves' destructive force by listing past death tolls or marking mass graves." People who endured those deaths tried to pass on knowledge to their descendants about the dangers of building closer to the coast than where the markers were placed. These markers were carved and erected recently enough that there was no uncertainty about the warning being communicated, and to whom

Stone tablet located in Aneyoshi, Japan, warning future residents not to build below the site of the stone (2011). Photo: Ko Sasaki.

they were being communicated. The text partly reads, "Do not build your homes below this point. . . . High dwellings insure the peace and happiness of our descendants." We don't ascribe such a relationship to those in the past: they didn't know what we know now. "Modern Japan, confident that advanced technology and higher seawalls would protect vulnerable areas, came to forget or ignore these ancient warnings, dooming it to repeat bitter experiences when the recent tsunami struck."[101]

Our strategies for nuclear markers, again, reveal delusions of competence. We imagine that we can succeed in communicating an appropriate sense of risk to future generations—that will compel what we define as appropriate behaviors on their part. In many ways our operating assumptions are that people in the future are likely to be irrational and we must strategize to make sure they construct the proper meaning out of the information-laden markers we have built for them. Strategies based on both emotion and information are built on the assumption that we need to control how people in the future think and feel. We imagine scaring them away from the bad thing we buried. We imagine creating mythologies to control how they narrate their world. We envision them as extensions of our own social needs and story, not as fully equal and autonomous human beings likely existing in social structures and in accordance with cultural narratives that they themselves find appropriate.

If the people in the future can be imagined as irrational, how can we pretend that we ourselves are different from them, that we are rational? If we have to give them an undergirding story, myth, or religion that will keep them from harming themselves with our legacy waste, how do we not grasp that we are the ones who have created a story, a myth, a religion to allow ourselves to believe we are behaving correctly? We flatter ourselves that the generations that follow us will look at our communication about the buried waste as sort of guide that will help them to "understand" the materials buried there. Is it not irrational to imagine that we are the kind teachers and they are the naive students? Our waste says far more than our markers can; our remedial communications are as graffiti on the walls of a garbage dump. The waste itself is the message. The waste says everything they need to know about us.

I propose that the most effective form of nuclear marker we can leave to future generations must be based on the concept of apology. We must apologize to them for putting them into relationship with the high-level nuclear waste in such immense quantities and for disposing of it in dozens of places. Fundamentally, we need to grasp for ourselves that what we did was wrong. With no clarity about how to dispose of the longest lived and most toxic waste ever produced, we manufactured hundreds of thousands of tons of this material anyway. We may have felt compelled to do so based on our visions of *survival*, or *energy efficiency*; it will not matter to our descendants what our rationale was. Failure to grasp that its production was an epic ecological catastrophe can easily lead to similarly dysfunctional notions of competently marking the waste sites and communicating anything coherent to anyone.

It may not be epidemiologically useful, but a message based on apology will be present with the situation they will inherit, and because of that it may endure longer than lecturing them or using some trickery or clever psychological manipulation. Honesty about the waste sites may lead to our communication having actual value. If we say, "We apologize for leaving this toxic mess for you to live with; please try to minimize the harm we have done to you," the message that "here lies something dangerous" is more likely to endure. Of course, even this will only last so long, and then our message will fade and the dangers that our descendants face will likely be unknown to them. We can only be incompetent about this. And that alone is a fundamental reason that we should apologize to them.

Temporal Violence

Will human beings manage the hundreds of thousands of tons of spent nuclear fuel in a way that adequately protects future generations from risks of sickness and mortality? As in most human endeavors, we will probably get it partly right and partly wrong. Compare this dilemma to the history of safely operating nuclear power plants. The consequences from catastrophic failures at nuclear power plants are almost impossible to overstate: millions may be exposed to radiation, centuries will be required to remove and contain melted nuclear fuel, radionuclides will embed into downwind ecosystems for hundreds or thousands of years. All nuclear power plants are operated with a commitment to safety procedures. Most are built with redundant safety systems to ensure adequate operation in an emergency. Looking back on the first eight decades with nuclear power, we can conclusively say: humans got it partly right and partly wrong. For the overwhelming majority of the days of operation, there were not serious accidents or radioactive leaks. But still, some have occurred. We have had a partial fuel meltdown at least once per decade. Two events have been classified at the highest magnitude on the International Nuclear and Radiological Event Scale, as designated by the IAEA, in less than a half century.[102] This cannot be called a total success. We got it partly right and partly wrong.

Almost certainly this will be an apt description of our capacity to safely dispose of spent nuclear fuel and protect future generations from their toxicity. Similar to operating nuclear power plants, this would be a catastrophic legacy. If our DGRs leak, or if some of them leak, we may contaminate ecosystems for thousands of generations of future biota. People living in those times may have very limited understandings of the specific risks that they would face from exposure to highly radioactive material, and may have no technological capacity to mediate any breaches or leaks.

Our waste may be permanently exposed for tens or hundreds of thousands of years. This would be the most catastrophic ecological event in Earth's history. Even simple leakage of radiological materials in groundwater could have devastating impacts on generations of future human beings and other creatures. All of the 177 underground tanks built for the storage of high-level radioactive and chemical liquid waste at the Hanford Site are leaking.[103] Not one or two—all of them. This is not to suggest that all DRGs will leak, but rather that systems designed and built with high technology and competent engineering at the time of their construction do not necessarily eventuate as intended. The future unfolds separately from our dreams of it unfolding.

Nuclear power plants typically operate for about two generations of human beings, but they leave behind highly toxic waste that will present a risk for at least three thousand generations of human beings. This is an essential component of our legacy to the thousands of generations of human beings that will follow us on the planet. No benefit, only burden. Only risk. And so we take it upon ourselves to alleviate that risk by competently disposing of the waste of these nuclear power plants deep underground—a project so unprecedented, so remote from us in time, with the most toxic materials ever made at the center of the project. Doing it perfectly is an unlikely outcome.

Rob Nixon has written about "slow violence," which he defines as "a violence that occurs gradually and out of sight, a violence of delayed destruction that is dispersed across time and space, an attritional violence that is typically not viewed as violence at all."[104] We are placing tons of the most dangerous materials ever created, which will remain dangerous for millennia, into deep holes in multiple places around the world. This is temporal violence. While normally we understand violence to be exerted over space, across battlefields, from territory to territory, this violence is being exerted over time, over generations, against people in future centuries. It is temporal violence: violence toward our descendants.

This is not something that we might do, like we might have a nuclear war; this is something that we have already done: the waste is here.[105] We have to move forward as competently as we can. We must assume that we will at least partly fail in our efforts to protect future generations from this waste. We should stop pretending that we can exert unwavering competence and control into deep time. Instead, we should work to mitigate that failure as much as possible; we must commit all of the funding required over the next four to five generations to design and execute deep geological storage of this waste. "It would be more than naïve to assume that Finland's nuclear waste experts painted perfect portraits of worlds for millennia to come. They did not," writes Ialenti. "In fact, many of them emphasized that their own computer models of the future, however sophisticated, were but highly educated guesses. They saw them as merely the most credible future-gazing strategies they could concoct using the best science and technology available to them at the time."[106] This self-awareness in the face of immense unknowns is wise and essential to achieving our best strategies; however, one of the most fundamental things we can do to protect future generations is to stop generating more nuclear waste immediately.

Although the vast majority of nuclear reactors have been used as power plants to generate electricity, they all manufacture plutonium as a by-product of their operation. Before nuclear reactors were invented, so little natural plutonium existed that scientists were unaware of it: they believed they invented it in Berkeley in 1940.[107] Virtually all of the plutonium on Earth was created inside nuclear reactors. That plutonium will remain viable for use in nuclear weapons for thousands of years.[108] Some military or government, distant in time, can dig up our deep geological repositories and separate the plutonium for use in nuclear weaponry. All that would be stopping them is the technology (technology we currently possess) and the fact that our warning signs tell them that "this place is not a place of honor."[109] Our peaceful use of nuclear power has given us electricity, but it has left behind thousands of metric tons of fissile material that will remain militarily viable for millennia. We made that. Maybe we made that for them.

Conclusion

Roughly one hundred thousand years ago a migration of *Homo sapiens* left Africa and eventually populated the entire planet with modern human beings. It took almost seventy thousand years before humans reached the Americas. This unthinkable period of time is how long we intend to isolate the waste of our nuclear power plants, used to make either nuclear weapons or electricity, from harming human beings and ecosystems in the future. A mere one hundred and fifty years ago humans began to generate electricity that could reach homes and businesses, transforming modern society. Nuclear power plants came online about halfway into our short history of electrification, yet in that brief moment we have generated hundreds of thousands of metric tons of radioactive heavy metals. These, less than eighty years' worth of high-level nuclear waste, will pose a catastrophic risk to future human beings for a time equal to the millennia our species has existed. "It must be seen as a novel feature of this point in history. Novel, because it represents a *new form* of waste," observes Peter van Wyck. "It really is matter *without* a place. A kind of waste that resists its own containment. A kind of waste that operates in a radically different temporality; it is material whose toxicity requires a different conception of history and time."[110]

Twentieth- and twenty-first-century human beings have cast a large, toxic shadow on the future of the Earth. Many scientists and policymakers assure us that maintaining control and facilitating millennia-long understandings of this threat will be something we can accomplish. It is easy to imagine catastrophes that will render our efforts at containment ineffective: a comet or meteorite may hit a bull's-eye right upon the site of one of our DGRs; a massive solar flare may halt the operation of our electrical grids and make technological society, and its memory culture, obsolete; a tyrannical form of government may arrive in some region of the Earth where rulers force people to dig up the marked DGRs to obtain transuranics for weapons. A lot can happen in one hundred thousand years. Just look at the last one hundred thousand years: most of what happened during that period is unknown to us. Still, we toil on, comforted in our irrational beliefs that we can contain this waste into infinity. These self-assurances may not succeed in alleviating future suffering, but they will likely alleviate our sense of shame and guilt over the mess we have made.

During the Cold War each of the protagonists had multiple thermonuclear weapons aimed at the plutonium production sites of their adversary. These sites contained most of the spent fuel from the decades of plutonium production as well as the nuclear reactors that were the engines of production. While these sites—Hanford, Mayak, Tomsk in the USSR, and Savannah River—were all deeply contaminated with radionuclides from the decades of production, had those thermonuclear weapons been launched at these sites, much of that waste would have been aerosolized and spread throughout the regions, along with the massive deposition of fallout radionuclides from the weapon detonations. Of course, had this war occurred, the distribution of fallout across the Soviet Union and the United States would have been sufficient to kill their entire populations as well as many beyond their borders.[111] This deeply strategized nuclear war would likely have occurred over the course of hours, days, or perhaps a week or two. Thousands, potentially tens of thousands, of nuclear weapons would have detonated in the course of this short war. What would be left behind would be, as Major General A. J. Old, then director of SAC Operations, told a military audience about the US attack plan in March 1954 (less than two weeks after the Bravo test), "a smoking radiating ruin." All of this would happen, according to Old, in "two hours." Death, sickness, radiological contamination.[112]

The deep-time presence of hundreds of thousands of tons of radioactive waste that will remain dangerous for over one hundred thousand years presents a different kind of violence. With dozens of DGRs, many with differing designs and materials, all intended to provide complete containment of this immense high-level waste burden, what may unfold is a slow-motion nuclear war. One in which the outcomes would have similarities—deaths, illnesses, toxic contamination, radiological migration—but the process of emergence would differ. Rather than a massive global contamination that happens in "two hours," or over a week, this outcome would unfold at different times in different places—sometimes separated by centuries and continents—yet cumulatively presenting a similar eventuation: a radiologically compromised planet and ecosystem. This is the ticking time bomb we will bury beneath the Earth and from which we will walk away, muttering about competence.

Afterword: Opening Our Eyes

Like the first stalkers entering one of the "Visitation Zones" in the Russian science fiction story *Roadside Picnic,* we are yet unable to grasp the complex alterations to our ecosystem left in the wake of our Cold War technologies.[1] We can measure the levels of radiation in places, track its migration, catalog mutations in insect samples, but we have yet to grasp the world we unfold around ourselves, the ecosystem we are crafting for our descendants to inhabit.

Where I live, in Hiroshima, there is anxiety about how to preserve the memory of the nuclear attack here seventy-seven years ago. Within this anxiety is both an obligation to remember those who suffered and died, and a vigilance to ensure that such an attack never happens to people in the future. My academic home, the Hiroshima Peace Institute, was born of this anxiety at the fiftieth anniversary. Here, and in Nagasaki, work on preserving memory preoccupies the cities. Increasingly it occurs to me that human civilization will actually never be able to forget our advent of nuclear technologies in the twentieth century. Long after our cities and monuments have crumbled, long after our languages are unspoken, long after our gods have turned away, our nuclear waste will endure. It is how our descendants will know us.

During the Cold War years, we anticipated an event—World War III— that would destroy the developed world and spread fallout clouds into the rest. Instead, we experienced a toxic seepage: the dispersal of radioactive fallout particles around the world and the mass production of highly toxic and long-lived radioactive waste in factories designed to manufacture plutonium with a by-product of heat. Not explosions, but contaminations. As I write, more than a dozen countries are mining uranium from beneath the Earth's surface; in several dozens of countries nuclear fuel produces both heat and plutonium. The heat will slowly abate, but the plutonium will endure.

We stand on the precipice of the Anthropocene, an epoch pegged to our invention, production, and global distribution of transuranic elements, born in the Manhattan Project, barely able to see beyond our eyelids.[2] Our understanding of the past is as fogged as our vision of the future. Much of the violence we unleashed in the Cold War was against noncombatants on the conflict's periphery—and future generations. Still, we struggle to see past the imaginary long peace we invoke like a bedtime fable.

To see clearly into the future, even the near-term future, we must focus on what we have done. If we can begin to grapple with those who endured the limited nuclear war of the Cold War, and sincerely remediate what we did to their—and our—bodies and lands, we may also be able to turn and see the rough beast we are midwifing into the future. Interventions now, such as halting the production of additional metric tons of spent nuclear fuel, are the most powerful means we have to caretake our descendants and nurture the gardens that will produce their food.[3]

Notes

Introduction

1. An A-bomb (fission weapon) fundamentally differs from a thermonuclear H-bomb (fusion weapon). The former releases energy by splitting the nuclei of heavy atoms. The latter releases energy by fusing together the nuclei of two lighter atoms. Fusion reactions power the sun and other stars. On 1 November 1952 the United States had achieved a fusion reaction and explosion: the Mike test at Enewetak Atoll was the first thermonuclear reaction achieved on Earth, but it had required an entire building to house the weapon's structural components.

2. Richard G. Hewlett and Jack M. Holl, *Atoms for Peace and War, 1953–1961: Eisenhower and the Atomic Energy Commission* (Berkeley: University of California Press, 1989): 173, 179, 182.

3. Hewlett and Holl, *Atoms for Peace and War:* 181.

4. *Operation Castle: Commander's Report* (USAF, Lookout Mountain Laboratory, 1954).

5. David Rosenberg, "The Origins of Overkill: Nuclear Weapons and American Strategy, 1945–1960," *International Security* 7:4 (1983): 34. JCS is the Joint Chiefs of Staff, the leaders of the various service branches of the US military.

6. NSC 162/2 Basic National Security Policy, 30 October 1953, Record Group 273, Archives of the National Security Council (National Archives of the United States): 4.

7. *Operation Castle*. The residents of Utirik were actually evacuated on the third day after the shot. The atolls range from 152 km to 187 km downwind. Martha Smith-Norris points out that residents had been evacuated from these same atolls prior to the 1946 Bikini Crossroads tests of weapons a thousand times smaller than the Bravo weapon. See Martha Smith-Norris, *Domination and Resistance: The United States and the Marshall Islands During the Cold War* (Honolulu: University of Hawai'i Press, 2016): 77–80.

8. George W. Albin, "Reports on Evacuation of Natives and Surveys of Several Marshall Island Atolls," 9 April 1954: https://web.archive.org/web/20110721041118 /http://www.hss.energy.gov/HealthSafety/IHS/marshall/collection/data /ihp1d/122574e.pdf (accessed 22 July 2020).

9. Atolls are subcircles of coral reef that remain after an island or underwater volcano recedes. They consist of a series of reefs and small islands around a central lagoon. They are desirable locations in the mid-ocean because the lagoon provides an easy place to fish and to come and go by canoe from the sea. See Patrick D. Nunn, "Atoll," in Andrew S. Goudie, *Encyclopedia of Geomorphology* (London: Routledge, 2004): 39–41.

10. Victor P. Bond et al., *Medical Examination of Rongelap People Six Months After Exposure to Fallout* (Bethesda, MD: Naval Medical Research Institute, 1955): 11, 14.

11. *Atoms for Peace and War:* 173–175. People who were 2.4 km from the epicenter in Hiroshima are officially recognized as hibakusha (survivors of the nuclear attacks on Hiroshima and Nagasaki) by the Japanese state, which provides them with monthly allowances and enhanced medical care. See "Relief for A-bomb Victims," *Japan Times,* 15 August 2007: https://www.japantimes.co.jp/opinion/2007/08/15 /editorials/relief-for-a-bomb-victims/ (accessed 13 August 2020).

12. Steven L. Simon et al., "Radiation Doses and Cancer Risks in the Marshall Islands Associated with Exposure to Radioactive Fallout from Bikini and Enewetak Nuclear Weapons Tests: Summary," *Health Physics* 99:2 (2010): 105–123.

13. Christopher Robert Hill has written that "The scientific objectivity that seemed to legitimize testing overseas was shaken by the *Lucky Dragon* incident." Christopher Robert Hill, "Britain, West Africa and 'The New Nuclear Imperialism': Decolonisation and Development During French Tests," *Contemporary British History* 33:2 (2018): 277. *Daigo Fukuryu Maru* is translated into English as *Lucky Dragon*.

14. Aya Homei, "The Contentious Death of Mr. Kuboyama: Science as Politics in the 1954 Lucky Dragon Incident," *Japan Forum* 25:2 (2013): 212–232.

15. The memorial day has previously been called "Nuclear Victims' Day" and "Nuclear Survivors' Day." See Ron Tanner, "Remembrance Day," Marshall Island Story Project: http://mistories.org/remembrance.php (accessed 20 August 2020).

16. "Nuclear Testing Tally, 1945–2017," Arms Control Association: https://www .armscontrol.org/factsheets/nucleartesttally (accessed 14 August 2020).

17. There is some suspicion that Israel did conduct one test in 1979, the so-called Vela Incident. See Lars-Erik de Geer and Christopher M. Wright, "The 22 September 1979 Vela Incident: Radionuclide and Hydroacoustic Evidence for a Nuclear Explosion," *Science & Global Security* 26:1 (2018): 20–54.

18. Yasuyuki Taira et al., "Current Concentration of Artificial Radionuclides and Estimated Radiation Doses from 137Cs Around the Chernobyl Nuclear Power Plant, the Semipalatinsk Nuclear Testing Site, and in Nagasaki," *Journal of Radiation Research* 52 (2011): 88.

19. See Steven L. Simon and André Bouvielle, "Radiation Doses to Local Populations Near Nuclear Weapons Test Sites Worldwide," *Health Physics* 82:5 (2002): 706–725.

20. Wright H. Langham, quoted from "Proceedings: Second Interdisciplinary Conference on Selected Effects of a General War," *DASIAC Special Report 95* (Santa Barbara, CA: Defense Atomic Support Agency, 1969): 45.

21. "The Active Straw," *Newsweek*, 12 November 1945: 50.

22. Roger A. Meade and Linda S. Meade, *"The World, We Think She Start Over Again": Nuclear Testing and the Marshall Islands, 1946–1958,* internal distribution report no. LA-UR-18-30848 (Los Alamos, NM: LANL, 2018): 116. I would add that within the Marshall Islands, the radiological disaster of the Crossroads Baker test in 1946 forced testing from Bikini Atoll to Enewetak Atoll until the Bravo test. For details, see chapter 2.

23. Justin McCurry, "Fukushima Grapples with Toxic Soil That No One Wants," *The Guardian,* 11 March 2019: https://www.theguardian.com/world/2019/mar/11/fukushima-toxic-soil-disaster-radioactive (accessed 21 July 2020).

24. See Robert Jacobs, "Born Violent: The Birth of Nuclear Power," *Asian Journal of Peacebuilding* 7:1 (July 2019): 9–29.

25. See Gabrielle Hecht, *Being Nuclear: Africans and the Global Uranium Trade* (Cambridge, MA: MIT Press, 2012).

26. Dana Kennedy, "Chernobyl Cleanup Survivor's Message for Japan: 'Run Away as Quickly as Possible,'" *AOL News,* 22 March 2011, quoted in Robert Jacobs, "Social Fallout: Marginalization After the Fukushima Nuclear Meltdown," *Asia-Pacific Journal* 9:28 (2011): https://apjjf.org/2011/9/28/Robert-Jacobs/3562/article.html (accessed 10 December 2020). Natasha Zaretsky links late–Cold War tropes of radiation-affected bodies with the concurrent collapse of trust in government and expert authorities. See Natasha Zaretsky, "Radiation Suffering and Patriotic Body Politics in the 1970s and 1980s," *Journal of Social History* 48:3 (2015): 487–510.

27. The United States during Operation Crossroads in 1946 and the Soviet Union during the Totskoye test in 1954.

28. Gilbert W. Beebe, Morihiro Ishida, and Seymour Jablon, *Life Span Study, Report Number 1: Description of Study Mortality in the Medical Subsample October 1950–June 1958,* Technical Report 05-61 (Hiroshima: Atomic Bomb Casualty Commission, 1961): 1–6.

29. Of note is a 2018 study that utilized ESR (electron spin resonance) to measure the received dose of radiation in a fragment taken from a human jawbone found 1.5 km from the epicenter in Hiroshima. The authors describe how the study determined the actual dose absorbed by the jawbone of this human being, while earlier studies relied on "estimating the doses received by the victims." See Angela Kinoshita, Oswaldo Baffa, and Sérgio Mascarenhas, "Electron Spin Resonance (ESR) Dose Measurement in Bone of Hiroshima A-bomb Victim," *PLOS ONE* 13:2 (2018): https://journals.plos.org/plosone/article?id=10.1371/journal.pone.0192444 (accessed 28 August 2020).

30. Adri De La Bruheze, "Radiological Weapons and Radioactive Waste in the United States: Insiders' and Outsiders' Views, 1941–55," *British Journal for the History of Science* 25:2 (1992): 207–227.

31. Lisa Onaga, "Reconstructing the Linear No-Threshold Model in Japan: A Historical Perspective on the Technics of Evaluating Radiation Exposure," *Technology and Culture* 58:1 (2017): 194–205.

32. John Lewis Gaddis, "The Long Peace: Elements of Stability in the Postwar International System," *International Security* 10:4 (1986): 99–142.

33. Zaretsky, "Radiation Suffering and Patriotic Body Politics": 487–510.

34. See Robert Jacobs, "Nuclear Conquistadors: Military Colonialism in Nuclear Test Site Selection During the Cold War," *Asian Journal of Peacebuilding* 1:2 (2013): 157–177: http://tongil.snu.ac.kr/ajp_pdf/201311/02_Robert%20Jacobs.pdf (accessed 21 August 2020): doi: 10.18588/201311.000011.

35. "Nuclear Testing Tally, 1945–2017."

36. B. H. Liddell Hart, *The Strategy of Indirect Approach* (London: Faber and Faber, 1941). See also Zygmunt Bauman, *Wasted Lives: Modernity and Its Outcasts* (Cambridge: Polity Press, 2004).

37. Hiroaki Koide, "The Fukushima Nuclear Disaster and the Tokyo Olympics," trans. Norma Field, *Asia-Pacific Journal* 17:5 (1 March 2019): https://apjjf .org/2019/05/Koide-Field.html (accessed 20 August 2020).

38. Jacobs, "Born Violent."

39. International Atomic Energy Agency, *Status and Trends in Spent Fuel and Radioactive Waste Management* (Vienna: IAEA, 2018): 35; Office of Scientific and Technical Information, *Categorization of Used Nuclear Fuel Inventory in Support of a Comprehensive National Nuclear Fuel Cycle Strategy* (Oak Ridge, TN: Oak Ridge National Laboratory, 2012).

40. Kate Brown, "The Last Sink: The Human Body as the Ultimate Radioactive Storage Site," in Christoph Mauch, ed., *Out of Sight, Out of Mind: The Politics and Culture of Waste* (Munich: RCC Perspectives, 2016): 45.

1. Hypocenter

1. The classic work on the effects of nuclear detonations is Samuel Glasstone and Philip J. Dolan, *The Effects of Nuclear Weapons*, 3rd ed. (Washington, DC: Government Printing Office, 1977).

2. A. B. Pittock et al., *Environmental Consequences of Nuclear War* (Scope 28), vol. 1: *Physical and Atmospheric Effects*, 2nd ed. (Chichester, UK: John Wiley & Sons, 1989): 5.

3. Hiroshima International Council for Medical Care of the Radiation-Exposed, *A-Bomb Radiation Effects Digest* (Tokyo: Bunkodo, 1995): 6. Gy denotes a "gray," which is a measure of the absorption of radiation by matter.

4. Letter to the author from Jeffrey Weiss, 16 December 2018.

5. Ulf Tveten, *Radionuclide Behavior in the Environment* (Albuquerque, NM: Sadia National Laboratory, 1991): 4.

6. When a nuclear weapon detonates, the fissile material releases energy in a short amount of time and then distributes as fission particles. At Fukushima the source of gamma radiation is the melted fuel, which continuously emits gamma radiation from the fuel indefinitely.

7. Councilman Morgan, "Hiroshima, Nagasaki, and the RERF," *American Journal of Pathology* 98:3 (March 1980): 844.

8. Doran M. Christensen et al., "Management of Ionizing Radiation Injuries and Illnesses, Part 1: Physics, Radiation Protection, and Radiation Instrumentation," *Clinical Practice* 114:3 (2014): 189.

9. *Report on a Feasibility Study of the Health Consequences to the American Population from Nuclear Weapons Tests Conducted by the United States and Other Nations,* vol. 1: *Technical Report* (Washington, DC: National Academies Press, May 2005), sec. 2.1.3: 19.

10. Brooke Buddemeier, *Nuclear Detonation Fallout: Key Considerations for Internal Exposure and Population Monitoring* (6 July 2018), LLNL-TR-754319: 1.

11. Glasstone and Dolan, *The Effects of Nuclear Weapons:* 418–19; J. Rosinski and J. Stockham, *Preliminary Studies of Scavenging Systems Related to Radioactive Fallout,* Armour Research Foundation of the Illinois Institute of Technology Report No. 3127-6 (30 April 1959).

12. Hiroko Takahashi, "One Minute After the Detonation of the Atomic Bomb: The Erased Effects of Residual Radiation," *Historia Scientiarum* 19:2 (2009): 147; Ritsu Sakata et al., "Long-term Effects of the Rain Exposure Shortly After the Atomic Bombings in Hiroshima and Nagasaki," *Radiation Research* 182:6 (2014): 599–606.

13. For an example of recent research demonstrating this dynamic, see Kazuko Shichijo et al., "Impact of Local High Doses of Radiation by Neutron Activated Mn Dioxide Powder in Rat Lungs: Protracted Pathologic Damage Initiated by Internal Exposure," *Biomedicines* 8:6 (2020): doi:10.3390/biomedicines8060171 (accessed 24 January 2021).

14. These figures are assessments made by the city governments. The actual numbers cannot be precisely known, as it is impossible to know where an entire urban population was located at the moment of attack.

15. Glasstone and Dolan, *The Effects of Nuclear Weapons:* 544.

16. Hiroshima International Council, *A-Bomb Radiation Effects Digest:* 5.

17. The Committee for the Compilation of Materials on Damage Caused by the Atomic Bombs in Hiroshima and Nagasaki, *Hiroshima and Nagasaki: The Physical, Medical, and Social Effects of the Atomic Bombings* (New York: Basic Books, 1981): 114.

18. Richard B. Frank, *Downfall: The End of the Imperial Japanese Empire* (New York: Random House, 1999): 3–19.

19. Warren Kozak, *LeMay: The Life and Wars of General Curtis LeMay* (Washington, DC: Regnery Publishing, 2009): x.

20. See, e.g., Howard W. Blakeslee, "Power of Atom Likened to Sun's," *New York Times,* 7 August 1945: 5; "A New Era: The Secrets of Science," *Newsweek,* 20 August 1945: 34–40. The book *The Atomic Age Opens,* which was a reprint of numerous newspaper and magazine articles rushed into print before the end of August 1945, dedicated 96 of 252 pages to explanations of the science of nuclear weapons; Donald Porter Geddes, *The Atomic Age Opens* (New York: Pocket Books, 1945): 62–158.

21. "No Slow Death at Hiroshima, Yank Reports," *Chicago Tribune,* 13 September 1945: 3. Leslie Groves famously testified before Congress about death from

radiation sickness: "As I understand it from the doctors it is a very pleasant way to die."
See Leslie Groves discussing radiation deaths in Hiroshima and Nagasaki before
Congress in 1946. US Congress, Senate, Special Committee on Atomic Energy, 79th
Congress, 1945–1946, Hearings (Washington, DC: Government Printing Office,
1946): 37.

22. Tetsuji Imanaka, "Radiation Survey Activities in the Early Stages After the
Atomic Bombing in Hiroshima," *Revisit the Hiroshima A-Bomb with Database* (Hiroshima: Hiroshima City, 2013): 69–81; Judith Anzures-Cabrera and Jane L. Hutton,
"Competing Risks, Left Truncation and Late Entry Effect in A-Bomb Survivors Cohort," *Journal of Applied Statistics* 37:5 (2010): 821–831.

23. Sadako Sasaki would become the most famous hibakusha from Hiroshima and
is often the only actual person that many Americans can name who was harmed by the
US attack. Her story and the story of her efforts to fold one thousand paper cranes to
obtain one wish would become one of the most frequently told stories of Hiroshima
and would also form the basis of a powerful peace education effort, as many thousands
of children around the world would read her story and learn to fold paper cranes. See
Robert Jungk, *Strahlen aus der Asche* (Bern: Scherz, 1959); Karl Bruckner, *Sadako will
leben!* (Vienna: Taschenbuch, 1961); Eleanor Coerr, *Sadako and the Thousand Paper
Cranes* (New York: Putnam Juvenile, 1977); Masamoto Nasu, *Children of the Paper
Crane: The Story of Sadako Sasaki and Her Struggle with the A-Bomb Disease* (Armonk:
M.E. Sharpe, 1996).

24. M. Susan Lindee, *Suffering Made Real: American Science and the Survivors at
Hiroshima* (Chicago: University of Chicago Press, 1994): 21.

25. Lindee, *Suffering Made Real:* 17–32.

26. Lindee, *Suffering Made Real:* 32.

27. Lewis H. Weed, "Forward," *General Report, Atomic Bomb Casualty Commission*
(Washington, DC: National Research Council, January 1947): ii.

28. Harry S. Truman, "Directive to the National Academy of Sciences," 26 November 1946, repr. in George D. Kerr, Tadashi Hashizume, and Charles W. Edington,
"Historical Review," *US-Japan Joint Reassessment of Atomic Bomb Radiation Dosimetry in Hiroshima and Nagasaki: Final Report*, vol. 1, DS86-vol. 1 (1986): 2.

29. Robert E. Anderson, "Establishment of ABCC," *Human Pathology* 2:4 (December 1971): 485.

30. Kenneth Osgood, *Total Cold War: Eisenhower's Secret Propaganda Battle at
Home and Abroad* (Lawrence: University Press of Kansas, 2006): 1.

31. Quoted in William L. Laurence, *Dawn over Zero: The Story of the Atomic Bomb*
(New York: Alfred A. Knopf, 1946): 11.

32. Holmes quoted in Lindee, *Suffering Made Real:* 5.

33. Spencer Weart, *Nuclear Fear: A History of Images* (Cambridge, MA: Harvard
University Press, 1988): 36.

34. Rachel Pressdee, "The Radium Girls: A Tale of Workplace Safety," *Juris
Magazine*, 1 December 2019: http://sites.law.duq.edu/juris/2019/12/01/the
-radium-girls-a-tale-of-workplace-safety/ (accessed 18 April 2020); Kate Moore, *The

Radium Girls: The Dark Story of America's Shining Women (Naperville, IL: Sourcebooks, 2017).

35. Hermann J. Muller, "Artificial Transmutation of the Gene," *Science* 66:1699 (22 July 1927): 84–87; Hermann J. Muller, *Studies in Genetics* (Bloomington: Indiana University Press, 1962).

36. Hermann J. Muller, "The Production of Mutations" (Nobel Lecture, 12 December 1946): https://www.nobelprize.org/nobel_prizes/medicine/laureates/1946/muller-lecture.html (accessed 25 July 2018).

37. Radiation Effects Research Foundation, *History of ABCC-RERF*, 2004: https://www.rerf.or.jp/uploads/2017/09/rerf30the-1.pdf (accessed 25 March 2018).

38. Lindee, *Suffering Made Real*: 245–251; Sue Rabbitt Roff, *Hotspots: The Legacy of Hiroshima and Nagasaki* (London: Cassell, 1995: 143–144; William J. Schull, *Effects of Atomic Radiation: A Half-Century of Studies from Hiroshima and Nagasaki* (New York: John Wiley & Sons, 1995): 72–79.

39. The ABCC was reestablished as the Radiation Effects Research Foundation under joint US and Japanese administration in April 1975. See *History of the ABCC-RERF*: https://www.rerf.or.jp/uploads/2017/09/rerf30the-1.pdf (accessed 25 July 2018).

40. Gilbert W. Beebe, Morihiro Ishida, and Seymour Jablon, *Life Span Study, Report Number 1: Description of Study Mortality in the Medical Subsample October 1950–June 1958,* Technical Report 05-61 (Hiroshima, Atomic Bomb Casualty Commission, 1961): 1.

41. Beebe, Ishida, and Jablon, *Life Span Study:* 2.

42. Schull, *Effects of Atomic Radiation:* 78.

43. Beebe, Ishida, and Jablon, *Life Span Study:* 3.

44. Beebe, Ishida, and Jablon, *Life Span Study:* 4–5.

45. Schull, *Effects of Atomic Radiation:* 80.

46. Jeffry A. Siegel, Charles W. Pennington, and Bill Sacks, "Subjecting Radiological Imaging to the Linear No-Threshold Hypothesis: A Non Sequitur of Non-Trivial Proportion," *Journal of Nuclear Medicine* 58:1 (January 2017): 2.

47. John Phair, "The Atomic Bomb Casualty Commission," *Northwestern Medicine*: http://www.publichealth.northwestern.edu/nphr/2014-v2i1/Phair.html (accessed 26 March 2018).

48. Leslie Roberts, "Atomic Bomb Doses Reassessed," *Science* 238:4834 (18 December 1987): 1649.

49. National Research Council of the National Academies, *Health Risks from Exposures to Low-Levels of Ionizing Radiation,* BEIR VII Phase 2 (Washington, DC: National Academies Press, 2006): 143.

50. In science studies, the two key works that inform this perspective are Thomas Kuhn, *The Structure of Scientific Revolutions* (Chicago: University of Chicago Press, 1962), and Bruno Latour and Steve Woolgar, *Laboratory Life: The Social Construction of Scientific Facts* (Los Angeles: Sage, 1979).

51. "Radiation epidemiology seeks to describe and quantify the risk of health effects, often cancer, in populations exposed to ionizing and non-ionizing radiation. To do so, it is important to estimate organ or tissue doses for large numbers of exposed individuals with a moderate to high degree of certainty." See Steven L. Simon et al., "Uses of Dosimetry in Radiation Epidemiology," *Radiation Research* 166 (2006): 125–127.

52. Beebe, Ishida, and Jablon, *Life Span Study:* 5.

53. *Life Span Study:* 4.

54. John A. Auxier, *Ichiban: Radiation Dosimetry and the Survivors of the Bombings of Hiroshima and Nagasaki* (Springfield, VA: National Technical Information Service, 1977): 6.

55. Kerr, Hashizume, and Edington, "Historical Review": 3.

56. There is a clear delineation of the factors contributing to illness from internalized exposures, discussed in a 1949 US Atomic Energy Commission handbook on the handling of radioactive waste. See Atomic Energy Commission, *Handling Radioactive Wastes in the Atomic Energy Program* (Washington, DC: Government Printing Office, 1949): 7.

57. Ralph E. Lapp, "Survey of Nucleonics Instrumentation Industry," *Nucleonics* 4:5 (1949): 100–104.

58. James B. Conant, Arthur H. Compton, and Harold C. Urey, "The Use of Radioactive Material as a Military Weapon," *Report of Subcommittee of the S-1 Committee,* 11 October 1943: 3–4. The document is included in the source material for the Advisory Committee on Human Radiation established by the US government in 1994 and archived at Georgetown University: https://bioethicsarchive.georgetown.edu/achre/commeet/meet3/brief3.gfr/tab_f/br3f1f.html (accessed 22 August 2020). Prior to their participation on the committee, Conant had been the president of Harvard University, and Compton and Urey had each won a Nobel Prize in Chemistry. See also Barton J. Bernstein, "Radiological Warfare: The Path Not Taken," *Bulletin of the Atomic Scientists* 41:7 (1985): 44–49; Sean Malloy, " 'A Very Pleasant Way to Die': Radiation Effects and the Decision to Use the Atomic Bomb Against Japan," *Diplomatic History* 36:3 (2012): 515–545; Janet Farrell Brodie, "Contested Knowledge: The Trinity Test Radiation Studies," in Brinda Sarathy, Vivian Hamilton, and Janet Farrell Brodie, *Inevitably Toxic: Historical Perspectives on Contamination, Exposure, and Expertise* (Pittsburgh: University of Pittsburgh Press, 2018): 50–73.

59. This was clearly a concern among Japanese researchers in the early years of the LSS. See Shizuyo Sutou, "Rediscovery of an Old Article Reporting That the Area Around the Epicenter in Hiroshima Was Heavily Contaminated with Residual Radiation, Indicating that Exposure Doses of A-Bomb Survivors Were Largely Underestimated," *Journal of Radiation Research* 58:5 (2017): 745–754.

60. Beebe, Ishida, and Jablon, *Life Span Study:* 5–6. See also Lindee, *Suffering Made Real:* 28.

61. Acknowledgment of the health damage to "early entrants" would unfold over the ensuing decades. See Janet Farrell Brodie, "Radiation Secrecy and Censorship After Hiroshima and Nagasaki," *Journal of Social History* 48:4 (2015): 842–864.

62. James V. Neel and William J. Schull, *The Children of Atomic Bomb Survivors: A Genetic Study* (Washington, DC: National Academies Press, 1991): 62. Also, "by and large individuals exposed to the effects of atomic bombs tend to leave the area as rapidly as possible" is an utterly subjective statement to offer as a basis for decisions on experimental design. This applied only to the people in Hiroshima and Nagasaki, and vast numbers remained near their homes and searched for or tended to loved ones.

63. Lindee writes that "the residents of Hiroshima and Nagasaki endured what many came to fear would be the fate of humanity as a whole," in *Suffering Made Real*: 4.

64. Shizuyo Sutou, "A Message to Fukushima: Nothing to Fear but Fear Itself," *Genes and Environment* 38 (2016): 12: https://genesenvironment.biomedcentral.com/articles/10.1186/s41021-016-0039-7 (accessed 1 February 2018). Sutou is actually a proponent of hormesis, or the belief that exposure to low-levels of external radiation actually *improves* health. A 2008 study questioned the veracity of the correlation of a low dose to the absence of health outcomes by documenting a higher cancer incidence in those the LSS found to have "very low dose" exposures. See Tomoyuki Watanabe et al., "Hiroshima Survivors Exposed to Very Low Doses of A-Bomb Primary Radiation Showed a High Risk for Cancers," *Environmental Health and Preventative Medicine* 13:5 (2008): 264–270.

65. Klervi Leuraud et al., "Ionising Radiation and Risk of Death from Leukaemia and Lymphoma in Radiation-Monitored Workers (INWORKS): An International Cohort Study," *Lancet Haemotology* 2:7 (21 June 2015): https://www.thelancet.com/action/showPdf?pii=S2352-3026%2815%2900094-0 (accessed 17 August 2018).

66. "ORAU History—1964," Oak Ridge Associate Universities: https://www.orau.org/about-orau/history/1964.aspx (accessed 4 April 2018).

67. R. L. Gotchy and R. J. Bores, *The Public Whole Body Counting Program Following the Three Mile Island Accident* (Washington, DC: US Nuclear Regulatory Commission, 1980).

68. Kazuko Shichijo et al., "Autoradiographic Analysis of Internal Plutonium Radiation Exposure in Nagasaki Atomic Bomb Victims," *Heliyon* 4:6 (2018): https://www.cell.com/heliyon/fulltext/S2405-8440(18)31775-4 (accessed 14 May 2020).

2. The Particles That Remain

1. Willard K. Libby, "Radioactive Fallout and Radioactive Strontium," *Science* 123:3199 (1956): 657–660.

2. Libby, "Radioactive Fallout and Radioactive Strontium": 658–659.

3. Libby, "Radioactive Fallout and Radioactive Strontium": 657–658.

4. T. F. Hamilton, J. C. Millies-Lacrox, and G. H. Hong, "^{137}Cs (^{90}Sr) and PU Isotopes in the Pacific Ocean: Sources & Trends" (paper presented at the Radionuclides in the Oceans: Inputs and Inventories, international symposium of radionuclides in the Ocean, 7–11 October 1996, Cherbourg-Octeville, France): 2. Preprint available at https://digital.library.unt.edu/ark:/67531/metadc690481/m2/1/high_res_d/621645.pdf (accessed 22 January 2021); United Nations Scientific Committee on the Effects of Atomic Radiation, *Sources and Effects of Ionizing Radiation*, vol. 1: *Sources* (New York: United Nations, 2000): 159.

5. Noriyuki Kawano and Megu Ohtaki, "Remarkable Experiences of the Nuclear Tests in Residents Near the Semipalatinsk Nuclear Test Site: Analysis Based on the Questionnaire Surveys," *Journal of Radiation Research* 47 (2006): A203–A204.

6. Andrei Sakharov, *Memoirs* (London: Hutchinson, 1990): 192. Also see David Holloway, *Stalin & the Bomb* (New Haven, CT: Yale University Press, 1984): 315–316.

7. Togzhan Kassenova, "Banning Nuclear Testing: Lessons from the Semipalatinsk Nuclear Testing Site," *Nonproliferation Review* 23:3–4 (2016): 332.

8. "22 November 1955—RDS-37," Comprehensive Nuclear-Test-Ban Treaty Organization: https://www.ctbto.org/specials/testing-times/22-november-1955-rds-37 (accessed 13 April 2018).

9. For Charles Perrow's survey of the systemic dismissal of health effects from radiation exposures, see Charles Perrow, "Nuclear Denial: From Hiroshima to Fukushima," *Bulletin of the Atomic Scientists* 69:5 (2013): 56–67.

10. The radiation measurement of "rem" stands for "Roentgen equivalent man," and was an early metric of radiation absorbed by the human body. The US Nuclear Regulatory Commission advises that "a single dose of 100 rem may cause a person to experience nausea or skin reddening (although recovery is likely), and about 25 rem can cause temporary sterility in men." See "High Radiation Doses," USNRC, 2 October 2017: https://www.nrc.gov/about-nrc/radiation/health-effects/high-rad-doses .html (accessed 21 July 2020). The use of the adult male body as standard in radiation protection has been criticized. See Arjun Makhijani, *The Use of Reference Man in Radiation Protection Standards and Guidance with Recommendations for Change* (Takoma Park, MD: Institute for Energy and Environmental Research, 2008).

11. K. Gordeev et al., "Fallout from Nuclear Tests: Dosimetry in Kazakhstan," *Radiation and Environmental Biophysics* 41 (2002): 65.

12. Cynthia Werner and Kathleen Purvis-Roberts, *Unraveling the Secrets of the Past: Contested Versions of Nuclear Testing in the Soviet Republic of Kazakhstan* (Washington, DC: National Council for Eurasian and East European Research, 2005): 26–27.

13. Werner and Purvis-Roberts, *Unraveling the Secrets of the Past:* 26.

14. Fred Hiatt, "Survivors Tell of '54 Soviet A-Blast," *Washington Post,* 15 September 1994: https://www.washingtonpost.com/archive/politics/1994/09/15 /survivors-tell-of-54-soviet-a-blast/d5e9e540-78a1-4e75-8e85-09708a5fbf05/ (accessed 10 June 2021).

15. Elizabeth Bryant, "Algeria: 60 Years on, French Nuclear Tests Leave Bitter Fallout," Deutsche Welle, 13 February 2020: https://www.dw.com/en/algeria-60-years-on-french-nuclear-tests-leave-bitter-fallout/a-52354351 (accessed 5 May 2020).

16. Jake Wallis Simons, "Forgotten Victims of Britain's Nuclear Tests on Christmas Islands," *The Telegraph,* 2 February 2014: https://www.telegraph.co.uk/finance /newsbysector/industry/defence/10611985/Forgotten-victims-of-Britains-nuclear -tests-on-Christmas-Island.html (accessed 13 April 2018).

17. Steve Bogan, "Radiation from 1960s Nuclear Tests Is Still Hurting My Family," *The Times,* 27 April 2009: https://www.thetimes.co.uk/article/radiation

-from-1960s-nuclear-tests-is-still-hurting-my-family-gvwx6gbs8wd (accessed 13 April 2018).

18. Robert MacKenzie, interviewed by Mary Palevsky, 1 January 2005, transcript, Nevada Test Site Oral History Project, University of Nevada at Las Vegas Special Collections (Las Vegas, Nev.): 42–43.

19. Kate Brown, "The Last Sink: The Human Body as the Ultimate Radioactive Storage Site," in Christoph Mauch, ed., *Out of Sight, Out of Mind: The Politics and Culture of Waste* (Munich: Rachel Carson Center Perspectives, 2016): 45–46. Susan Lindee places our orientation to global hibakusha in the context many policymakers strategize, describing them as "the ultimate consumers of nuclear weapons." See Susan Lindee, *Rational Fog: Science and Technology in Modern War* (Cambridge, MA: Harvard University Press, 2020): 6.

20. S. M. Loyland Asbury, S. P. Lamont, and S. B. Clark, "Plutonium Partitioning to Colloidal and Particulate Matter in an Acidic, Sandy Sediment: Implications for Remediation Alternatives and Plutonium Migration," *Environmental Science and Technology* 35:11 (2001): 2295. Those 73 million cubic meters of soil are roughly equivalent to 868 square miles of soil at 1 inch. By contrast, the city of Los Angeles occupies 469 square miles of area.

21. Mark Oliphant to Hedley Marston, 9 September 1956, repr. in Roger Cross, *Fallout: Hedley Marston and the Atomic Bomb Tests in Australia* (Kent Town, AU: Wakefield Press, 2011): 53–54.

22. William R. Kennedy Jr., *Fallout Forecasting—1945 through 1962*, Los Alamos Report No. LA-1605-RS (March 1988): 2.

23. Wright H. Langham, quoted from "Proceedings: Second Interdisciplinary Conference on Selected Effects of a General War," *DASIAC Special Report 95* (Santa Barbara, CA: Defense Atomic Support Agency, 1969): 45. See also Kenneth Bainbridge, *Trinity*, Los Alamos Report No. LA-6300-H (May 1976): 31–36. In fact, according to Janet Brodie, the exposures of this family were very concerning in Warren's own report to General Groves: "the map of radiation at what he labeled the 'Hot Canyon' included his handwritten note that 'House (with family)' located 0.9 miles beyond the canyon received 'an accumulated total dose of 57–60r.'" See Janet Farrell Brodie, "Contested Knowledge: The Trinity Test Radiation Studies," in Brinda Sarathy, Vivian Hamilton, and Janet Farrell Brodie, *Inevitably Toxic: Historical Perspectives on Contamination, Exposure, and Expertise* (Pittsburgh: University of Pittsburgh Press, 2018): 54.

24. William A. Shurcliff, *Bombs at Bikini: The Official Report of Operation Crossroads* (New York: Wm. H. Wise, 1947): 2.

25. See Atsuko Shigesawa, *Demystifying the Atomic Bomb: The U.S. Strategic Bombing Survey Goes to Hiroshima and Nagasaki* (PhD diss., Hiroshima City University, 2019).

26. Kenneth O. Emery, J. I. Tracey Jr., and H. S. Ladd, *Geology of Bikini and Nearby Atolls* (Washington, DC: Government Printing Office, 1954).

27. Emery, Tracey, and Ladd, *Geology of Bikini and Nearby Atolls*: 2, 14, 186–191.

28. Nuell W. Paschal, "Voices from Nuclear Hell," in *National Association of Atomic Victims Newsletter*, March 2010: 10.

29. L. Berkhouse et al., *Operation Crossroads—1946*, DNA 6032F (Washington, DC: Defense Nuclear Agency, 1984): 1.

30. Jonathan M. Weisgall, *Operation Crossroads: The Atomic Tests at Bikini Atoll* (Annapolis, MD: Naval Institute Press, 1994): 227.

31. Weisgall, *Operation Crossroads:* 229–230.

32. Shurcliff, *Bombs at Bikini:* 198–199.

33. Weisgall, *Operation Crossroads:* 260.

34. Salvatore Lazzari, *Loss-of-Use Damages from U.S. Nuclear Testing in the Marshall Islands: Technical Analysis of the Nuclear Claims Tribunal's Methodology and Alternative Estimates* (Washington, DC: Congressional Research Service, 2005): 1.

35. "The Evaluation of the Atomic Bomb as a Military Weapon," *The Final Report of the Joint Chiefs of Staff Evaluation Board for Operation Crossroads,* Enclosure "A," JCS 1691/3 (30 June 1947): 23.

36. The use of fallout radiation as a primary lethal effect in nuclear war planning will be explored in much greater detail in chapter 6.

37. See Carl Maag et al., *Shot Hood: A Test of the PLUMBOB Series* (Washington, DC: Defense Nuclear Agency, 1983): 1.

38. Maag et al., *Shot Hood*: 22. See also Harvey Wasserman and Norman Solomon, *Killing Our Own: The Disaster of America's Experience with Atomic Radiation* (New York: Dell, 1982): 54–72.

39. Jacques J. Richardson, "Serious Misapplications of Military Research: Dysfunction Between Conception and Implementation," *Science and Engineering Ethics* 7 (2001): 352.

40. Marlise Simons, "Soviet Atom Test Used Thousands as Guinea Pigs, Archives Show," *New York Times*, 7 November 1993: 1, 20.

41. Sergei Zelentsov, Vadim Logachev, and Anatoliy Matushchenko, "The 1954 Nuclear Exercise at the Totskoye Test Range: How Is This 'Radiation Legacy' Dangerous?," in *Second Russian National Dialogue on Energy, Society and Security: 21–22 April 2008* (St. Petersburg: Green Cross Russia, 2008): 279.

42. I. Kirov, "A Time to Scatter Stones," *Environmental Issues*, United States Joint Publications Research Service—TEN-91-012 (26 June 1991): 69.

43. Nikolay Leonov, quoted in *The Red Bomb: In the Name of Peace*, DVD directed and produced by Jamie Doran (1994; Silver Springs, MD: Discovery Channel).

44. A. A. Romanyukhaf et al., "The Distance Effect on Individual Exposures Evaluated from the Soviet Nuclear Bomb Test at Totskoye Test Site in 1954," *Radiation Protection Dosimetry* 86:1 (1999): 54.

45. Noriyuki Kawano et al., "Human Suffering Effects of Nuclear Tests at Semipalatinsk, Kazakhstan: Established on the Basis of Questionnaire Surveys," *Journal of Radiation Research* 47 (2006): A209.

46. Steven L. Simon and André Bouville, "Health Effects of Nuclear Weapons Testing," *The Lancet* 386 (1 August 2015): 407–408.

47. Ralph Lapp, "Radioactive Fall-Out III," *Bulletin of the Atomic Scientists* 11:6 (1955): 206.

48. Luigi Monte, "Modelling Multiple Dispersion of Radionuclides Through the Environment," *Journal of Environmental Radioactivity* 101 (2010): 134.

49. *Bikini: Radiobiological Laboratory* (AEC, Lookout Mountain Laboratory, 1949). See also Ronald Rainger, "Science at the Crossroads: The Navy, Bikini Atoll, and American Oceanography in the 1940s," *Historical Studies in the Physical and Biological Sciences* 30:2 (2000): 349–371; Jacob D. Hamblin, *Oceanographers and the Cold War: Disciples of Marine Science* (Seattle: University of Washington Press, 2005): 3–31.

50. S. M. Greenfield, "Rain Scavenging of Radioactive Particulate Matter from the Atmosphere," *Journal of Meteorology* 14 (April 1957): 115–125.

51. C. L. Comar, "Movement of Fallout Radionuclides Through the Biosphere and Man," *Annual Review of Nuclear Science* 15 (1965): 176. Recent research has also revealed that electrical charges released by atmospheric nuclear weapon tests altered rainfall patterns over Scotland. See R. Giles Harrison et al., "Precipitation Modification by Ionization," *Physical Review Letters* 124 (2020): https://journals.aps.org/prl/abstract/10.1103/PhysRevLett.124.198701 (accessed 2 July 2020).

52. The Division of Nuclear Safety and Security, *Nuclear Explosions in the USSR: The North Test Site* (Vienna: IAEA, 2004): 123.

53. M. H. Frere et al., *The Behavior of Radioactive Fallout in Soils and Plants* (Washington, DC: National Academy of Sciences / National Research Council, 1963): 1.

54. Division of Nuclear Safety and Security, *Nuclear Explosions in the USSR*: 176–177. See also Marko Štrok, Borut Smolduš, and Klemen Eler, "Natural Radionuclides in Trees Grown on a Uranium Mill Tailings Waste Pile," *Environmental Science and Pollution Research* 18 (2011): 819–826.

55. Thomas K. Riesen, "Radiocaesium in Forests—a Review on Most Recent Research," *Environmental Reviews* 10:2 (2002): 79. See also John Dighton, Tatanya Tuga, and Nelli Zhdanova, "Fungi and Ionizing Radiation from Radionuclides," *FEMS Microbiology Letters* 281 (2008): 109–120.

56. Sergey Mamikhin, F. A. Tikhomirov, and A. I. Shcheglov, "Dynamics of Cs-137 in the Forests of the 30-km Zone Around the Chernobyl Nuclear Power Plant," *Science in the Total Environment* 193 (1997): 169.

57. Cynthia Dion-Schwarz et al., *Technological Lessons from the Fukushima Dai-Ichi Accident* (Santa Monica, CA: RAND Corporation, 2016): 34.

58. Olivier Evrard, J. Patrick Laceby, and Atsushi Nakao, "Effectiveness of Landscape Decontamination Following the Fukushima Nuclear Accident: A Review," *Soil* 5 (2019): 333.

59. UNSCEAR, "Annex B—Exposures from Man-Made Sources of Radiation," *Report to the General Assembly*, 1993: 96.

60. Hugh D. Livingston and Pavel P. Povinec, "A Millennium Perspective on the Contribution of Global Fallout Radionuclides to Ocean Science," *Health Physics* 82:5 (May 2002):656.

61. Asker Aarkrog, "Input of Anthropogenic Radionuclides into the World Ocean," *Deep-Sea Research Part II* 50 (2003): 2597.

62. R. Hill et al., "Sea to Land Transfer of Anthropogenic Radionuclides to the North Wales Coast, Part II: Aerial Modelling and Radiological Assessment," *Journal of Environmental Radioactivity* 99 (2008): 21.

63. Aarkrog, "Input of Anthropogenic Radionuclides into the World Ocean": 2604–2605.

64. Ning Wang et al., "Penetration of Bomb ^{14}C into Deepest Ocean Trench," *Geophysical Research Letters* 46 (2019): 5413–5419.

65. Gordon Edwards, "Responding to the Nuclear Present II," (lecture presented at An Atomic Symposium, Toronto, Canada, 15 June 2014).

66. "Over 9 Million Bags of Nuclear Cleanup Waste Piled up Across Fukushima Pref.," *Mainichi Shimbun*, 10 December 2015: https://mainichi.jp/english/articles /20151210/p2a/00m/0na/020000c (accessed 12 December 2019).

67. Eric J. Grant et al., "Solid Cancer Incidence Among the Life Span Study of Atomic Bomb Survivors: 1958–2009," *Radiation Research Society* 187:5 (2017): 513–537; Tetsuji Imanaka, "Casualties and Radiation Dosimetry of the Atomic Bombings on Hiroshima and Nagasaki," in Arrigo A. Cigna and Marco Durante, eds., *Radiation Risk Estimates in Normal and Emergency Situations* (Dordrecht, Netherlands: Springer, 2005): 149–156.

68. See Harold L. Beck et al., "Review of Methods of Dose Estimation for Epidemiological Studies of the Radiological Impact of Nevada Test Site and Global Fallout," *Radiation Research* 166 (2006): 209–218.

69. Committee on the Biological Effects of Ionizing Radiation, *Health Risks of Radon and Other Internally Deposited Alpha-Emitters* BEIR IV (Washington, DC: National Academies Press, 1988): 2, 17. You should look up the reticuloendothelium system yourself!

70. Michael Shellenberger, "It Sounds Crazy, but Fukushima, Chernobyl, and Three Mile Island Show Why Nuclear Is Inherently Safe," *Forbes*, 11 March 2019: https://www.forbes.com/sites/michaelshellenberger/2019/03/11/it-sounds-crazy -but-fukushima-chernobyl-and-three-mile-island-show-why-nuclear-is-inherently -safe/#13f993381688 (accessed 6 August 2020). Shellenberger has also advocated (on the anniversary of the nuclear attack on Hiroshima) for the proliferation of nuclear weapons; see Michael Shellenberger, "Who Are We to Deny Weak Nations the Nuclear Weapons They Need for Self-Defense?" *Forbes*, 6 August 2018: https://www.forbes .com/sites/michaelshellenberger/2018/08/06/who-are-we-to-deny-weak-nations -the-nuclear-weapons-they-need-for-self-defense/?sh=631765d5522f (accessed 11 March 2021). Discussion of differing numbers of deaths from Chernobyl is further discussed in the next chapter.

71. "The Active Straw," *Newsweek*, 12 November 1945: 50; J. H. Webb, "The Fogging of Photographic Film by Radioactive Contaminants in Cardboard Packaging Materials," *Physical Review* 76:3 (August 1949): 375–380; "Oral History of Merril Eisenbud" (conducted 26 January 1995), *Human Radiation Studies, Remembering the Early Years* (US Department of Energy, Office of Human Radiation Experiments, May 1995): 33: http://www.eh.doe.gov/ohre/roadmap/histories/0456/0456toc .html (accessed 21 June 2018).

72. David Bradley, *No Place to Hide* (Boston: Little, Brown, 1948). The book was a Book-of-the-Month Club selection in 1948.

73. US General Accounting Office, *Operation Crossroads: Personnel Radiation Exposure Estimates Should Be Improved,* GAO/RCED-8615 (November 1985): 4.

74. For details on human experimentation funded by the US government, see "Executive summary," *The Final Report of the Advisory Committee on Human Radiation Experiments,* 21 October 1994: https://ehss.energy.gov/ohre/roadmap/achre/summary.html (accessed 9 August 2020).

75. "Oral History of Merril Eisenbud": 61. For more on Project GABRIEL, see Jacob Darwin Hamblin, *Arming Mother Nature: The Birth of Catastrophic Environmentalism* (Oxford: Oxford University Press, 2013): 113–117; Lisa Martino-Taylor, *Behind the Fog: How the U.S. Cold War Radiological Weapons Program Exposed Innocent Americans* (New York: Routledge, 2018): 109–117; Toshihiro Higuchi, *Political Fallout: Nuclear Weapons Testing and the Making of a Global Environmental Crisis* (Stanford, CA: Stanford University Press, 2020): 48–49.

76. Australian Radiation Protection and Nuclear Safety Agency Report, *Australian Strontium 90 Testing Program 1957–1978,* attachment A, 12: http://www.nuclearfiles.org/menu/key-issues/nuclear-weapons/issues/testing/PDFs/sr90pubrep[1].pdf (accessed 14 June 2018).

77. Richard G. Hewlett and Jack M. Holl, *Atoms for Peace and War, 1953–1961: Eisenhower and the Atomic Energy Commission* (Berkeley: University of California Press, 1989): 265–266.

78. *Report on Project Gabriel,* July 1954, in folder 26–2, box 3363, DOE Historian Record Group, DOE Archives, United States National Archives and Records Administration.

79. See Sue Rabbitt Roff, "Project Sunshine and the Slippery Slope," *Medicine, Conflict and Survival* 18:3 (2002): 299–310; Keith Schneider, "Stillborns' Ashes Used in Studies of Radiation, Documents Show," *New York Times,* 5 May 1994: B12; Warren E. Leary, "In 1950s, U.S. Collected Human Tissue to Monitor Atomic Tests," *New York Times,* 21 June 1995: B8; Eddie Goncalves, "Britain Snatched Babies' Bodies for Nuclear Labs," *The Guardian,* 3 June 2001: https://www.theguardian.com/uk/2001/jun/03/highereducation.research (accessed 14 June 2018).

80. *Worldwide Effects of Atomic Weapons: Project Sunshine* (Santa Monica, CA: RAND Corporation, 1953): 7.

81. Caroline Jack and Stephanie Steinhardt, "Atomic Anxiety and the Tooth Fairy: Citizen Science in the Midcentury Midwest," *The Appendix,* 26 November 2014: https://theappendix.net/issues/2014/10/atomic-anxiety-and-the-tooth-fairy-citizen-science-in-the-midcentury-midwest (accessed 25 June 2018).

82. Louise Zibold Reiss, "Strontium-90 Absorption by Deciduous Teeth," *Science* 134:3491 (24 November 1961): 1669–1673. See also W. K. Wyant Jr., "50,000 Baby Teeth," *The Nation,* 13 June 1959: 535–537.

83. "Strontium-90 in Teeth," *British Medical Journal* 2:5422 (5 December 1964): 1411; Australian Radiation Protection and Nuclear Safety Agency, "Australian Strontium 90 Testing Program 1957–1978," ARPANSA Report (September 2011):

http://www.nuclearfiles.org/menu/key-issues/nuclear-weapons/issues/testing /PDFs/sr90pubrep[1].pdf (accessed 10 June 2021).

84. Hewlett and Holl, *Atoms for Peace and War:* 376–377.

85. Neal O. Hines, *Proving Ground: An Account of the Radiobiological Studies in the Pacific, 1946–1961* (Seattle: University of Washington Press, 1962): 45. The AFL would conduct the primary marine radiobiological studies for the AEC during the period of atmospheric nuclear testing. See also Ronald Rainger, "Science at the Crossroads: The Navy, Bikini Atoll, and American Oceanography in the 1940s," *Historical Studies in the Physical and Biological Sciences* 30:2 (2000): 349–371.

86. Toshihiro Higuchi, "Atmospheric Nuclear Weapons Testing and the Debate on Risk Knowledge in Cold War America, 1945–1963," in J. R. McNeill and Corinna R. Unger, eds., *Environmental Histories of the Cold War* (Cambridge: Cambridge University Press, 2010): 303.

87. Neal O. Hines, *Atoms, Nature, and Man: Man-Made Radioactivity in the Environment* (Oak Ridge, TN: US AEC Division of Technical Information Extension, 1966): 20.

88. See Laura A. Bruno, "Bequest of the Nuclear Battlefield: Science, Nature, and the Atom During the First Decade of the Cold War," *Historical Studies in the Physical and Biological Sciences* 33:2 (2003): 237–260.

89. Lester Machta, "Meteorological Benefits from Atmospheric Nuclear Tests," *Health Physics* 82:5 (May 2002): 635, 641.

90. A. H. Seymour, "Introduction," *Radioactivity in the Marine Environment* (Washington, DC: National Academy of Sciences, 1971): 1.

91. E. Jerry Jessee, "A Heightened Controversy: Nuclear Weapons Testing, Radioactive Tracers, and the Dynamic Stratosphere," in James Roger Fleming and Ann Johnson, eds., *Toxic Airs: Body, Place, Planet in Historical Perspective* (Pittsburgh: University of Pittsburgh Press, 2014): 155. See also Lester Machta, "Discussion of Meteorological Factors and Fallout Distribution," in *Environmental Contamination from Weapon Tests* (Oak Ridge, TN: US AEC, Division of Technical Information Extension, October 1958): 310–325.

92. J. A. Corcho et al., "Anthropogenic Radionuclides in Atmospheric Air Over Switzerland During the Last Few Decades," *Nature Communications* 5:3030 (2014): 2. For more on SNAP-9A, see E. P. Hardy, E. W. Krey, and H. L. Volchock, *Global Inventory and Distribution of PU-239 from SNAP-9A,* Atomic Energy Commission Report No. HASL-250, 1 March 1972.

93. D. Tsumune et al., "Calculation of Artificial Radionuclides in the Ocean by an Ocean General Circulation Model," *Journal of Radioanalytical and Nuclear Chemistry,* 248:3 (2001): 777.

94. D. R. Walling and S. B. Bradley, "Some Applications of Caesium-137 Measurements in the Study of Erosion, Transport and Deposition," *Erosion, Transport and Deposition Processes* (Proceedings of the International Association of Hydrological Sciences 189, 1990): 179.

95. Emil Fulajtar et al., *Use of 137Cs for Soil Erosion Assessment* (Rome: Food and Agriculture Organization of the UN IAEA, 2017): v. See also Walling and Bradley, "Some Applications of Caesium-137 Measurements": 179–203.

96. Michaela Čadová et al., "Radioactivity in Mushrooms from Selected Locations in the Bohemian Forest, Czech Republic," *Radiation and Environmental Biophysics* 56 (2017): 167. At the Hanford Site, radionuclides were transported from contaminated locations by swallows. See Annette Cary, "Birds at Hanford Vit Plant Spread Contaminated Waste," *Tri-City Herald*, 25 June 2013: https://www.tri-cityherald.com/news/local/hanford/article32128425.html (accessed 5 July 2020).

97. R. M. Alexakhin et al., "Chernobyl Radionuclide Distribution, Migration, and Environmental and Agricultural Impacts," *Health Physics* 93:5 (November 2007): 418, 420.

98. Jussi Paatero et al., "Airborne and Deposited Radioactivity from the Chernobyl Accident—A Review of Investigations in Finland," *Boreal Environment Research* 15: 19, 29–30.

99. J. S. Chaplow, N. A. Beresford, and C. L. Barnett, "Post-Chernobyl Surveys of Radiocaesium in Soil, Vegetation, Wildlife and Fungi in Great Britain," *Earth System Science Data*, 2015: https://www.earth-syst-sci-data.net/7/215/2015/essd-7-215-2015.pdf (accessed 23 June 2018).

100. National Council on Radiation Protection and Measurements, *Cesium-137 from the Environment to Man: Metabolism and Dose* (Washington, DC: NCRP, 1977).

101. Petr Dvořák, Petr Snášel, and Katarína Beňová, "Transfer of Radiocesium into Wild Boar Meat," *Acta Veterinaria Brno* 79 (2010): S85. See also Verena Schmitt-Roschmann, "Radioactive Boars on the Rise in Germany," *Phys.org*, 20 August 2010: https://phys.org/news/2010-08-radioactive-boars-germany.html (accessed 25 June 2018); "Radioactive Boar Shot Dead in Sweden—31 Years After Chernobyl Disaster," *The Local*, 5 October 2017: https://www.thelocal.se/20171005/radioactive-boar-shot-dead-in-sweden-31-years-after-chernobyl-disaster (accessed 25 June 2018).

102. Yasuyuki Taira et al., "Current Concentration of Artificial Radionuclides and Estimated Radiation Doses from 137Cs Around the Chernobyl Nuclear Power Plant, the Semipalatinsk Nuclear Testing Site, and in Nagasaki," *Journal of Radiation Research* 52 (2011): 89.

103. Georg Steinhauser and Paul R. J. Saey, "137Cs in the Meat of Wild Boars: A Comparison of the Impacts of Chernobyl and Fukushima," *Journal of Radioanalytical and Nuclear Chemistry* 307 (2016): 1801.

104. Serge Schmemann, "Soviet Announces Nuclear Accident at Electric Plant," *New York Times*, 29 April 1986: A1.

105. Lavrans Skuterud and Helge Hansen, "Managing Radiocaesium Contamination in Norwegian Reindeer Twenty-Two Years after the Chernobyl Accident—The Need for a New Regulation" (paper presented to the International Conference on Radioecology & Environmental Radioactivity held in Bergen, Norway, 2008): http://radioecology.info/Bergen2008/proceedings/138.%20Skuterud%20and%20Hansen%20O.pdf (accessed 25 June 2018).

106. Jussi Paatero and Timo Jaakkola, "Transfer of Plutonium, Americium and Curium from Fallout into Reindeer After the Chernobyl Accident," *Boreal Environmental Research* 3 (1998): 181–182. See also Bernt-E. V. Jones, Olof Eriksson, and Magnus Nordkvist, "Radiocesium Metabolism in Reindeer," *Rangifer* 3 (1990): 45–48. See also Merril Eisenbud, *Environment, Technology, and Health: Human*

Ecology in Historical Perspective (New York: New York University Press, 1978): 56; Swedish National Institute of Radiation Protection, *Chernobyl—Its Impact on Sweden,* SSI-rapport 86-12 (1986): 7.

107. "Slow Decline of Caesium-137 in Wild Reindeer," *Norwegian Radiation Protection Authority,* 22 December 2017: http://www.environment.no/goals/4 .-pollution/target-4.2/levels-of-selected-radioactive-substances-in-the-environment /slow-decline-of-caesium-137-in-wild-reindeer/ (accessed 25 June 2018).

108. C. E. Miller, H. A. May, and Leonidas D. Marinelli, *The Use of Low Level Scintillation Spectroscopy in the Evaluation of Radioactive Contamination of the Human Body* (Lemont, IL: Argonne National Laboratory, 1958). This included the "Radium Girls."

109. "News of Tennessee Science," *Journal of the Tennessee Academy of Science* 39:2 (1964): 42–43.

110. Frederick W. Lengemann and John H. Woodburn, *Whole Body Counters* (Washington, DC: US Atomic Energy Commission, 1964): 13–14.

111. Daniel Allen Butler, *Distant Victory: The Battle of Jutland and the Allied Triumph in the First World War* (Santa Barbara, CA: Greenwood, 2006): 229.

112. Jehani Ragai, *The Scientist and the Forger: Insights into the Scientific Detection of Forgery in Paintings* (London: Imperial College Press, 2015): 69–70. See also Edwin Cartlidge, "Nuclear Fallout Used to Spot Fake Art," *Physics World,* 4 July 2008: https:// physicsworld.com/a/nuclear-fallout-used-to-spot-fake-art/ (accessed 13 June 2018).

113. Rami Tzabar, "Wine Makers Crack Open Hi-Tech Tricks," *BBC Science,* 28 November 2008: http://news.bbc.co.uk/2/hi/science/nature/7755014.stm (accessed 13 June 2018). See also Frances Robinson, "Chernobyl Fallout Levels Can Date Wine," *Decanter,* 16 June 2005: http://www.decanter.com/wine-news/chernobyl-fallout-levels -can-date-wine-97133/ (accessed 13 June 2018). See also Michael S. Pravikoff, Christine Marquet, and Philippe Hubert, "Dating of Wines with Cesium-137: Fukushima's Imprint," arXiv, 2018: https://arxiv.org/pdf/1807.04340.pdf (accessed 21 December 2020).

114. Gordon T. Cook et al., "Using Carbon Isotopes to Fight the Rise in Fraudulent Whisky," *Radiocarbon* 62:1 (2020): 51–62.

115. Kyle S. Van Houtan et al., "Time in Tortoiseshell: A Bomb Radiocarbon-Validated Chronology in Sea Turtle Scutes," *Proceedings of the Royal Society B* 283 (2016): http://rspb.royalsocietypublishing.org/content/royprsb/283/1822/20152220.full .pdf (accessed 25 June 2018).

116. Joyce J. L. Ong et al., "Annual Bands in Vertebrae Validated by Bomb Radiocarbon Assays Provide Estimates of Age and Growth of Whale Sharks," *Frontiers in Marine Science* 7:188 (2020): https://www.frontiersin.org/articles/10.3389 /fmars.2020.00188/full (accessed 7 July 2020).

117. Simon L. Lewis and Mark A. Maslin, *The Human Planet: How We Created the Anthropocene* (London: Penguin Books, 2018): 309–310.

118. Ronald K. Chesser and Robert J. Baker, "Growing Up with Chernobyl: Working in a Radioactive Zone, Two Scientists Learn Tough Lessons About Politics, Bias and the Challenges of Doing Good Science," *American Scientist* 94 (2006): 542.

119. Chesser and Baker, "Growing Up with Chernobyl": 544.

120. Pritikin herself grew up in Richland, Washington, during the years of peak plutonium production at Hanford. Trisha Pritikin, *The Hanford Plaintiffs: Voices from the Fight for Atomic Justice* (Lawrence: University Press of Kansas, 2020): 26.

3. Falling Apart Inside

1. Natalie Wolchover, "Timeline of Events at Japan's Fukushima Nuclear Reactors," *LiveScience*, 17 March 2011: https://www.livescience.com/13294-timeline-events-japan-fukushima-nuclear-reactors.html (accessed 7 January 2021).

2. "Hot particle" has been frequently used in recent discourse to refer to radionuclides.

3. This story was told to me during an interview with journalist Akiko Kamemoto on 9 April 2015 in Fukushima City, Japan. (Pseudonyms used to protect interviewee and the subjects' privacy.)

4. Arnfinn Tønnessen, Bertil Mårdberg, and Lars Weisæth, "Silent Disaster: A European Perspective on Threat Perception from Chernobyl Far Field Fallout," *Journal of Traumatic Stress* 15:6 (2002): 453–454. The "scientific uncertainties" refer to disagreements about the effects of long-term external exposure to low-dose radiation and the relationship of internalized particles to radiogenic diseases. See also Richard E. Adams et al., "Stress and Well-Being in Mothers of Young Children 11 Years after the Chornobyl Nuclear Power Plant Accident," *Psychological Medicine* 32:1 (2002): 143–156.

5. Studies of survivors in Hiroshima and Nagasaki showed that those with more serious radiogenic illnesses also experienced heightened anxiety. See Michiko Yamada and Shizue Izumi, "Psychiatric Sequelae in Atomic Bomb Survivors in Hiroshima and Nagasaki Two Decades After the Explosions," *Social Psychiatry and Psychiatric Epidemiology* 37:9 (2002): 409–415.

6. This book uses the term "indigenous" as defined by Workshop on Data Collection and Disaggregation for Indigenous Peoples (2004): "Indigenous communities, peoples and nations are those which, having a historical continuity with pre-invasion and pre-colonial societies that developed on their territories, consider themselves distinct from other sectors of the societies now prevailing on those territories, or parts of them. They form at present non-dominant sectors of society and are determined to preserve, develop and transmit to future generations their ancestral territories, and their ethnic identity, as the basis of their continued existence as peoples, in accordance with their own cultural patterns, social institutions and legal system." See "Workshop on Data Collection and Disaggregation for Indigenous Peoples," in *The Concept of Indigenous Peoples* (New York: United Nations Department of Economic and Social Affairs, 2004): 2.

7. Siw Ellen Jakobsen, "Surprisingly High Levels of Radioactivity in Norwegian Reindeer and Sheep," *ScienceNorway.no*, 8 October 2014: https://sciencenorway.no/chernobyl-forskningno-nature-conservation/surprisingly-high-levels-of-radioactivity-in-norwegian-reindeer-and-sheep/1408148 (accessed 4 October 2019); Nuclear Energy Agency, *Chernobyl: Assessment of Radiological and Health Impact, 2002 Update of Chernobyl: Ten Years On* (Paris: Organization for Economic Co-operation and Development, 2002): 102.

8. Sharon Stevens, "Chernobyl Fallout: A Hard Rain for the Sámi," *Cultural Survival*, June 1987: https://www.culturalsurvival.org/publications/cultural-survival -quarterly/chernobyl-fallout-hard-rain-Sámi (accessed 4 October 2019).

9. Stevens, "Chernobyl Fallout."

10. Robert Paine, " 'Chernobyl' Reaches Norway: The Accident, Science, and the Threat to Cultural Knowledge," *Public Understanding of Science* 1 (1992): 268. Snåsa is a municipality in Norway central to a community of Southern Sámi people.

11. Svetlana Alexievich, *Voices from Chernobyl: Chronicle of the Future* (London: Aurum Press, 1999): 20.

12. Sharon Stephens, "Physical and Cultural Reproduction in a Post-Chernobyl Norwegian Sami Community," in Faye D. Ginsburg and Rayna Rap, eds., *Conceiving the New World Order: The Global Politics of Reproduction* (Berkeley: University of California Press, 1995): 276.

13. Stephens, "Physical and Cultural Reproduction": 275.

14. Tønnessen, Mårdberg, and Weisæth, "Silent Disaster": 453.

15. The only reported perceptibility of radiation is a vague "metallic taste" in the mouth when exposed to high levels of external radiation, or fallout containing irradiated metallic particles: "Downwind residents intermittently experienced a metallic taste in the air due to the incineration, and irradiation, of the bomb casing, shed enclosing the bomb, and tower upon which both were perched as well as reported ailments symptomatic of radiation sickness," from James Rice and Susan Steinkopf Rice, " 'Radiation Is Not New to Our Lives': The U.S. Atomic Energy Commission, Continental Atmospheric Weapons Testing, and Discursive Hegemony in the Downwind Communities," *Journal of Historical Sociology* 28:4 (2014): 21. This phenomenon has also been reported by military personnel participating in nuclear tests, as well as by people downwind of Chernobyl and Fukushima. See George Mace, "An Atomic Bomb Test Veteran Remembers," in Walter E. Venator Jr., ed., *Where the Boys Were: Nuclear Testing at Eniwetok Atoll in 1958: Atomic Veteran Stories* (Scribd: 2010): 26: https://www .scribd.com/document/47828956/Where-the-Boys-Were (accessed 25 May 2018).

16. Kai Erikson, "Radiation's Lingering Dread," *Bulletin of the Atomic Scientists* 47:2 (1991): 36.

17. Katie Koralis, "Meetings for 'Downwinders' as Compensation Program Set to Come to an End," ABC4, 9 May 2019: https://www.abc4.com/news/meetings-for-downwinders-as-compensation-program-set-to-come-to-an-end/ (accessed 30 ay 2020).

18. Lucinda Dillon, "Toxic Utah: Ghosts in the Wind," *Deseret News*, 15 February 2001: https://www.deseret.com/2001/2/15/19781193/toxic-utah-ghosts-in-the -wind#blaine-johnson-talks-about-a-1980-life-magazine-story-on-the-downwinders -of-southern-utah-johnsons-daughter-sybil-died-at-age-12-from-cancer-likely-caused -by-nevada-nuclear-tests (accessed 30 March 2022).

19. Interview with Claudia Peterson conducted by Robert Jacobs and Mick Broderick on 7 February 2012 in St. George, Utah.

20. Atmospheric testing ended at the NTS in 1963 but continued globally until the last atmospheric test in China in 1980. See details in chapter 5.

21. Quoted in Carole Gallagher, *American Ground Zero: The Secret Nuclear War* (New York: Random House, 1993): 147.

22. Hiroaki Katayamam et al., "An Attempt to Develop a Database for Epidemiological Research in Semipalatinsk," *Journal of Radiation Research* 47 (2006): 189–197.

23. K. N. Apsalikov et al., "The State Scientific Automated Medical Registry, Kazakhstan: An Important Resource for Low-Dose Radiation Health Research," *Radiation and Environmental Biophysics* 58 (2019): 2.

24. See Nuclear Energy Agency, *Radioactive Waste in Perspective* (Paris: Organization for Economic Co-operation and Development, 2010).

25. Mary Hudetz, "US Official: Research Finds Uranium in Navajo Women, Babies," Associated Press, 8 October 2019: https://apnews.com/334124280ace4b36 beb6b8d58c328ae3 (accessed 9 October 2019).

26. Garet Bleir, "Havasupai Prayer Gathering: Indigenous Nations Unite Against Nuclear Colonialism," *Intercontinentalcry.org,* 8 March 2018: https://intercontinentalcry.org/havasupai-prayer-gathering-indigenous-nations-unite-nuclear-colonialism/ (accessed 8 May 2019).

27. Dan Frosh, "Amid Toxic Waste, a Navajo Village Could Lose its Land," *New York Times.* 19 February 2014: http://www.nytimes.com/2014/02/20/us/nestled-amid-toxic-waste-a-navajo-village-faces-losing-its-land-forever.html (accessed 17 October 2018).

28. Chapter 5 outlines the basis of this choice.

29. See Robert Jacobs, "Post-nuclear/Post-colonial Challenges to Democratization in the Pacific," in Narayanan Ganesan, ed., *International Perspectives on Democratization and Peace* (London: Emerald Publishing, 2020): 27–42.

30. Quoted in Kevin Rafferty, Jayne Loader, and Pierce Rafferty, *Atomic Cafe: The Book of the Film* (New York: Bantam Books, 1982): 29.

31. Jonathan Weisgal, "The Nuclear Nomads of Bikini," *Foreign Policy* 39 (Summer 1980): 79. See also Francis X. Hezel, *Strangers in Their Own Land: A Century of Colonial Rule in the Caroline and Marshall Islands* (Honolulu: University of Hawai'i Press, 1995): 271.

32. Martha Smith-Norris, "American Cold War Policies and the Enewetaks," *Journal of the Canadian Historical Association* 22:2 (2011): 201. Lagoons are the essential features making atolls desirable places to live.

33. Quotation reproduced in Smith-Norris, "American Cold War Policies and the Enewetaks": 201, from: National Archives and Records of the United States, National Archives Gift Collection, Trust Territories of the Pacific Islands, RG200, box 27, roll 323, Petition of the People of Enewetak to the Trusteeship Council of the United Nations, May 1981: 3–4.

34. Jack Adair Tobin, "Land Tenure in the Marshall Islands," *Atoll Research Bulletin,* no. 11 (Washington, DC: Pacific Science Board, National Research Council, 1952): 2.

35. Satoe Nakahara, "Perceptions of the Radiation Disaster from H-Bomb Testing: Subsistence Economy, Knowledge and Network Among the People of Rongelap in the Marshall Islands," *Sociology and Anthropology* 6:1 (2018): 183.

36. Salvatore Lazzari, *Loss-of-Use Damages from U.S. Nuclear Testing in the Marshall Islands: Technical Analysis of the Nuclear Claims Tribunal's Methodology and Alternative Estimates* (Washington, DC: Congressional Research Service, 2005): 1–2.

37. Holly M. Barker, *Bravo for the Marshallese: Regaining Control in a Post-Nuclear, Post-Colonial World* (Belmont, CA: Wadsworth/Thompson Learning, 2004): 61.

38. Lazzari, *Loss-of-Use Damages:* 2.

39. Richard J. Hewlett and Jack M. Holl, *Atoms for Peace and War, 1953–1961: Eisenhower and the Atomic Energy Commission* (Berkeley: University of California Press, 1989): 173–175.

40. Holly M. Barker and Barbara Rose Johnston, "Seeking Compensation for Radiation Survivors in the Marshall Islands: The Contributions of Anthropology," *Cultural Survival* 24:1 (March 2000): 48–50.

41. Barker, *Bravo for the Marshallese:* 60.

42. Minutes of the 56th meeting of the Advisory Committee for Biology and Medicine, held at the Atomic Energy Commission, Washington, DC, 26–27 May 1956: 21.

43. Whittie B. McCool, "Return of Rongelapese to Their Home Island," Atomic Energy Commission Note by the Secretary AEC 125/30 (6 February 1957): 2.

44. Robert Conard et al., "Medical Survey of Rongelap and Utirik People Three Years After Exposure to Radioactive Fallout," ACHRE No. DOE-033195-B (March 1957): 22.

45. Robert A. Conrad et al., *Medical Survey of Rongelap People Seven Years After Exposure to Radioactive Fallout* (Upton, NY: Brookhaven National Laboratory, 1962): 5.

46. "Residents Evacuate Atomic-Test Atoll, *Washington Post,* 22 May 1985: A-2. See also David Robie, *Eyes of Fire: The Last Voyage of the Rainbow Warrior* (Philadelphia: New Society Publishers, 1987): 47–63. After this mission the *Rainbow Warrior* docked in Auckland Harbor in New Zealand to prepare to disrupt upcoming nuclear tests by the French in French Polynesia. French intelligence agents carried out a terror attack, detonating two bombs on board the boat and killing one crew member on the evening of 10–11 July; see Robie, *Eyes of Fire:* 92–104.

47. Barker and Johnson, "Seeking Compensation for Radiation Survivors": 48–50.

48. Song sung during the commemoration event held as part of the official commemoration of the sixtieth anniversary of the Bravo test by the Bikini Atoll community at Majuro Atoll on 25 February 2014. Words taken from a video filmed by the author.

49. Philip Brasor and Masako Tsubuku, "Temporary Disaster Housing Has an Unforeseen Permanence," *Japan Times,* 2 April 2017: https://www.japantimes.co.jp/community/2017/04/02/how-tos/temporary-disaster-housing-unforeseen-permanence/#.W8WC-aeB2Cc (accessed 16 October 2018).

50. Brasor and Tsubuku, "Temporary Disaster Housing."

51. "Fukushima Evacuation Split 50% of Families: Survey," *Japan Times,* 4 May 2014: https://www.japantimes.co.jp/news/2014/05/04/national/fukushima-evacuation-split-50-of-families-survey/#.W8WUrqeB2Cc (accessed 16 October 2018).

52. Sarai Flores, "Five Years On, Fukushima Evacuees Voice Lingering Anger, Fear and Distrust," *Japan Times,* 9 March 2016: https://www.japantimes.co.jp/community /2016/03/09/voices/five-years-fukushima-evacuees-voice-lingering-anger-fear -distrust/ (accessed 4 July 2020).

53. Rupert Wingfield-Hayes, "Is Fukushima's Exclusion Zone Doing More Harm than Radiation?" BBC, 10 March 2016: https://www.bbc.com/news/world -asia-35761136 (accessed 11 April 2019).

54. "No-Go Zones Keep Kin from Burying Deceased Fukushima Evacuees at Ancestral Gravesites," *Japan Times,* 24 August 2017: https://www.japantimes.co.jp /news/2017/08/24/national/no-go-zones-keep-kin-burying-deceased-fukushima -evacuees-ancestral-gravesites/#.W8bQDqeB2Cc (accessed 16 October 2018).

55. See Robert E. Buswell Jr. and Donald S. Lopez Jr., *The Princeton Dictionary of Buddhism* (Princeton, NJ: Princeton University Press, 2014): 936.

56. Yosuke Fukudome, "New Cemetery in Futaba Offers Chance to Honor Spirits for Obon," *Asahi Shimbun,* 13 August 2019: http://www.asahi.com/ajw/articles /AJ201908130050.html (accessed 30 September 2019).

57. "Disaster-Hit Fukushima Shrines Eye Consolidation as Key to Survival," *Kyodo News,* 15 May 2019: https://english.kyodonews.net/news/2019/05/7e46cb6a34e5 -disaster-hit-fukushima-shrines-eye-consolidation-as-key-to-survival.html (accessed 19 June 2020).

58. George Curran et al., "Central Australian Aboriginal Songs and Biocultural Knowledge: Evidence from Women's Ceremonies Relating to Edible Seeds," *Journal of Ethnobiology* 39:3 (2019): 354. See also Clint Bracknell, "Rebuilding as Research: Noongar Song Language and Ways of Knowing," *Journal of Australian Studies* 44:2 (2020): 210–223.

59. Diana James, "Tjukurpa Time," in Ann McGrath and Mary Ann Jebb, eds., *Long History, Deep Time: Deepening Histories of Place* (Canberra: Australian National University Press, 2015): 34.

60. Peter N. Grabosky, *Wayward Governance: Illegality and its Control in the Public Sector* (Australia: Australian Institute of Criminology, 1989): https://aic.gov .au/publications/lcj/wayward/chapter-16-toxic-legacy-british-nuclear-weapons -testing-australia (accessed 8 May 2020).

61. Kingsley Palmer, "Dealing with the Legacy of the Past: Aborigines and Atomic Testing in South Australia," *Aboriginal History* 14:2 (1990): 199–200.

62. Satoe Nakahara, "Perceptions of the Radiation Disaster from H-Bomb Testing: Subsistence Economy, Knowledge and Network Among the People of Rongelap in the Marshall Islands," *Sociology and Anthropology* 6:1 (2018): 183.

63. Jessica A. Schwartz, " 'Between Death and Life': Mobility, War, and Marshallese Women's Songs of Survival," *Women & Music* 16 (2012): 54–55.

64. Mary X. Mitchell, "Offshoring American Environmental Law: Land, Culture, and Marshall Islanders' Struggles for Self-determination During the 1970s," *Environmental History* 22 (2017): 222.

65. Interview with Roland Oldham in Papeete, French Polynesia, 16 February 2017.

66. United Nations Scientific Committee on the Effects of Atomic Radiation, *Sources and Effects of Ionizing Radiation,* vol. 2 (New York: United Nations, 2011): 64–65. See also Mark Peplow, "Counting the Dead," *Nature* 440 (20 April 2006): 982–983. The same script is being run at Fukushima. Two days before the tenth anniversary of the disaster, UNSCEAR released a report declaring there would likely be no disease impact or legacy of the multiple nuclear meltdowns and contaminations. See United Nations Scientific Committee on the Effects of Atomic Radiation, *UNSCEAR 2020 Report: Sources, Effects and Risks of Ionizing Radiation,* 2021: https://www.unscear.org/docs/publications/2020/UNSCEAR_2020_AnnexB_Advance-Copy.pdf (accessed 10 March 2021); "Fukushima Radiation Unlikely to Raise Cancer Rates, U.N. Experts Say," Reuters, 9 March 2021: https://www.reuters.com/article/japan-fukushima-radiation/fukushima-radiation-unlikely-to-raise-cancer-rates-un-experts-say-idUSL8N2L64MJ (accessed 9 March 2021).

67. Hannah Ritchie, "What Was the Death Toll from Chernobyl and Fukushima?" Our World in Data, 24 July 2017: https://ourworldindata.org/what-was-the-death-toll-from-chernobyl-and-fukushima (accessed 29 April 2020). Our World in Data is a collaborative statistical project of the University of Oxford and the Global Change Data Lab. Yablokov and fellow researchers have placed the death toll at almost one million people. See Alexey V. Yablokov, Vassily B. Nesterenko, and Alexey V. Nesterenko, *Chernobyl: Consequences of the Catastrophe for People and Environment* (Boston: Blackwell, 2009). Kate Brown has pointed out that thirty-five thousand women were awarded compensation for husbands having died of "Chernobyl-related health problems" in Ukraine alone. See Kate Brown, *Manual for Survival: A Chernobyl Guide to the Future* (New York: Allen Lane, 2019): 3, 310.

68. UNSCEAR, *Sources and Effects of Ionizing Radiation*: 57.

69. Burton Bennett, Michael Repacholi, and Zhanat Carr, eds., *Health Effects of the Chernobyl Accident and Special Health Care Programmes* (Geneva: World Health Organization, 2006): 93–94.

70. Magdalena E. Stawkowski, "Radiophobia Had to Be Reinvented," *Culture, Theory and Critique* 58:4 (2017): 360. See also Aliaksandr Novikau, "What Is 'Chernobyl Syndrome'? The Use of Radiophobia in Nuclear Communications," *Environmental Communication* 11 (2017): 800–809.

71. Klaus Becker, "Radiophobia: A Serious but Curable Mental Disorder," *Proceedings of 3rd International Symposium on Radiation Education,* 2005: https://inis.iaea.org/collection/NCLCollectionStore/_Public/36/113/36113743.pdf (accessed on the International Atomic Energy Agency website on 20 December 2020).

72. Yehoshua Socol, "Reconsidering Health Consequences of the Chernobyl Accident," *Dose Response* 13:1 (2014): 3. Radiation hormesis is the theory that exposure to low doses of radiation, above background levels, is beneficial to the health of living creatures.

73. David Ropeik, "Fear of Radiation Is More Dangerous Than Radiation Itself," *Aeon,* 5 July 2017: https://aeon.co/ideas/fear-of-radiation-is-more-dangerous-than-radiation-itself (accessed 29 April 2020).

74. Michael Edwards, "Stories from Experience: Using the Phenomenological Psychological Method to Understand the Needs of Victims of the Fukushima Nuclear Accident," *Asian Perspectives* 37:4 (2013): 617. Experts trained in psychology or psychiatry rarely use the term "radiophobia" or make a diagnosis of it. Its promoters tend to be in the hard sciences or public relations.

75. See Novikau, "What Is 'Chernobyl Syndrome'?": 800–809.

76. Shunichi Yamashita speaking in Fukushima City, 21 March 2011, quoted in Gayle Green, "Science with a Skew: The Nuclear Power Industry after Chernobyl and Fukushima," *Asia-Pacific Journal* 10:1 (2011): https://apjjf.org/2012/10/1/Gayle -Greene/3672/article.html (accessed 4 February 2021).

77. Aya Hirata Kimura, *Radiation Brain Moms and Citizen Scientists: The Gender Politics of Food Contamination after Fukushima* (Durham, NC: Duke University Press, 2016): 28.

78. Radionuclides distributed throughout the region have been widely documented by researchers. See, e.g., Marco Kaltofen and Arnie Gundersen, "Radioactively-Hot Particles Detected in Dusts and Soils from Northern Japan by Combination of Gamma Spectrometry, Autoradiography, and SEM/EDS Analysis and Implications in Radiation Risk Assessment," *Science of the Total Environment* 607–608 (2017): 1065–1072; Asumi Ochiai et al., "Uranium Dioxides and Debris Fragments Released to the Environment with Cesium-Rich Microparticles from the Fukushima Daiichi Nuclear Power Plant," *Environmental Science & Technology* 52:2 (2018): 2586–2594.

79. Ross H. Pastel, "Radiophobia: Long-Term Psychological Consequences of Chernobyl," *Military Medicine* 167:2 (2002): 134–135. Even before the Chernobyl disaster, similar diagnoses had been made of US service personnel who became anxious about their exposures to radiation during nuclear tests. See Henry M. Vyner, "The Psychological Effects of Ionizing Radiation," *Culture, Medicine, and Psychiatry* 7 (1983): 241–261. Pripyat is the city adjacent to the Chernobyl reactor complex where most of the employees lived prior to the disaster.

80. Johan M. Havenaar, G. M. Rumiantseva, and Jan van den Bout, "Mental Health Problems in the Chernobyl Area," *Russian Social Science Review* 35:4 (1994): 89.

81. Adolph Kharash, "A Voice from Dead Pripyat," website of Prof. Paul Brians: https://brians.wsu.edu/2016/12/05/a-voice-from-dead-pripyat/ (accessed 29 April 2020).

82. Geoff Brumfiel, "Fallout of Fear," *Nature* 493 (2013): 291, 293. The article quotes extensively from Dr. Shunichi Yamashita from Nagasaki University, widely criticized in the prefecture for telling citizens that smiling "deflects" radiation and advising against the distribution of potassium iodine pills to children to prevent their absorption of iodine 131 in the immediate aftermath of the disaster. See Majia Holmer Nadesan, "Nuclear Governmentality: Governing Nuclear Security and Radiation Risk in Post-Fukushima Japan," *Security Dialogue* 50:6 (2019): 512–530. Additionally, many of the people Yabe describes as having "fled from the nuclear disaster" were compelled to leave their homes by mandatory evacuation orders.

83. Shusuki Murai, "Fukushima No. 1 Cleanup Continues but Radioactive Water, and Rumors, also Prove Toxic," *Japan Times*, 9 March 2018: https://www.japantimes.co.jp/news/2018/03/09/national/fukushima-no-1-cleanup-continues-radioactive-water-rumors-also-prove-toxic/ (accessed 1 May 2020).

84. Justin McCurry, "Fukushima's Children at Centre of Debate over Rates of Thyroid Cancer," *The Guardian*, 9 March 2014: https://www.theguardian.com/world/2014/mar/09/fukushima-children-debate-thyroid-cancer-japan-disaster-nuclear-radiation (accessed 2 May 2020).

85. Reiko Hasegawa, "Disaster Evacuation from Japan's 2011 Tsunami Disaster and the Fukushima Nuclear Accident," Study No. 5, *IDDRI* (Institute for Sustainable Development and International Relations): *Sciences Po*, 13 May 2013: 31.

86. See "Fukushima Residents Demand Stricter Decontamination to Enable Safe Return," *Japan Times*, 22 January 2021: https://www.japantimes.co.jp/news/2021/01/22/national/fukushima-decontaminating-town/ (accessed 23 January 2021). The article was first published in Japanese in the *Fukushima Minpo*, the largest local newspaper in the region, and was reprinted in *Japan Times*.

87. Justin McCurry, "Fukushima Disaster: First Residents Return to Town Next to Nuclear Plant," *The Guardian*, 10 April 2019: https://www.theguardian.com/world/2019/apr/10/fukushima-disaster-first-residents-return-to-town-next-to-nuclear-plant (accessed 2 May 2020).

88. Stephens, "Physical and Cultural Reproduction": 278.

89. Michael Edwards, "Stories from Experience: Using the Phenomenological Psychological Method to Understand the Needs of Victims of the Fukushima Nuclear Accident," *Asian Perspectives* 37:4 (2013): 618.

4. Cloaking Contamination

1. Beverley Ann Deepe Kever, *News Zero: The New York Times and the Bomb* (Monroe, ME: Common Courage Press, 2004): 53–54. See also William L. Laurence, *Dawn Over Zero: The Story of the Atomic Bomb* (New York: Alfred A. Knopf, 1946).

2. William L. Laurence, "Drama of the Atomic Bomb Found Climax in July 16 Test," *New York Times*, 26 September 1945: 16.

3. Wilfred Burchett, "The Atomic Plague," *London Daily Mail*, 5 September 1945: 1, repr. in George Burchett and Nick Shimmin, eds., *Rebel Journalism: The Writings of Wilfred Burchett* (Cambridge: Cambridge University Press, 2009): 5.

4. Amy Goodman and David Goodman, *The Exception to the Rulers: Exposing Oily Politicians, War Profiteers, and the Media That Love Them* (New York: Hyperion Books, 2004): 296.

5. William L. Laurence, "U.S. Atom Bomb Site Belies Tokyo Tales: Tests on New Mexico Range Confirm That Blast, and Not Radiation, Took Toll," *New York Times*, 12 September 1945: 1. See also Janet Farrell Brodie, "Radiation Secrecy and Censorship after Hiroshima and Nagasaki," *Journal of Social History* 48:4 (2015): 842–864.

6. Amy Goodman and David Goodman, "The Hiroshima Cover-up," *Baltimore Sun*, 5 August 2005: https://www.baltimoresun.com/news/bs-xpm-2005-08-05-0508050019-story.html# (accessed 4 September 2019).

7. See Robert Jacobs, "Good Bomb / Bad Bomb: Talking About Atomic Tests in Nevada," *Interdisciplinary Humanities* 24:1 (Spring 2007): 65–82.

8. International Physicians for the Prevention of Nuclear War, *Radioactive Heaven and Earth: The Health and Environmental Effects of Nuclear Weapon Testing in, on, and Above the Earth* (New York: Apex Press, 1991): 57.

9. Richard L. Miller, *Under the Cloud: The Decades of Nuclear Testing* (The Woodlands, TX: Two-Sixty Press, 1991): 180.

10. Quoted in Philip L. Fradkin, *Fallout: An American Nuclear Tragedy* (Tucson: University of Arizona Press, 1989): 22. Butrico's statement was made during court testimony given in the case, *Irene Allen v. the United States of America*, filed in 1979 on behalf of downwind claimants.

11. Richard J. Hewlett and Jack M. Holl, *Atoms for Peace and War, 1953–1961: Eisenhower and the Atomic Energy Commission* (Berkeley: University of California Press, 1989): 450. Strauss took over the chairmanship from Dean in July 1953.

12. Jim Kichas, "Downwind in Utah," Utah State Archives and Record Service, 26 June 2015: https://archivesnews.utah.gov/2015/06/26/downwind-in-utah / (accessed 23 July 2018).

13. Sarah Alisabeth Fox, *Downwind: A People's History of the Nuclear West* (Lincoln: University of Nebraska Press, 2014): 55–56. A 1982 court decision found that AEC scientists had used deception in their denials of fallout playing a role in the sheep deaths. See R. Jeffrey Smith, "Scientists Implicated in Atom Test Deception," *Science* 218:4572 (1982): 545–547.

14. Bengt Danielsson, "Under a Cloud of Secrecy: The French Nuclear Tests in the Southeastern Pacific," *Ambio* 13:5/6 (1984): 338.

15. Angelique Chrisafis, "French Nuclear Tests 'Showered Vast Area of Polynesia with Radioactivity,'" *The Guardian*, 3 July 2013: https://www.theguardian.com /world/2013/jul/03/french-nuclear-tests-polynesia-declassified (accessed 10 September 2019); see also "Essais nucléaires: Ce que l'Etat a caché pendant 50 ans," *Le Parisien*, 3 July 2013: http://www.leparisien.fr/archives/essais-nucleaires-ce-que-l-etat-a-cache -pendant-50-ans-03-07-2013-2949221.php (accessed 10 September 2019).

16. Walter Zweifel, "For 30 Years We Lied About the Nuclear Tests, Says Tahiti's Fritch," Radio New Zealand, 21 November 2018: https://www.rnz.co.nz/international /pacific-news/376391/for-30-years-we-lied-about-the-nuclear-tests-says-tahiti-s-fritch (accessed 10 September 2019).

17. "Le document choc sur la bombe A en Algérie," *Le Parisien*, 14 February 2014: http://www.leparisien.fr/faits-divers/le-document-choc-sur-la-bombe-a-en -algerie-14-02-2014-3590523.php (accessed 7 May 2020); "Fallout from 1960s French Nuclear Test Reached Sicily, Chad," Radio France Internationale, 14 February 2014: http://www.rfi.fr/en/africa/20140214-fallout-1960s-french-nuclear-test- reached-sicily-chad (accessed 7 May 2020). See also Roxanne Panchasi, "'No Hiroshima in Africa': The Algerian War and the Question of French Nuclear Tests in the Sahara," *History of the Present: A Journal of Critical History* 9:1 (2019): 84–112; Karena Kalmbach, "Radiation and Borders: Chernobyl as a National and Transnational Site of Memory," *Global Environment* 11 (2013): 130–159.

18. Ralph Lapp, *The Voyage of the Lucky Dragon* (New York: Harper and Brothers, 1957): 33–34. The Metallurgical Laboratory was a primary Manhattan Project lab center located at the University of Chicago.

19. Matashichi Ōishi, *The Day the Sun Rose in the West: Bikini, the Lucky Dragon, and I*, trans. by Richard H. Minear (Honolulu: University of Hawai'i Press, 2011): 5.

20. John C. Bugher and Merril Eisenbud, "Contamination of the Fukuryu Maru and Associated Problems in Japan: Preliminary Report," 1954: 6, https://www.osti.gov/opennet/servlets/purl/16005492.pdf (accessed 2 December 2020).

21. It is rare to find the word "fallout" in a public document before the Bravo test. After the nuclear attacks on Hiroshima and Nagasaki, many publications used the phrase "residual radiation" to refer to fallout, but it was not often discussed. After the Bravo test, the word "fallout" can be found in thousands of newspaper and magazine articles every year. See "The Bravo Test and the Death and Life of the Global Ecosystem in the Early Anthropocene," *Asia-Pacific Journal* 13:29 (July 20, 2015): http://japanfocus.org/-Robert-Jacobs/4343/article.html (accessed 28 August 2019).

22. Eiichiro Ochiai, *Hiroshima to Fukushima: Biohazards of Radiation* (Heidelberg, Germany: Springer, 2014): v.

23. James C. Hagerty, in Robert H. Ferrell, ed., *The Diary of James C. Hagerty* (Bloomington: Indiana University Press, 1983): 40–42. See also Robert A. Divine, *Blowing on the Wind: The Nuclear Test Ban Debate, 1954–1960* (Oxford: Oxford University Press, 1978): 8, 18–19.

24. Lewis Strauss, quoted in "Copy of Aide-Memoire Prepared by the Embassy of Japan to the United States (April 12, 1954)," *Castle Series 1954*, DNA 60354 (1982): 469.

25. Dwight D. Eisenhower, "The President's News Conference of June 26, 1957," *Public Papers of the Presidents of the United States, Dwight D. Eisenhower, 1957* (Washington, DC: Government Printing Office, 1958): 499–500. See also Robert A. Divine, *Eisenhower and the Cold War* (Oxford: Oxford University Press, 1981): 124–125; Toshihiro Higuchi, "'Clean' Bombs: Nuclear Technology and Nuclear Strategy in the 1950s," *Journal of Strategic Studies* 29:1 (2006): 83–116.

26. Dwight D. Eisenhower, "The President's News Conference of July 3, 1957," *Public Papers of the Presidents of the United States, Dwight D. Eisenhower, 1957* (Washington, DC: Government Printing Office, 1958): 519–520. This reference to nuclear geoforming, formalized in the US as Project Plowshare, would ultimately prove disastrous, with two tests at the NTS and the Polygon leaving behind two of the most highly radioactive legacy sites at either test site, the Sedan Crater (NTS) and Lake Chagan (Polygon). See Bernd Grosche et al., "Studies of Health Effects from Nuclear Testing near the Semipalatinsk Nuclear Test Site, Kazakhstan," *Central Asian Journal of Global Health* 4:1 (2015): https://www.ncbi.nlm.nih.gov/pmc/articles/PMC5661192/ (accessed 4 June 2020); Scott Kaufman, *Project Plowshare: The Peaceful Use of Nuclear Explosives in Cold War America* (Ithaca, NY: Cornell University Press, 2013): 102–117.

27. US Department of Energy, *Advisory Committee on Human Radiation Experiments (ACHRE) Final Report* (Washington, DC: Government Printing Office,

1995), chap. 11: https://ehss.energy.gov/ohre/roadmap/achre/chap11_4.html (accessed 4 September 2019). In fact, the public remained unaware of this event and exposure until the US Nuclear Regulatory Commission made documents of Hanford's operation available to journalists and litigants in 1986. See Trisha T. Pritikin, *The Hanford Plaintiffs: Voices from the Fight for Atomic Justice* (Lawrence: University Press of Kansas, 2020): 204.

28. See Robert Jacobs, "Born Violent: The Origins of Nuclear Power," *Asian Journal of Peacebuilding* 7:1 (2019): 9–29. See also Lester Machta, "Finding the Site of the First Soviet Nuclear Test in 1949," *Bulletin of the American Meteorological Society* 73:11 (1992): 1797–1806.

29. D. E. Jenne and J. W. Healy, *Dissolving of Twenty Day Metal at Hanford,* HW-17381-DEL (Richland, WA: General Electric Company, 1950): 60–32.

30. Thomas E. Marceau et al., *History of the Plutonium Production Facilities as the Hanford Site Historic District, 1943–1990* (Richland, WA: US Department of Energy, 2002): 2–7.11. Italics added.

31. The first reporting on the disaster was by Karen Dorn Steele, "In 1949 Study Hanford Allowed Radioactive Iodine into Area Air," *Spokesman-Review,* 6 March 1986: 6. See also Karen Dorn Steele, "Hanford's Bitter Legacy," *Bulletin of the Atomic Scientists* 44:1 (1988): 17–23.

32. The Technical Steering Panel, *The Green Run* (Richland WA: Hanford Environmental Dose Reconstruction Project, 1992): 4.

33. Kate Brown, *Plutopia: Nuclear Families, Atomic Cities, and the Great Soviet and American Plutonium Disasters* (Oxford: Oxford University Press, 2013): 235.

34. Serhii Plokhy, *Chernobyl: History of a Tragedy* (London: Allen Lane, 2018): 174.

35. Lorna Arnold, *Windscale 1957: Anatomy of a Nuclear Accident,* 2nd ed. (London: Macmillan, 1995): 48–49, 54.

36. Steve Jones, "Health Effects of the Windscale Pile Fire," *Journal of Radiological Protection* 36 (2016): E23–E25.

37. William Penney et al., *Report on the Accident at Windscale No. 1 Pile on 10 October 1957,* republished in the *Journal of Radiological Protection* 37 (2017): 780.

38. Ray McGrath and Paul Nolan, "Revisiting the 1957 Windscale Nuclear Accident Using Atmospheric Reanalysis Data," Irish Centre for High-End Computing, 24 November 2017: https://www.ichec.ie/news/revisiting-1957-windscale-nuclear-accident-using-atmospheric-reanalysis-data (accessed 16 May 2020).

39. Arnfinn Tønnessen, Bertil Mårdberg, and Lars Weisæth, "Silent Disaster: A European Perspective on Threat Perception from Chernobyl Far Field Fallout," *Journal of Traumatic Stress* 15:6 (2002): 453.

40. Kristin Shrader-Frechette, "Rights to Know and the Fukushima, Chernobyl, and Three Mile Island Accidents," in Behnam Taebi and Sabine Roeser, eds., *The Ethics of Nuclear Energy: Risk, Justice, and Democracy in the Post-Fukushima Era* (Cambridge: Cambridge University Press, 2015): 53. See also Yuki Shimada, "Truth and Truth-Telling in Dietrich Bonhoeffer: Reconsidered After 3.11 and 'Fukushima,'" *Theology Today* 71:1 (2014): 121–131.

41. Olga Kuchinskaya, *The Politics of Invisibility: Public Knowledge about Radiation Health Effects after Chernobyl* (Cambridge, MA: MIT Press, 2014): 2.

42. Interview with Ludmila Dyatlova in Kyiv, Ukraine, 26 February 2019.

43. Edward Geist, "Political Fallout: The Failure of Emergency Management at Chernobyl," *Slavic Review* 74:1 (2015): 106, 124.

44. Kate Brown, *Manual for Survival: An Environmental History of the Chernobyl Disaster* (New York: W. W. Norton, 2019): 1–2.

45. David Lochbaum, Edwin Lyman, and Susan Q. Stranahan, *Fukushima: The Story of a Nuclear Disaster* (New York: The New Press, 2014): 109. See also Jinbong Choi and Seohyeon Lee, "Managing a Crisis: A Framing Analysis of Press Releases Dealing with the Fukushima Nuclear Power Station Crisis," *Public Relations Review* 43:5 (2017): 1016–1024.

46. "Cold Shutdown," US NRC, 2020: https://www.nrc.gov/reading-rm/basic -ref/glossary/cold-shutdown.html (accessed 13 June 2020).

47. Shrader-Frechette, "Rights to Know and the Fukushima, Chernobyl, and Three Mile Island Accidents": 56. See also Geoff Brumfiel, "Fukushima Reaches Cold Shut-down: But Milestone Is More Symbolic than Real," *Nature,* 16 December 2011: https://www.nature.com/news/fukushima-reaches-cold-shutdown-1.9674 (accessed 13 June 2020).

48. Ryoko Ando, "Trust—What Connects Science to Daily Life," *Health Physics* 115:5 (2018): 581, 588.

49. "Japanese Prosecutor Suspends Contempt Proceedings Against Journalist," *Reporters Without Borders,* 30 May 2014: https://rsf.org/en/news/japanese-prose-cutor-suspends-contempt-proceedings-against-journalist (accessed 3 September 2019).

50. Sezin Topçu, "Chernobyl Empowerment? Exporting 'Participatory Governance' to Contaminated Territories," in Sevin Boudia and Nathalie Jas, eds., *Toxicants, Health and Regulation since 1945* (London: Pickering & Chatto, 2013): 136, 144.

51. Maxime Polleri, "Being Clear-Eyed about Citizen Science in the Age of COVID-19," *Sapiens,* 15 July 2020: https://www.sapiens.org/culture/fukushima-citizen -science/ (accessed 22 July 2020).

52. Maxime Polleri, "Indeterminate Life: Dealing with Radioactive Contamination as a Voluntary Evacuee Mother," in Alys Einion and Jen Rinaldi, eds., *Bearing the Weight of the World: Exploring Maternal Embodiment* (Ontario: Demeter Press, 2018): 165.

53. Aya Hirata Kimura, *Radiation Brain Moms and Citizen Scientists* (Durham, NC: Duke University Press, 2016): 1. See also Sasha Davis and Jessica Hayes-Conroy, "Living with Contamination: Alternative Lessons and Perspectives from the Marshall Islands," in Mitsuo Yamakawa and Daisaku Yamamoto, eds., *Unravelling the Fukushima Disaster* (London: Routledge, 2017): 118–135.

54. "Is Radiation Safe?" World Nuclear Association: https://world-nuclear.org /nuclear-essentials/is-radiation-safe.aspx (accessed 13 June 2020).

55. See Jim Green, "The Banana Equivalent Dose of Catastrophic Nuclear Acci-dents," *Nuclear Monitor* 855:4694 (2017): https://www.wiseinternational.org

/nuclear-monitor/855/banana-equivalent-dose-catastrophic-nuclear-accidents (accessed 13 June 2020).

56. Polleri, "Being Clear-Eyed About Citizen Science in the Age of COVID-19."

57. Sven Ove Hansson, "Nuclear Energy and the Ethics of Radiation Protection," in Taebi and Roeser, *The Ethics of Nuclear Energy:* 31.

58. Maria Varenikova, "Chernobyl Wildfires Reignite, Stirring Up Radiation," *New York Times,* 11 April 2020: A19.

59. Andrew Roth, "Ukraine: Wildfires Draw Dangerously Close to Chernobyl Site," *The Guardian,* 13 April 2020: https://www.theguardian.com/environment/2020/apr/13/ukraine-wildfires-close-chernobyl-nuclear-site (accessed 10 May 2020).

60. François Murphy, "Fires Near Chernobyl Pose 'No Risk to Human Health,' IAEA Says," Reuters, 25 April 2020: https://www.reuters.com/article/us-ukraine-chernobyl-fire-iaea/fires-near-chernobyl-pose-no-risk-to-human-health-iaea-says-idUSKCN2262YU (accessed 12 May 2020).

61. Klaas Buijs et al., "The Dispersion of Radioactive Aerosols in Fires," *Journal of Nuclear Materials* 166 (1989): 199–207. See also Nikolaos Evangeliou and Sabine Eckhardt, "Uncovering Transport, Deposition and Impact of Radionuclides Released after the Early Spring 2020 Wildfires in the Chernobyl Exclusion Zone," *Scientific Reports* 10:10655 (2020): https://doi.org/10.1038/s41598-020-67620-3 (accessed 9 January 2021); E. Evangeliou et al., "Fire Evolution in the Radioactive Forests of Ukraine and Belarus: Future Risks for the Population and the Environment," *Ecological Monographs* 85:1 (2015): 49–72; Fernando P. Carvalho, Joao M. Oliviera, and Margarida Malta, "Forest Fires and Resuspension of Radionuclides into the Atmosphere," *American Journal of Environmental Sciences* 8:1 (2012): 1–4.

62. Richard Knox and Andrew Price, "Early Radiation Data from Near Plant Ease Health Fears," NPR, 18 March 2011: https://www.npr.org/sections/health-shots/2011/03/20/134658088/radiation-data-near-nuclear-plant-offers-little-cause-for-concern (accessed 18 September 2019).

63. Denise Grady, "Radiation Is Everywhere, but How to Rate Harm?" *New York Times,* 4 April 2011: https://www.nytimes.com/2011/04/05/health/05radiation.html#story-continues-2 (accessed 18 September 2019).

64. Beate Ritz et al., "The Effects of Internal Radiation Exposure on Cancer Mortality in Nuclear Workers at Rocketdyne/Atomics International," *Environmental Health Perspectives* 108:8 (2000): 743, 749.

65. Magdalena E. Stawkowski, "Radiophobia Had to Be Reinvented," *Culture, Theory and Critique* 58:4 (2017): 361. Brown points out that "consultants from UN agencies dismissed the findings of scientists in Ukraine and Belarus" for similar reasons or because they were seen as remedial. See Brown, *Manual for Survival:* 308.

66. Susan Thaul et al., *Mortality of Military Personnel Present at Atmospheric Tests of Nuclear Weapons* (Washington, DC: National Academies Press, 2000): 1, 3, 8. Italics added.

67. J. Goetz et al., *Analysis of Radiation Exposure for Troop Observers, Exercise Desert Rock V, Operation Upshot-Knothole,* DNA5247F (McLean, VA: Science

Applications, 1981): 80–81. Many troops felt betrayed to have been irradiated by the government they served and that was charged with protecting them. See Betty Garcia, "Social-Psychological Dilemmas and Coping of Atomic Veterans," *American Journal of Orthopsychiatry* 64:4 (1994): 651–655.

68. Gayle Green, "Science with a Skew: The Nuclear Power Industry after Chernobyl and Fukushima," *Asia-Pacific Journal* 10:1 (2 January 2012): https://apjjf .org/-Gayle-Greene/3672/article.pdf (accessed 18 September 2019).

69. Kate Brown, "The Last Sink: The Human Body as the Ultimate Radioactive Storage Site," *RCC Perspectives* 1 (2016): 44.

70. Peter Stegnar and Tony Wrixon, "Semipalatinsk Revisited: Radiological Evaluation of the Former Test Site," *IAEA Bulletin* 40:4 (1988): 14.

71. Tami Freeman, "Job-Exposure Matrix Sheds Light on Plutonium Workers' Radiation Exposure," *Physics World*, 22 May 2019: https://physicsworld.com/a /job-exposure-matrix-sheds-light-on-plutonium-workers-radiation-exposure/ (accessed 14 May 2020).

72. Krupar develops this idea extensively in Shiloh R. Krupar, *Hot Spotter's Report: Military Fables of Toxic Waste* (Minneapolis: University of Minnesota Press, 2013).

73. Keith B. Noble, "U.S., for Decades, Let Uranium Leak at Weapon Plant," *New York Times*, 15 October 1988: 1.

74. The DOE and its predecessor organizations, on the other hand, did bear responsibility for the ongoing contamination. "Transition to Cleanup: A New Beginning," Fernald Closure Project website, US DOE: https://www.lm.doe.gov/land/sites/oh /fernald_orig/50th/clean.htm (accessed 18 May 2020).

75. *Fernald Preserve, Ohio, Site Fact Sheet* (Washington, DC: US DOE Legacy Management, 2018): 1.

76. Brown, *Manual for Survival:* 10.

77. Jenny Wohlfarth, "What Lies Beneath the Fernald Preserve," *Cincinnati Magazine*, 7 June 2019: https://www.cincinnatimagazine.com/citywiseblog/what -lies-beneath-the-fernald-preserve/ (accessed 18 May 2020). Italics added.

78. Krupar, *Hot Spotter's Report*: 6.

79. "2020 National Federal Facility Excellence in Site Reuse Awards," US EPA, 2020: https://www.epa.gov/fedfac/2020-national-federal-facility-excellence-site- reuse-awards (accessed 16 June 2020).

80. *Weldon Spring Site Disposal Facility (Cell) Fact Sheet*, US DOE Legacy Management, 12 December 2011: https://firstsecretcity.files.wordpress.com/2015/09 /disposal_cell.pdf (accessed 25 May 2020).

81. *Historical Land Use at the Weldon Spring Site*, US DOE Legacy Management brochure, August 2015.

82. "Koeberg Nature Reserve," West Coast Way: https://www.westcoastway.co.za /koeberg-nature-reserve/ accessed 24 May 2020).

83. Lei Yuan et al., "Wild Camels in the Lop Nur Nature Reserve," *Journal of Camel Practice and Research* 21:2 (2014): 137–144.

84. Thomas Nilsen, "From Nuclear Tests to Polar Bears Reserve," *Barents Observer,* 16 June 2009: https://barentsobserver.com/en/node/18542 (accessed 18 May 2020). In 2019, polar bears "invaded" the closed military town located on the peninsula; see Isaac Stanley-Becker, "A 'Mass Invasion' of Polar Bears Is Terrorizing an Island Town. Climate Change Is to Blame," *Washington Post,* 11 February 2019: https://www.washingtonpost.com/nation/2019/02/11/mass-invasion-polar-bears-is-terrorizing-an-island-town-climate-change-is-blame/ (accessed 18 May 2020); Eugenio Luciano, "The Bears' Famous Invasion of Novaya Zemlya," *Arcadia* 41 (2019): http://www.environmentandsociety.org/node/8924 (accessed 21 June 2020).

85. Inna V. Molchanova et al., "Radioactive Inventories Within the East-Ural Radioactive State Reserve on the Southern Urals," *Radioprotection* 44:5 (2009): 747–757.

86. Douglas D. Kautz et al., "The Pit Production Story," *Los Alamos Science* 28 (2003): 58; David E. Hunter et al., *Independent Assessment of the Two-Site Pit Production Decision: Executive Summary* (Alexandria, VA: Institute for Defense Analysis, 2019).

87. Len Ackland, "Rocky Flats: Expect a Fire, but Produce," *Montana: The Magazine of Western History* 50:2 (2000): 36.

88. Len Ackland, "Rocky Flats: Closing in on Closure," *Bulletin of the Atomic Scientists* 57:6 (2001): 55.

89. The report can be read at *Colorado Federal District Court Report of the Federal District Special Grand Jury 89–2 January 24, 1992:* https://constitution.org/jury/gj/rocky_flats/rocky-flats-grand-jury-report.htm (accessed 13 June 2020). See also Wes McKinley and Caron Balkany, *The Ambushed Grand Jury: How the Justice Department Covered Up Government Nuclear Crimes and How We Caught Them Red Handed* (New York: The Apex Press, 2004).

90. *Rocky Flats Fact Sheet* (Westminster, CO: US DOE, 2018): 2.

91. "Rocky Flats: Facts at a Glance," Colorado Department of Public Health & Environment, 2019: https://www.colorado.gov/pacific/cdphe/rocky-flats-facts-glance (accessed 20 May 2020).

92. John Aguilar, "Nearly 300,000 Colorado Public School Students Now Barred from Making Field Trips to Rocky Flats," *Denver Post,* 29 April 2018: https://www.denverpost.com/2018/04/29/rocky-flats-school-field-trips-ban/ (accessed 20 May 2020).

93. John Aguilar, "Potential Plutonium Hot Spot Found on Eastern Edge of Rocky Flats," *Denver Post,* 16 August 2019: https://www.denverpost.com/2019/08/16/rocky-flats-plutonium-hot-spot-jefferson-parkway/ (accessed 20 May 2020).

94. Michael Ketterer and Scott Szechenyi, "Interim Report: PuO_2 Particles in the Indiana St. Corridor," 10 September 2019: https://4500b39c-9ca0-4e86-aee9-0ef1f6556788.filesusr.com/ugd/76432a_63a1809d88e6458caa043fbd65e53aeb.pdf (accessed 20 May 2020). See also M. P. Johansen et al., "Plutonium in Wildlife and Soils at the Maralinga Legacy Site: Persistence over Decadal Time Scales," *Journal of Environmental Radioactivity* 131 (2014): 72–80.

95. Shannon Cram, "Wild and Scenic Wasteland: Conservation Politics in the Nuclear Wilderness," *Environmental Humanities* 7 (2015): 89, 103.

96. Nikolaus Evangeliou et al., "Resuspension and Atmospheric Transport of Radionuclides Due to Wildfires Near the Chernobyl Nuclear Power Plant in 2015: An Impact Assessment," *Nature: Scientific Reports* 6:26062 (2016): 2.

97. Georgia Paliouris et al., "Fire as an Agent in Redistributing Fallout ^{137}Cs in the Canadian Boreal Forest," *Science of the Total Environment* 160–161 (1995): 153–166.

98. Nikolaos Evangeliou et al., "Wildfires in Chernobyl-Contaminated Forests and Risks to the Population and the Environment: A New Nuclear Disaster About to Happen?" *Environment International* 73 (2014): 346.

99. Department of Energy, *Type B Accident Investigation, Response to the 24 Command Wildland Fire on the Hanford Site, June 27–July 1, 2000,* 23 October 2000, DOE /RL-2000-63: 3–24.

100. Eliot Marshall, "Hanford's Radioactive Tumbleweed," *Science* 236 (1987): 1616.

101. "Lightning Strikes Ignite 9,000-Acre Fire on Hanford Nuclear Reservation Mountain," *Tri-City Herald,* 1 June 2020: https://www.tri-cityherald.com/news /local/hanford/article243170681.html (accessed 14 February 2021).

102. See John E. McCoy II, *The Department of Energy's Wildland Fire Prevention Efforts at the Los Alamos National Laboratory* (Washington, DC: Department of Energy, 2021); John M. Volkerding, "Comparison of the Radiological Dose from the Cerro Grande Fire to a Natural Wildfire," *Environment International* 29 (2003): 987–993.

103. Magdalena E. Stawkowski, "'I Am a Radioactive Mutant': Emergent Biological Subjectivities at Kazakhstan's Semipalatinsk Nuclear Test Site," *American Ethnologist* 43:1 (2016): 147.

104. Fred Pearce, "Rocky Flats: A Wildlife Refuge Confronts Its Radioactive Past," *Yale Environment 360,* 16 August 2016: https://e360.yale.edu/features/rocky_flats _wildlife_refuge_confronts_radioactive_past (accessed 21 May 2020).

105. Robert Jacobs, *The Dragon's Tail: Americans Face the Atomic Age* (Amherst: University of Massachusetts Press, 2010): 23–28.

106. This was most profoundly expressed by the Chernobyl Forum group in 2006; see Chernobyl Forum Expert Group "Environment," *Environmental Consequences of the Chernobyl Accident and Their Remediation: Twenty Years of Experience* (Vienna: IAEA, 2006): 137.

107. See Nicholas A. Beresford and David Copplestone, "Effects of Ionizing Radiation on Wildlife: What Knowledge Have We Gained Between the Chernobyl and Fukushima Accidents?" *Integrated Environmental Assessment and Management* 7:3 (2011): 371–373. The article was first submitted to the journal on the day of the 3/11 earthquake and tsunami; Tatiyana G. Derbyabina et al., "Long-Term Census Data Reveal Abundant Wildlife Populations at Chernobyl," *Current Biology* 25 (2015): R824–R826; Mike Wood and Nicholas A. Beresford, "The Wildlife of Chernobyl: 30 Years Without Man," *The Biologist* 63:2 (2016): 16–19. Such articles also followed at Fukushima. See

Shaena Montinari, "In the Wake of the Fukushima Nuclear Disaster, Some Animals are Thriving," *The Hill,* 23 January 2020: https://thehill.com/changing-america /sustainability/environment/479544-in-the-wake-of-the-fukushima-nuclear-disaster (accessed 22 February 2020).

108. Laura Helmuth, "Chernobyl's Wildlife Survivors: The Radioactive Fallout Zone has Turned into a Refuge," *Slate,* 21 January 2013: https://slate.com/technology /2013/01/chernobyl-wildlife-the-radioactive-fallout-zone-is-a-wildlife-refuge-photos .html (accessed 21 May 2020).

109. Barry Starr, "The Benefits of Radioactive Fallout," KQED, 9 January 2012: https://www.kqed.org/quest/29086/the-benefits-of-radioactive-fallout (accessed 8 June 2021); Tania Rabesandratana, "Humans Are Worse Than Radiation for Chernobyl Animals: Study Finds," *Science,* 5 October 2015: https://www.sciencemag.org /news/2015/10/humans-are-worse-radiation-chernobyl-animals-study-finds (accessed 23 May 2020).

110. Timothy Mousseau, "Ecology in Fukushima: What Does a Decade Tell Us?" (presented to the International Physicians for the Prevention of Nuclear War-Symposium, 10 Years Living with Fukushima, 27 February 2021): https://www.youtube.com /watch?v=3nDKJdkq390 (accessed 27 March 2021). Mousseau found that only ten of the five hundred Fukushima top articles on the Web of Science website related to actual biological effects.

111. Karine Beaugelin-Seiller et al., "Dose Reconstruction Supports the Interpretation of Decreased Abundance of Mammals in the Chernobyl Exclusion Zone," *Scientific Reports* 10:14083 (2020): 1.

112. Mayumi Itoh, *Animals and the Fukushima Nuclear Disaster* (London: Palgrave Macmillan, 2018): 178–179. The film *Radioactive Wolves* aired on the PBS show *Nature* on 18 October 2011.

113. Michael E. Byrne et al., "Evidence of Long-Distance Dispersal of a Gray Wolf from the Chernobyl Exclusion Zone," *European Journal of Wildlife Research* 64:4 (2018): https://link.springer.com/article/10.1007/s10344-018-1201-2 (accessed 24 July 2020).

114. Timothy A. Mousseau and Anders P. Møller, "Genetic and Ecological Studies of Animals in Chernobyl and Fukushima," *Journal of Heredity* 105:5 (2014): 704. See also Anders P. Møller and Timothy A. Mousseau, "Are Organisms Adapting to Ionizing Radiation at Chernobyl?" *Trends in Ecology & Evolution* 31:4 (2016): 281–289; Beaugelin-Seiller, "Dose Reconstruction"; Jacqueline Garnier-Laplace et al., "Radiological Dose Reconstruction for Birds Reconciles Outcomes of Fukushima with Knowledge of Dose-Effect Relationships," *Scientific Reports* 5:16594 (2015): https:// doi.org/10.1038/srep16594 (accessed 19 November 2020).

115. Narration from the film *Radioactive Wolves.*

116. Volodymyr Tykhyy, "Solving the Social Problems Caused by the Chernobyl Catastrophe: 20 Years Is Not Enough," in Testsuji Imanaka, ed., *Multi-side Approach to the Realities of the Chernobyl NPP Accident: Summing-up of the Consequences of the Accident Twenty Years After (II)* (Tokyo: Toyota Foundation, 2008): 199.

117. Hiroyuki Kaneko, "Radioactive Contamination of Forest Commons: Impairment of Minor Subsistence Practices as an Overlooked Obstacle to Recovery in the Evacuated Areas," in Mitsuo Yamakawa and Daisaku Yamamoto, eds., *Unravelling the Fukushima Disaster* (London: Routledge, 2017): 139, 146.

118. Masaharu Tsubokura et al., "Reduction of High Levels of Internal Radio-Contamination by Dietary Intervention in Residents of Areas Affected by the Fukushima Daiichi Nuclear Plant Disaster: A Case Series," *PLOS ONE* 9:6 (2014): https://journals.plos.org/plosone/article/file?id=10.1371/journal.pone.0100302&type=printable (accessed 1 May 2020).

119. Sharon Stephens, "Physical and Cultural Reproduction in a Post-Chernobyl Norwegian Sami Community," in Faye D. Ginsburg and Rayna Rap, eds., *Conceiving the New World Order: The Global Politics of Reproduction* (Berkeley: University of California Press, 1995): 272, 278.

120. Gerd Persson, quoted in Stephens, "Physical and Cultural Reproduction": 277.

121. Stephens, "Physical and Cultural Reproduction": 272.

122. Bruno Latour, *On the Modern Cult of the Factish Gods* (Durham, NC: Duke University Press, 2010): 110–113.

123. Stephens, "Physical and Cultural Reproduction": 271.

124. Stephens, "Physical and Cultural Reproduction": 276.

125. Douglas Almond, Lena Edlund, and Mårten Palme, "Chernobyl's Subclinical Legacy: Prenatal Exposure to Radioactive Fallout and School Outcomes in Sweden" (Working Paper 13347, Cambridge, MA: National Bureau of Economic Research, 2007): 2.

126. See Masanori Otake, Hiroshi Yoshimaru, and Willam Shull, "Prenatal Exposure to Atomic Radiation and Brain Damage," *Congenital Anomalies* 29 (1989): 309–320; William Shull and Masanori Otake, "Learning Disabilities in Individuals Exposed Prenatally to Ionizing Radiation: The Hiroshima and Nagasaki Experiences," *Advances in Space Research* 6:11 (1986): 223–232; William Shull and Masanori Otake, "Cognitive Functioning and Prenatal Exposure to Ionising Radiation," *Teratology* 59 (1999): 222–226; and Masanori Otake, "Review: Radiation-Related Brain Damage and Growth Retardation among the Prenatally Exposed Atomic Bomb Survivors," *International Journal of Radiation Biology* 74 (1998): 159–171.

127. Interview with Roland Oldham in Papeete, French Polynesia, 16 February 2017.

128. Holly M. Barker and Barbara Rose Johnston, "Seeking Compensation for Radiation Survivors in the Marshall Islands: The Contributions of Anthropology," *Cultural Survival* 24:1 (March 2000): 48–50.

129. Mary X. Mitchell, "Offshoring American Environmental Law: Land, Culture, and Marshall Islanders' Struggles for Self-Determination During the 1970s," *Environmental History* 22 (2017): 215.

130. Jane Diblin, *Day of Two Suns: U.S. Nuclear Testing and the Pacific Islanders* (New York: New Amsterdam Books, 1988): 23. See also Martha Smith-Norris,

"American Cold War Policies and the Enewetaks," *Journal of the Canadian Historical Association* 22:2 (2011): 195–236.

131. Quoted in *Marshall Islands, a Chronology 1944–1981,* 2nd ed. (Honolulu: Micronesia Support Committee, 1981): 11.

132. Sasha Davis, *The Empires' Edge: Militarization, Resistance, and Transcending Hegemony in the Pacific* (Athens: University of Georgia Press, 2015): 40–41. See also Lauren Hirshberg, "Nuclear Families: (Re)producing 1950s Suburban America in the Marshall Islands," *OAH Magazine of History* 26:4 (2012): 39–43.

133. Seiji Yamada, "Cancer, Reproductive Abnormalities, and Diabetes in Micronesia: The Effect of Nuclear Testing," *Pacific Health Dialogue* 11:2 (2004): 218–219.

134. Magdalena E. Stawkowski, "Everyday Radioactive Goods? Economic Development at Semipalatinsk, Kazakhstan," *Journal of Asian Studies* 76:2 (2017): 432–433.

135. Stawkowski, " 'I am a Radioactive Mutant' ": 145, 155. See also Joseph Masco, "Mutant Ecologies: Radioactive Life in Post–Cold War New Mexico," *Cultural Anthropology* 19:4 (2004): 517–550.

136. Stawkowski, " 'I am a Radioactive Mutant' ": 149–150.

137. Lamine Chikhi, "French Nuclear Tests in Algeria Leave Toxic Legacy," Reuters, 4 March 2010: https://uk.reuters.com/article/algeria-france-nuclear/french-nuclear-tests-in-algeria-leave-toxic-legacy-idUKCHI23393320100304 (accessed 5 May 2020). See also Pier R. Danesi, "Residual Radionuclide Concentrations and Estimated Radiation Doses at the Former French Nuclear Weapons Test Sites in Algeria," *Applied Radiation and Isotopes* 66 (2008): 1671–1674.

138. Quoted in Johnny Magdaleno, "Algerians Suffering from French Atomic Legacy, 55 Years After Nuke Tests," Al Jazeera America, 1 March 2015: http://america.aljazeera.com/articles/2015/3/1/algerians-suffering-from-french-atomic-legacy-55-years-after-nuclear-tests.html (accessed 5 May 2020). See also IAEA, *Radiological Conditions at the Former French Nuclear Test Sites in Algeria: Preliminary Assessment and Recommendations* (Vienna: International Atomic Energy Association, 2005): 32.

139. Sarah Elizabeth Fox, *Downwind: A People's History of the Nuclear West* (Lincoln: University of Nebraska Press, 2014): 150.

140. Harvey Wasserman and Norman Solomon, *Killing Our Own: The Disaster of America's Experience with Atomic Radiation* (New York: Dell, 1982): 56.

141. Fox, *Downwind*: 149.

142. Lois Gibbs, *Love Canal: My Story* (Albany: State University of New York Press, 1982): 66; see also Richard S. Newman, *Love Canal: A Toxic History from Colonial Times to the Present* (Oxford: Oxford University Press, 2016): 151–160. Mapping of disease occurrence has a long history among epidemiologists as well, such as Dr. John Snow's maps of the locations of victims of a cholera epidemic in London in 1854. See Tom Koch, *Cartographies of Disease: Maps, Mapping, and Medicine* (Redlands, CA: Esri Press, 2005); Sandra Hempel, *The Atlas of Disease: Mapping Deadly Epidemics and Contagion from the Plague to the Zika Virus* (London: White Lion, 2018).

143. Karen Dorn Steele, "Introduction," in Trisha T. Pritikin, *The Hanford Plaintiffs: Voices from the Fight for Atomic Justice* (Lawrence: University Press of Kansas, 2020): 4.

144. Marco Kaltofen, Robert Alvarez, and Lucas W. Hixson, "Forensic Micro-analysis of Manhattan Project Legacy Radioactive Wastes in St. Louis," *Applied Radiation and Isotopes* 136 (2018): 143–149. See also Lisa Martino-Taylor, *Behind the Fog: How the U.S. Cold War Radiological Weapons Program Exposed Innocent Americans* (New York: Routledge, 2018); Agencies for Toxic Substances and Disease Registry, *Evaluation of Community Exposures Related to Coldwater Creek* (Atlanta: US Department of Health and Human Services, 2019).

145. Marko P. J. Kaltofen et al., "Tracking Legacy Radionuclides in St. Louis, Missouri, Via Unsupported 210Pb," *Journal of Environmental Radioactivity* 153 (2016): 105.

146. Bridjes O'Neil, "Documents Sought on Radioactive Landfill Fire," *St. Louis American,* 23 October 2013: http://www.stlamerican.com/news/local_news/article_8b8dbf58-3c49-11e3-bd9b-001a4bcf887a.html (accessed 23 July 2020).

147. Robert Alvarez, *The Westlake Landfill: A Radioactive Legacy of the Nuclear Arms Race* (Washington, DC: Institute for Policy Studies, 2013): 4. Alvarez is referring to the amount of energy being radiated from the particles in his comparison of the thorium and uranium elements.

148. Quote taken from *Atomic Homefront,* directed by Rebecca Cammisa (2017; New York: HBO Documentary Films).

149. Gibson, the preeminent cyberpunk author, can't recall the first time he constructed this idea. For an inquiry into its iterations, see "The Future Has Arrived—It's Just Not Evenly Distributed Yet," Quote Investigator, 2012: https://quoteinvestigator.com/2012/01/24/future-has-arrived/ (accessed 13 June 2020).

150. *Declaration of the Indigenous World Uranium Summit Window Rock, Navajo Nation, USA,* The Indigenous World Uranium Summit, 2 December 2 006: http://swuraniumimpacts.org/wp-content/uploads/2010/06/IWUS-Declaration-Final-2.pdf (accessed 25 October 2018).

5. Selecting the Irradiated

1. Jonathan M. Weisgall, *Operation Crossroads: The Atomic Tests at Bikini Atoll* (Annapolis, MD: Naval Institute Press, 1994): 32.

2. Jane Dibblin, *Day of Two Suns: U.S. Nuclear Testing and the Pacific Islanders* (New York: New Amsterdam Books, 1988): 20.

3. Office of the Historian, Joint Task Force One, *Operation Crossroads: The Official Pictorial Record* (New York: Wm. H. Wise, 1946): 12.

4. Weisgall, *Operation Crossroads:* 31.

5. *Marshall Islands: A Chronology, 1944–1981,* vol. 2 (Honolulu: Micronesia Support Committee, 1981): 5.

6. Dorothy E. Richard, *United States Naval Administration of the Trust Territories of the Pacific Islands,* vol. 3 (Washington, DC: Government Printing Office, 1957): 510.

7. Jack Neidenthal, *For the Good of Mankind: A History of the People of Bikini and Their Islands* (Majuro, Marshall Islands: Bravo Publishers, 2001): 2.

8. Robert C. Kiste, *The Bikinians: A Study in Forced Migration* (Menlo Park, CA: Cummings, 1974): 28.

9. Steve Brown, "Poetics and Politics: Bikini Atoll and World Heritage Listing," in Sally Brockwell, Sue O'Connor, and Denis Byrne, eds., *Transcending the Culture—Nature Divide in Cultural Heritage: Views from the Asia-Pacific Region* (Canberra: Australian National University E Press, 2013): 38–39.

10. Quoted in Weisgall, *Operation Crossroads:* 114.

11. *Bikini—The Atom Island: A Carey Wilson Special Miniature* (1946; Beverly Hill, CA: MGM Studios). The short film was released on 15 June 1946.

12. Jeffrey Sasha Davis, "Representing Place: 'Deserted Isles' and the Reproduction of Bikini Atoll," *Annals of the Association of American Geographers* 95:3 (2005): 607–625.

13. Gabrielle Hecht, *Being Nuclear: Africans and the Global Uranium Trade* (Cambridge: MIT Press, 2012): 4–5. See also Gregory Hooks and Chad L. Smith, "The Treadmill of Destruction: National Sacrifice Areas and Native Americans," *American Sociological Review* 69:4 (2004): 558–575.

14. Achille Mbembe, "Necropolitics," *Public Culture* 15:1 (2003): 11–12.

15. Barbara Rose Johnston, "Half-Lives, Half-Truths, and Other Radioactive Legacies of the Cold War," in Barbara Rose Johnston, ed., *Half-Lives and Half-Truths: Confronting the Radioactive Legacies of the Cold War* (Santa Fe, NM: School for Advanced Research Press, 2007): 6.

16. Cited in Barbara Rose Johnston, " 'More Like Us Than Mice': Radiation Experiments with Indigenous Peoples," in Johnston, *Half-Lives and Half-Truths:* 25.

17. *Bikini: Radiobiological Laboratory* (AEC, Lookout Mountain Laboratory, 1949).

18. See Susan C. Schultz and Vincent Schultz, "Bikini and Enewetak Marshallese: Their Atolls and Nuclear Weapons Testing," *Critical Reviews in Environmental Science and Technology* 24:1 (1994): 33–118.

19. *Enewetak Radiological Support Project. Final Report* (Las Vegas, NV: Department of Energy, 1982): 5.

20. "Marshall Islands Nuclear Claims Tribunal: In the Matter of the People of Enewetak," *International Legal Materials* 39:5 (2000): 1214.

21. M. Lee Davisson, Terry F. Hamilton, and Andrew F. B. Tompson, "Radioactive Waste Buried Beneath Runit Dome on Enewetak Atoll, Marshall Islands," *International Journal of Environment and Pollution* 49:3–4 (2012): 161.

22. Ken Buesseler et al., "Lingering Radioactivity at the Bikini and Enewetak Atolls," *Science of the Total Environment* 621 (2018): 1185–1198.

23. Ken Buesseler, "The Isotopic Signature of Fallout Plutonium in the North Pacific," *Journal of Environmental Radioactivity* 36:1 (1997): 70.

24. Thomas B. Cochran, Robert S. Norris, and Oleg A. Bukharin, *Making the Russian Bomb: From Stalin to Yeltsin* (Boulder, CO: Westview Press, 1995).

25. Jerome Taylor, "The World's Worst Radiation Hotspot," *The Independent*, 10 September 2009: http://www.independent.co.uk/news/world/europe/the-worlds -worst-radiation-hotspot-1784502.html (accessed 14 March 2019).

26. David Holloway, *Stalin and the Bomb: The Soviet Union and Atomic Energy, 1939–1956* (New Haven, CT: Yale University Press, 1994): 213; Cochran, Norris, and Bukharin, *Making the Russian Bomb*: 11–12.

27. Bernd Grosche, "Semipalatinsk Test Site: Introduction," *Radiation and Environmental Biophysics* 41 (2002): 53.

28. Cynthia Werner and Kathleen Purvis-Roberts, "Unraveling the Secrets of the Past: Contested Versions of Nuclear Testing in the Soviet Republic of Kazakhstan," in Barbara Rose Johnston, ed., *Half-Lives and Half-Truths*: 277–98.

29. UNESCO, *International Memory of the World Register: Documents on Closure of Semipalatinsk Test Site (Kazakhstan)*: http://www.unesco.org/new/fileadmin /MULTIMEDIA/HQ/CI/CI/pdf/mow/nomination_forms/kazakhstan _semipalatinskonline.pdf (accessed 13 March 2019).

30. Togzhan Kassenova, "The Lasting Toll of Semipalatinsk's Nuclear Testing," *The Bulletin of the Atomic Scientists*, 28 September 2009: http://www.thebulletin .org/web-edition/ features/the-lasting-toll-of-semipalatinsks-nuclear-testing (accessed 14 March 2019).

31. Susanne Bauer et al., "Radiation Due to Local Fallout from Soviet Atmospheric Nuclear Weapon Testing in Kazakhstan: Solid Cancer Mortality in the Semipalatinsk Historical Cohort, 1960–1999," *Radiation Research* 164:4 (2004): 409–419.

32. Vitaly I. Khalturin et al., "A Review of Nuclear Testing by the Soviet Union at Novaya Zemlya, 1955–1990," *Science and Global Security* 13 (2005): 10.

33. Leonid Serebryanny, "The Colonization and People of Novaya Zemlya Then and Now," *Nationalities Papers* 25:2 (1997): 305–306.

34. The Division of Nuclear Safety and Security, *Nuclear Explosions in the USSR: The North Test Site* (Vienna: IAEA, 2004): 4–5.

35. Khalturin et al., "A Review of Nuclear Testing by the Soviet Union": 18–19.

36. Roger Took, *Running With Reindeer: Encounters In Russian Lapland* (London: Westview Press, 2004): 271.

37. Salve Dahle et al., "A Return to the Nuclear Waste Dumping Sites in the Bays of Novaya Zemlya," *Radioprotection* 44:5 (2009): 281. See also Thomas Nilsen, Igor Kudrik, and Alexandr Nikitin, "The Russian Northern Fleet: Sources of Radioactive Contamination," *Belona Report 2*, 28 August 1996: http://spb.org.ru/bellona /ehome/russia/nfl/index.htm (accessed 27 August 2020). See also Justin P. Gwyn et al., "Main Results of the 2012 Joint Norwegian-Russian Expedition to the Dumping Sites of the Nuclear Submarine K-27 and Solid Radioactive Waste in Stepovogo Fjord, Novaya Zemlya," *Journal of Environmental Radioactivity* 151 (2016): 417–426.

38. Thomas Nilsen, "Melting Glaciers at Novaya Zemlya Contain Radiation from Nuclear Bomb Tests," *Barents Observer*, 9 October 2018: https://thebarentsobserver .com/en/ecology/2018/10/melting-glaciers-novaya-zemlya-contain-radiation -nuclear-bomb-tests (accessed 5 July 2020).

39. The title of the US Department of Energy history of the years of atmospheric testing at the NTS explicitly calls it a "battlefield of the Cold War." See Thomas R. Fehner

and F. G. Gosling, *Battlefields of the Cold War, Nevada Test Site: Atmospheric Nuclear Weapons Testing, 1951–1963*, vol. 1 (Washington, DC: US DOE, 2006).

40. International Physicians for the Prevention of Nuclear War, *Radioactive Heaven and Earth: The Health and Environmental Effects of Nuclear Weapons Testing in, on, and Above the Earth* (New York: Apex Press, 1991): 56.

41. Constandina Titus, *Bombs in the Backyard: Atomic Testing and American Politics* (Reno: University of Nevada Press, 1986).

42. Eltona Henderson, testimony before the Senate Judiciary Committee, hearing on "Examining the Eligibility Requirements for the Radiation Exposure Compensation Program to Ensure all Downwinders Receive Coverage," 27 June 2018: https://www.judiciary.senate.gov/imo/media/doc/06-27-18%20Henderson%20Testimony.pdf (accessed 21 June 2020). See also Sarah Alisabeth Fox, *Downwind: A People's History of the Nuclear West* (Lincoln: University of Nebraska Press, 2014): 174; Carole Gallagher, "Nuclear Photography: Making the Invisible Visible," *Bulletin of the Atomic Scientists* 69:6 (2013): 43.

43. Richard L. Miller, *Under the Cloud: The Decades of Nuclear Testing* (The Woodlands, TX: Two-Sixty Press, 1991).

44. See Eric Frohmberg et al., "The Assessment of Radiation Exposures in Native American Communities from Nuclear Weapons Testing in Nevada," *Risk Analysis* 20:1 (2000): 101–111.

45. Harvey Wasserman and Norman Solomon, *Killing Our Own: The Disaster of America's Experience with Atomic Radiation* (New York: Dell Publishing, 1982): 89–90.

46. United States Atomic Energy Commission, *Assuring Public Safety in Continental Weapons Tests* (Washington, DC: Government Printing Office, 1953): 81.

47. United States Atomic Energy Commission, *Atomic Test Effects in the Nevada Test Site Region* (Washington, DC: Government Printing Office, 1955): 3.

48. Thomas H. Saffer, "Interview with Tom Saffer, U.S. Marine Corps," *People's Century: Fallout,* PBS, 15 June 1999: https://www.pbs.org/wgbh/peoplescentury/episodes/fallout/saffertranscript.html (accessed 22 July 2020); Mary Jo Viscuso et al., *Shot Priscilla: A Test of the Plumbbob Series* (Washington, DC: Defense Nuclear Agency, 1981): 20.

49. United States Department of Energy, *United States Nuclear Tests: July 1945 through September 1992* (Las Vegas, NV: US DOE, 1993): viii.

50. Centers for Disease Control, "Leukemia Among Persons Present at an Atmospheric Nuclear Test (SMOKY)," *Morbidity and Mortality Weekly Report* 28:31 (1979): 361–362; Ethel S. Gilbert et al., "Thyroid Cancer Rates and 131I Doses from Nevada Atmospheric Nuclear Bomb Tests: An Update," *Journal of the National Cancer Institute* 90:21 (1999): 1654–1660; L. E. Peterson and R. L. Miller, "Association Between Radioactive Fallout from 1951–1962 US Nuclear Tests at the Nevada Test Site and Cancer Mortality in Midwest US populations," *Russian Journal of Ecology* 39:7 (2008): 495–509.

51. Congressional Research Service, *The Radiation Exposure Compensation Act (RECA): Compensation Related to Exposure to Radiation from Atomic Weapons Testing and Uranium Mining* (Washington, DC: Congressional Research Service, 2019): 2.

52. Lorna Arnold and Mark Smith, *Britain, Australia and the Bomb: The Nuclear Tests and Their Aftermath,* 2nd ed. (London: Palgrave Macmillan, 2006): 1–16.

53. Arnold and Smith, *Britain, Australia and the Bomb:* 17–20.

54. "Key Events in the UK Atmospheric Nuclear Test Programme," UK Ministry of Defense, 14 February 2013: https://www.gov.uk/government/publications/key-events-in-the-uk-atmospheric-nuclear-test-programme (accessed 21 June 2020).

55. Kingsley Palmer, "Dealing with the Legacy of the Past: Aborigines and Atomic Testing in South Australia," *Aboriginal History* 14:2 (1990): 199.

56. James Robert McClelland, *The Report of the Royal Commission into British Nuclear Tests in Australia,* vol. 1 (Canberra: Australian Government Publishing Service, 1985): 308–309.

57. Tom Gara, "Walter MacDougall and the Emu and Maralinga Nuclear Tests," 1 May 2008: http://www.history.sa.gov.au/history/conference/Tom_Gara2.pdf (accessed 5 April 2019).

58. The Maralinga Rehabilitation Technical Advisory Committee, *Rehabilitation of Former Nuclear Test Sites at Emu and Maralinga* (Canberra: Commonwealth of Australia, 2002): 351–379. Glen Mitchell tracks the indoctrination around radiation issues of the British troops at these tests. See Glen Mitchell, "See an Atomic Blast and Spread the Word: Indoctrination at Ground Zero," in Jordan Goodman, Anthony McElligott, and Lara Marks, eds., *Useful Bodies: Humans in the Service of Medical Science in the Twentieth Century* (Baltimore, MD: Johns Hopkins University Press, 2003): 133–161.

59. Roger Cross, "British Nuclear Tests and the Indigenous People of Australia," in David Holdstock and Frank Barnaby, eds., *The British Nuclear Weapons Programme, 1952–2002* (London: Frank Cass, 2003): 85–86.

60. Arnold and Smith, *Britain, Australia and the Bomb:* 109–110.

61. Lorna Arnold, *Britain and the H-Bomb* (Baskingstoke, UK: Palgrave Macmillan, 2001): 95–107.

62. Anita Smith, "Colonialism and the Bomb in the Pacific," in John Schofield and Wayne Cockroft, eds., *Fearsome Heritage: Diverse Legacies of the Cold War* (Walnut Creek, CA: Left Coast Press, 2007): 58–59.

63. Quoted in International Physicians for the Prevention of Nuclear War, *Radioactive Heaven and Earth:* 124.

64. Nic Maclellan, "The Nuclear Age in the Pacific Islands." *Contemporary Pacific* 17:2 (2005): 363.

65. Nic Maclellan, *Grappling with the Bomb: Britain's Pacific H-Bomb Tests* (Canberra: Australian National University Press, 2017): xxiv, 174, 254.

66. Interview with Paul Ah Poy in Suva, Fiji, 17 January 2017. Paul Ah Poy is the president of the Fiji Nuclear Veterans Association.

67. Teeua Tetoa, interviewed by Mick Broderick on Christmas Island, Kiribati, 16 January 2017.

68. "Glossary," Comprehensive Nuclear-Test-Ban Treaty Organization: https://www.ctbto.org/index.php?id=280&no_cache=1&letter=u#uk-us-mutual-defense-agreement (accessed 22 June 2020).

69. See William E. Ogle, *An Account of the Return to Nuclear Weapons Testing by the United States After the Test Moratorium 1958–1961* (Las Vegas, NV: US DOE, 1985).

70. Tariq Rauf, "French Nuclear Testing: A Fool's Errand," *Nonproliferation Review* 3:1 (1995): 50.

71. Jean-Marc Regnault, "France's Search for Nuclear Test Sites, 1957–1963," *Journal of Military History* 67:4 (2003): 1223–1248.

72. Regnault, "France's Search for Nuclear Test Sites, 1957–1963": 1229–1230.

73. Ramesh Thakur, "The Last Bang Before a Total Ban: French Nuclear Testing in the Pacific," *International Journal* 51:3 (1996): 466–486.

74. Regnault, "France's Search for Nuclear Test Site, 1957–1963," 1223–1248.

75. See Mervyn O'Driscoll, "Explosive Challenge: Diplomatic Triangles, the United Nations, and the Problem of French Nuclear Testing, 1959–1960," *Journal of Cold War Studies* 11:1 (2009): 28–56.

76. For an exploration of the efforts by downwind nations to assess the radiological hazards transported by the winds to their own populations, see Abena Dove Osseo-Asare, *Atomic Junction: Nuclear Power in Africa After Independence* (Cambridge: Cambridge University Press, 2019): 28–39.

77. Johnny Magdaleno, "Algerians Suffering from French Atomic Legacy, 55 Years After Nuke Tests," Al Jazeera America, 1 March 2015: http://america.aljazeera.com/articles/2015/3/1/algerians-suffering-from-french-atomic-legacy-55-years-after-nuclear-tests.html (accessed 5 May 2020).

78. For an overview of problems in the French nuclear weapon program, see Benoît Pelopidas and Sébastien Phillipe, "Unfit for Purpose: Reassessing the Development and Deployment of French Nuclear Weapons (1956–1974)," *Cold War History*, 20 December 2020: https://doi.org/10.1080/14682745.2020.1832472 (accessed 19 February 2021).

79. Elizabeth Bryant, "Algeria: 60 Years On, French Nuclear Tests Leave Bitter Fallout," Deutsche Welle, 13 February 2020: https://www.dw.com/en/algeria-60-years-on-french-nuclear-tests-leave-bitter-fallout/a-52354351 (accessed 5 May 2020).

80. Lamine Chikhi, "French Nuclear Test in Algeria Leave Toxic Legacy," Reuters, 4 March 2010: http://in.reuters.com/article/2010/03/04/idINIndia-46657120100304 (accessed 13 March 2019).

81. Martin Evans, *Algeria: France's Undeclared War* (Oxford: Oxford University Press, 2012): 311. See also Vincent Crapanzano, "The Wound That Never Heals," *Alif: Journal of Comparative Poetics* 30 (2010): 57–84.

82. M'hamed Zengui, quoted in the film *Sandstorm: Sahara of Nuclear Testing*, directed by Larbi Benchiha (2008; Rennes, France: France 3 Corse), DVD. See P. R. Danesi et al., "Residual Radionuclide Concentrations and Estimated Radiation Doses at the Former French Nuclear Weapons Test Sites in Algeria," *Applied Radiation and Isotopes* 66 (2008): 1671–1674.

83. Regnault, "France's Search for Nuclear Test Sites, 1957–1963": 1234.

84. Regnault, "France's Search for Nuclear Test Sites, 1957–1963": 1241.

85. James W. Davidson, "French Polynesia and the French Nuclear Tests: The Submission of John Teariki," *Journal of Pacific History* 2:1 (1967): 149–154.

86. Stewart Firth, *Nuclear Playground* (Honolulu: University of Hawai'i Press, 1987): 95.

87. Bengt Danielsson, "Poisoned Pacific: The Legacy of French Nuclear Testing," *Bulletin of the Atomic Scientists* 46:2 (1990): 24.

88. Tilden Durden, "De Gaulle Sees French Nuclear Test in Pacific," *New York Times,* 12 September 1966: 1.

89. Danielsson, "Poisoned Pacific": 25.

90. Bengt Danielsson and Marie-Thérèse Danielsson, *Poisoned Reign: French Nuclear Colonialism in the Pacific* (New York: Penguin Books, 1977): 169.

91. Bengt Danielsson, "Under a Cloud of Secrecy: The French Nuclear Tests in the Southeast Pacific," *Ambio* 13:5/6 (1984): 338.

92. Interview with Yann Cambon in Bordeaux, France, 22 July 2013.

93. Interview with Tanemaruatoa Michel Arakino in Papeete, French Polynesia, 14 February 2017.

94. See Elizabeth Willis, "French Nuclear Tests in Polynesia," *Medicine, Conflict and Survival* 22:2 (2006): 159–165.

95. This work does not have sufficient time to review or assess the bombshell documents released just at the tenth anniversary of the Fukushima disaster. See Sébastien Philippe and Tomas Statius, *Toxique: Enquête sur les essais nucléaires français en polynésie* (Paris: Presses Universitaires de France, 2021); "Moruroa Files: Investigation into French Nuclear Tests in the Pacific," Moruroa Files, 10 March 2021: https://moruroa-files.org/en/investigation/moruroa-files (accessed 11 March 2021).

96. "Study Finds Cancer Prevalent among NZ Veterans Who Witnessed French Nuclear Explosions in 1973," 1 News, 22 May 2020: https://www.tvnz.co.nz/one-news/new-zealand/study-finds-cancer-prevalent-among-nz-veterans-witnessed-french-nuclear-explosions-in-1973 (accessed 25 May 2020). The *Otago* had been observing the 11 kt Euterpe test, and the *Canterbury* observed the 50 t Melpomène test.

97. "Moruroa Nuclear Site Could Collapse, MP Warns UN," Radio New Zealand, 10 October 2019: https://www.rnz.co.nz/international/pacific-news/400637/moruroa-nuclear-site-could-collapse-mp-warns-un (accessed 22 June 2020).

98. Laurence Cordonnery, "The Legacy of French Nuclear Testing in the Pacific," in David D. Caron and Harry N. Scheiber, eds., *Oceans in the Nuclear Age: Legacies and Risks* (Leiden: Brill / Nijhoff, 2014): 72.

99. Leo Yueh-Yun Liu, *China as a Nuclear Power in World Politics* (London: Palgrave Macmillan, 1972): 33–34. See also Liu Yanqiong and Liu Jifeng, "Analysis of Soviet Technology Transfer in the Development of China's Nuclear Weapons," *Comparative Technological Transfer and Society* 7:1 (2009): 66–110.

100. Robert Guillain, "Ten Years of Secrecy," *Bulletin of the Atomic Scientists* 21:2 (1965): 24.

101. John Wilson Lewis and Xue Litai, *China Builds the Bomb* (Stanford, CA: Stanford University Press, 1988): 111–112. For a detailed explanation of the various facilities, see Hui Zhang, "The History of Fissile Material Production in China," *Nonproliferation Review* 25:5–6 (2018): 477–499.

102. William Burr and Jeffrey T. Richelson, "A Chinese Puzzle," *Bulletin of the Atomic Scientists* 53:4 (1997): 42.

103. Chris Buckley and Adam Wu, "Where China Built its Bomb, Dark Memories Haunt the Ruins," *New York Times,* 20 January 2018: 6. See also Yin Shusheng, "The Pain of Jinyintan," *Yan Huang Chunqiu,* 2012: http://www.yhcqw.com/30/8715 .html (accessed 11 December 2019).

104. Lewis and Litai, *China Builds the Bomb*: 176–177. Robert Norris notes that large numbers of prisoners were among the tens of thousands of laborers who built the site. See Robert S. Norris, "French and Chinese Nuclear Weapon Testing," *Security Dialogue* 27:1 (1996): 48.

105. Kassym Zhumadilov et al., "The Influence of the Lop Nor Nuclear Weapons Test Base to the Population of the Republic of Kazakhstan," *Radiation Measurements* 46 (2011): 425. See also John R. Matzko, "Geology of the Chinese Nuclear Test Site Near Lop Nor, Xinjiang Uygur Autonomous Region, China," *Engineering Geology* 36 (1994): 73–181.

106. Zhihua Shen and Yafeng Xia, "Between Aid and Restriction: The Soviet Union's Changing Policies on China's Nuclear Weapon Program, 1954–1960," *Asian Perspectives* 36 (2012): 112.

107. Lewis and Litai, *China Builds the Bomb:* 190–218; See also "China's Third Nuclear Test," *Survival: Global Politics and Strategy* 8:7 (1966): 229.

108. Justin V. Hastings, "Charting the Course of Uyghur Unrest," *China Quarterly* 208 (2011): 893–912. See also Gardner Bovingdon, *The Uyghurs: Strangers in Their Own Land* (New York: Columbia University Press, 2010): 123.

109. Lawrence S. Wittner, *Confronting the Bomb: A Short History of the World Nuclear Disarmament Movement* (Stanford, CA: Stanford University Press, 2009): 206.

110. *China: "Where Are They?" Time for Answers About Mass Detentions in the Xinjiang Uighur Autonomous Region* (London: Amnesty International, 2018): https:// www.amnesty.org/download/Documents/ASA1791132018ENGLISH.PDF (accessed 31 July 2020).

111. David Lague, "China Now Pays Troops Involved in Nuclear Tests," *New York Times,* 28 January 2008: http://www.nytimes.com/2008/01/28/world/ asia/28china.html (accessed 7 April 2019).

112. Zeeya Merali, "Blasts from the Past," *Scientific American* 301 (2009): 16–20.

113. Wenting Bu et al., "Pu Isotopes in Soils Collected Downwind from Lop Nor: Regional Fallout vs. Global Fallout," *Scientific Reports* 5:12262 (17 July 2015): https://www.nature.com/articles/srep12262 (accessed 23 June 2020).

114. Center for Nonproliferation Studies, "China's Nuclear Tests: Dates, Yields, Types, Methods, and Comments," 2002: https://archive.vn/20131205083146/http:// cns.miis.edu/archive/country_china/coxrep/testlist.htm (accessed 31 July 2020).

115. Henry Kissinger discussing the Marshallese, quoted in Walter J. Hickel, *Who Owns America?* (Englewood Cliffs, NJ: Prentice Hall, 1971): 208.

116. Hecht, *Being Nuclear*: ix. See also Nelta Edwards, "Nuclear Colonialism and the Social Construction of Landscape in Alaska," *Environmental Justice* 4:2 (2011): 109–114.

117. Recent scholarship has shown that these dynamics were true for the production of nuclear materials as well as the testing of nuclear weapons. Gabrielle Hecht has written about colonialism in the mining of uranium in Africa: Gabrielle Hecht, ed., *Entangled Geographies: Empire and Technopolitics in the Global Cold War* (Cambridge, MA: MIT Press, 2012). Peter van Wyck has written about the legacies of uranium mining in native communities in Canada: Peter van Wyck, *The Highway of the Atom* (Montreal: McGill-Queen's University Press, 2010). Several writers have examined the impact of uranium mining on the native communities in the American Southwest: Stephanie A. Malin, *The Price of Nuclear Power: Uranium Communities and Environmental Justice* (New Brunswick, NJ: Rutgers University Press, 2015); Doug Brugge and Rob Goble, "The History of Uranium Mining and the Navaho People," *Public Health Then and Now* 92:9 (September 2002) 1410–1419; Michael A. Amundson, *Yellowcake Towns: Uranium Mining Communities in the American West* (Boulder: University of Colorado Press, 2002).

118. Robert Bauman, "Jim Crow in the Tri-Cities, 1943–1950," *Pacific Northwest Quarterly* 96:3 (2005): 124–131.

119. There is a rich scholarship on the siting of nuclear waste facilities on traditional lands. See Danielle Enders, "The Rhetoric of Nuclear Colonialism: Rhetorical Exclusion of American Indian Arguments in the Yucca Mountain Nuclear Waste Siting Decision," *Communication and Critical/Cultural Studies* 6:1 (2009): 39–60; Anne Sisson Runyan, "Disposable Waste, Lands and Bodies Under Canada's Gendered Nuclear Colonialism," *International Feminist Journal of Politics* 20:1 (2018): 24–38.

6. The Cold War Was a Limited Nuclear War

1. John F. Kennedy, "An Urgent Letter to All Americans from President Kennedy," *Life* 51:11 (15 September 1961): 95.

2. "A New Urgency, Big Things to Do—and What You Must Learn," *Life* 51:11 (15 September 1961): 96.

3. John F. Kennedy, "Letter to the Members of the Committee on Civil Defense of the Governors' Conference," 6 October 1961: https://www.presidency.ucsb.edu /documents/letter-the-members-the-committee-civil-defense-the-governors-conference (accessed 13 February 2021).

4. Terrence R. Fehner and F. G. Gosling, *Origins of the Nevada Test Site* (Washington, DC: US Department of Energy, 2002): 198–202.

5. Matthew Wald and Benjamin Zeimann, "Introduction: The Cold War as an Imaginary War," in Matthew Wald and Benjamin Zeimann, eds., *The Imaginary War: Culture, Thought and Nuclear Conflict, 1945–90* (Manchester, UK: Manchester University Press, 2016): 2. Italics added.

6. Jonathan Schell, *The Fate of the Earth* (New York: Alfred A. Knopf, 1982): 1.

7. See Barbara Rose Johnston, "Environmental Disaster and Resilience: The Marshall Islands Experience Continues to Unfold," *Cultural Survival,* September 2016: https://www.culturalsurvival.org/publications/cultural-survival-quarterly/environmental-disaster-and-resilience-marshall-islands-0 (accessed 1 July 2020); Martha Smith-Norris, "American Cold War Policies and the Enewetakese: Community Displacement, Environmental Degradation, and Indigenous Resistance in the Marshall Islands," *Journal of the Canadian Historical Association* 22:2 (2011): 195–236; Elizabeth DeLoughrey, Jill Didur, and Anthony Carrigan, "Introduction: A Postcolonial Environmental Humanities," in Elizabeth DeLoughrey, Jill Didur, and Anthony Carrigan, eds., *Introduction to Global Ecologies and the Environment: Postcolonial Approaches* (New York: Routledge, 2016): 1–32; Jessie Boylan, "Grievability and Nuclear Memory," *American Quarterly* 71:2 (2019): 379–388; Mick Broderick and Robert Jacobs, "The Global Hibakusha Project: Nuclear Post-colonialism and Its Intergenerational Legacy," *Unlikely: Journal for the Creative Arts* 5, 2018: http://unlikely.net.au/issue-05/the-global-hibakusha-project (accessed 4 July 2020); Satoe Nakahara, "Overcoming Nuclear Tragedy: The Case of the Rongelap People in the Marshall Islands Suffered from H-Bomb Test," *Japanese Review of Cultural Anthropology* 14 (2013): 73–93.

8. "Nuclear Testing Tally, 1945–2017," Arms Control Association: https://www.armscontrol.org/factsheets/nucleartesttally (accessed 14 August 2020).

9. This would have been "rung 44" in Kahn's ladder of escalation, the final rung. See Herman Kahn, *On Escalation: Metaphors and Scenarios* (New York: Fredrick A. Praeger, 1965): 194–195.

10. Quoted in Michael Howard, *The Franco-Prussian War: The German Invasion of France, 1870–1871* (London: Routledge Books, 1979): 301.

11. Mark Duffield, "Total War as Environmental Terror: Linking Liberalism, Resilience, and the Bunker," *South Atlantic Quarterly* 110:3 (2011): 757. Emmanuel Kreike writes that the United States engaged in environmental warfare against Native American tribes through the slaughter of the buffalo and winter attacks aimed at depriving entire communities of food and shelter. See Emmanuel Kreike, *Scorched Earth: Environmental Warfare as a Crime Against Humanity and Nature* (Princeton, NJ: Princeton University Press, 2021): 137–172.

12. David Alan Rosenberg, "The Origins of Overkill: Nuclear Weapons and American Strategy, 1945–1960," *International Security* 7:4 (Spring 1983): 3–71.

13. First widely popularized in the book by Paul R. Ehrlich et al., *The Cold and the Dark: The World after Nuclear War* (New York: W. W. Norton, 1985). See also Lawrence Badash, *A Nuclear Winter's Tale: Science and Politics in the 1980s* (Cambridge, MA: MIT Press, 2009).

14. Alan Robock, Luke Oman, and Georgiy L. Stenchikov, "Nuclear Winter Revisited with a Modern Climate Model and Current Nuclear Arsenals: Still Catastrophic Consequences," *Journal of Geophysical Research: Atmospheres* 112:D13107 (July 2007): 1–14; Owen B. Toon et al., "Atmospheric Effects and Societal Consequences of Regional

Scale Nuclear Conflicts and Acts of Individual Nuclear Terrorism," *Atmospheric Chemistry and Physics* 7 (2007): 1973–2002; Michael J. Mills, Owen B. Toon, Julia Lee-Taylor, and Alan Robock, "Multidecadal Global Cooling and Unprecedented Ozone Loss Following a Regional Nuclear Conflict," *Earth's Future* 2:4 (2014): 161–176.

15. For an examination of the effects of a limited nuclear war in South Asia on agricultural production in the United States, see David Pimentel and Michael Burgess, "Nuclear War Investigation Related to a Limited Nuclear Battle with Emphasis on Agricultural Impacts in the United States," *Ambio* 41 (2012): 894–899.

16. "Bizarro World" is an alternate world in the *Superman* comic series where everything is the direct opposite of the "normal" world in which we live. See Leonard Finkelman, "Superman and Man: What a Kryptonian Can Teach Us About Humanity," in Mark D. White, ed., *Superman and Philosophy: What Would the Man of Steel Do?* (Malden, MA: Wiley-Blackwell, 2013): 171–173.

17. Thomas Rabl, "The Nuclear Disaster of Kyshtym 1957 and the Politics of the Cold War," Environment & Society Portal, *Arcadia* 20 (2012), Rachel Carson Center for Environment and Society: http://www.environmentandsociety.org/arcadia/nuclear-disaster-kyshtym-1957-and-politics-cold-war (accessed 2 February 2019).

18. See Kate Brown, *Plutopia: Nuclear Families, Atomic Cities, and the Great Soviet and American Plutonium Disasters* (Oxford: Oxford University Press, 2013): 189–196.

19. A. A. Romanyukha et al., "The Distance Effect on Individual Exposures Evaluated from the Soviet Nuclear Bomb Test at Totskoye Test Site in 1954," *Radiation Protection Dosimetry* 86:1 (1999): 53–58.

20. Kate Brown, *Manual for Survival: A Chernobyl Guide to the Future* (London: Allen Lane, 2019): 42.

21. See Pavel Palazhchenko, George P. Schultz, and Kiron K. Skinner, *Turning Points in Ending the Cold War* (Stanford, CA: Hoover Institution Press, 2007).

22. Stephen Schwartz, ed., *Atomic Audit: The Costs and Consequences of U.S. Nuclear Weapons since 1940* (Washington, DC: Brookings Institution Press, 1998): 3 (original italics). I would add that Social Security is not properly a government expense but rather a trust fund paid into by American workers.

23. For an analysis of this discourse and the arguments, see Atsuko Shigesawa, *Demystifying the Atomic Bomb: The U.S. Strategic Bombing Survey Goes to Hiroshima and Nagasaki* (PhD diss., Hiroshima City University, 2019).

24. United States Department of Energy, *United States Nuclear Tests: July 1945 through September 1992* (Oak Ridge, TN: Office of Scientific and Technical Information, 1993).

25. Jai Prakash Agrawal and Robert D. Hodgson, *Organic Chemistry of Explosives* (Chichester: John Wiley and Sons, 2007): xxv–xxvii.

26. "The Evaluation of the Atomic Bomb as a Military Weapon," *The Final Report of the Joint Chiefs of Staff Evaluation Board for Operation Crossroads*, Enclosure "A," JCS 1691/3 (30 June 1947): 57–89, quoted in part in chap. 3.

27. RAND Corporation, "Conference on Methods for Studying the Psychological Effects of Unconventional Weapons," *Research Memorandum 120* (26–28 January 1949): 10.

28. Viktor P. Maslov and Vladimir S. Shpinel, "Claim for an Invention from V. Maslov and V. Shpinel, 'About Using of Uranium as an Explosive and Toxic Agent,'" 17 October 1940, doc. no. 75, *Atomic Project of USSR: Documents and Materials,* vol. 1, pt. 1: History and Public Policy Program Digital Archive: 193–196, https://digitalarchive.wilsoncenter.org/document/12163 (accessed 15 February 2021). See also Samuel Meyer, Sarah Bidgood, and William C. Potter, "Death Dust: The Little-Known Story of U.S. and Soviet Pursuit of Radiological Weapons," *International Security* 45:2 (2020): 78–82.

29. David Alan Rosenberg, "'A Smoking, Radiating Ruin at the End of Two Hours': Documents on American Plans for Nuclear War with the Soviet Union, 1954–1955," *International Security* 6:3 (Winter 1981/1982): 11.

30. Richard G. Hewlett and Jack M. Holl, *Atoms for Peace and War, 1953–1961* (Berkeley: University of California Press, 1989): 182.

31. Letter dated 1 July 1954 to Senator John W. Bricker, Chairman, Military Applications Subcommittee, Joint Committee on Atomic Energy, from Lewis L. Strauss, Chairman, US AEC, quoted in Chuck Hansen, *Swords of Armageddon,* version 2, vol. 4 (Sunnyvale, CA: Chukelea Publications, 2007): 15

32. *Report on Atomic Energy—1949–1954,* "Growth of Military Atomic Knowledge, Robert LeBaron, Chairman, Department of Defense Military Liaison Committee to the Atomic Energy Commission, Attachment to Memorandum for the Secretary of Defense and the Chairman, U.S. Atomic Energy Commission Dated 31 July 1954, from Robert LeBaron, Chairman, MLC," quoted in *Swords of Armageddon,* 21.

33. Letter dated 17 September 1954 to Lewis L. Strauss, Chairman, US AEC, from Donald A. Quarles, Assistant Secretary of Defense for Research and Development, Department of Defense, quoted in *Swords of Armageddon* 2.4, 22.

34. For an analysis of radioactive fallout in the ecosystem being understood as a political problem that required policy management, and specifically of the history of the United Nations Scientific Committee on the Effects of Radiation (UNSCEAR) in that policy management, see Toshihiro Higuchi, *Political Fallout: Nuclear Weapons Testing and the Making of a Global Environmental Crisis* (Stanford, CA: Stanford University Press, 2020): 109–135.

35. Nils-Olov Bergkvist and Ragnhild Ferm, *Nuclear Explosions 1945–1998* (Stockholm: Stockholm International Peace Research Institute, 2000).

36. George C. Reinhardt, *Nuclear Weapons and Limited Warfare: A Sketchbook History* (Santa Monica, CA: RAND Corporation, 1964): 10.

37. The UN Security Council's P5.

38. John Lewis Gaddis, "The Long Peace: Elements of Stability in the Postwar International System," *International Security* 10:4 (1986): 99–142. See also John Lewis Gaddis, *The Long Peace: Inquiry into the History of the Cold War* (Oxford: Oxford University Press, 1987).

39. Nina Tannenwald, *The Nuclear Taboo: The United States and the Non-Use of Nuclear Weapons since 1945* (Cambridge: Cambridge University Press, 2007): 1–2.

40. Odd Arne Wested, *The Global Cold War: Third World Interventions and the Making of our Times* (Cambridge: Cambridge University Press, 2007). See also Paul Thomas Chamberlin, *The Cold War's Killing Fields: Rethinking the Long Peace* (New York: HarperCollins, 2018).

41. Gaddis, "The Long Peace: Elements": 101.

42. Jacques Derrida, "No Apocalypse, Not Now (Full Speed Ahead, Seven Missiles, Seven Missives)," *Diacritics* 14:2 (Summer 1984): 23.

43. Robert Jacobs, "Nuclear Conquistadors: Military Colonialism in Nuclear Test Site Selection during the Cold War," *Asian Journal of Peacebuilding* 1:2 (2013): 157–177: http://tongil.snu.ac.kr/ajp_pdf/201311/02_Robert%20Jacobs.pdf (accessed 21 August 2020). See also Aliya Sartbayeva Peleo, "The Rights of the Wronged: Norms of Nuclearism, the Polygon and the Making of Waste-Life," *Contemporary Chinese Political Economy and Strategic Relations: An International Journal* 3:1 (2017): 285–330.

44. Gaddis, *The Long Peace: Inquiry:* 216.

45. Robert Jacobs, "Imagining a Nuclear World War Two in Europe: Preparing U.S. Troops for the Battlefield Use of Nuclear Weapons," *Estonian Yearbook of Military History* 7:13 (2018): 166–186.

46. See Lawrence Freedman, *The Evolution of Nuclear Strategy,* 3rd ed. (London: Palgrave Macmillan, 2003); Sharon Ghamari-Tabrizi, *The Worlds of Herman Kahn: The Intuitive Science of Thermonuclear War* (Cambridge, MA: Harvard University Press, 2005).

47. Robert E. Osgood, *Limited War: The Challenge to American Society* (Chicago: University of Chicago Press, 1957): 1–2.

48. Robert E. Osgood, *Limited War Revisited* (Boulder, CO: Westview Press, 1979): 11.

49. Jan Zalasiewicz et al., "When Did the Anthropocene Begin? A Mid-Twentieth Century Boundary Level is Stratigraphically Optimal," *Quaternary International* 383 (2015): 196.

50. "Nuclear Testing Tally, 1945–2017."

51. Hans M. Kristensen and Matt Korda, "Status of World Nuclear Forces," Federation of American Scientists: https://fas.org/issues/nuclear-weapons/status-world-nuclear-forces/ (accessed 14 August 2020).

52. Christopher E. Paine, Thomas B. Cochrane, and Robert S. Norris, *The Arsenals of Nuclear Weapon Powers: An Overview* (Washington, DC: Natural Resources Defense Council, 1996): 6.

53. Frans de Waal, "What Animals Can Teach Us About Politics," *The Guardian,* 12 March 2019: https://www.theguardian.com/science/2019/mar/12/what-animals-can-teach-us-about-politics (accessed 27 March 2019).

54. B. H. Liddell Hart, *The Strategy of Indirect Approach* (London: Faber and Faber, 1941): 190. Hart and other strategists constituted what was known as the British "graduated deterrence" school. See Lawrence Freedman and Jeffrey Michaels, *The Evolution of Nuclear Strategy,* 4th ed. (London: Palgrave Macmillan, 2019): 138–139.

55. Brian Holden Reid, "The Legacy of Liddell Hart: The Contrasting Responses

of Michael Howard and André Beaufre," *British Journal of Military History* 1:1 (October 2014): 68 (original italics).

56. Anthony Buzzard, "The H-Bomb: Massive Retaliation or Graduated Deterrence?" *International Affairs* 32:2 (1956): 154.

57. Amy Fass Emery, "The Zombie in/as the Text: Zora Neale Hurston's 'Tell My Horse,'" *African American Review* 39:3 (Fall 2005): 332.

58. Anne Harrington de Santana, "Nuclear Weapons as Currency of Power: Deconstructing the Fetishism of Force," *Nonproliferation Review* 16:3 (2009): 330.

59. The quote refers to the Marshallese exposed to fallout radiation from the Bravo test. Roy D. Maxwell et al., *Evaluation of Radioactive Fallout* (Washington, DC: Armed Forces Special Weapons Project, 1955): 88.

60. Kreike, *Scorched Earth:* 2–3.

7. The Slow-Motion Nuclear War

1. George F. Kennan, "The Nuclear Deterrent and the Principle of 'First Use,'" *The Nuclear Delusion: Early Reflections on the Atomic Bomb* (New York: Pantheon Books, 1983): 6. See also Andrew M. Johnston, *Hegemony and Culture in the Origins of NATO Nuclear First-Use, 1945–1955* (New York: Palgrave Macmillan, 2005).

2. Congressional Research Service, *U.S. Strategic Nuclear Forces: Background, Developments, and Issues* (Washington, DC: Government Printing Office, 2018): 2–3.

3. Andrew Brown and Lorna Arnold, "The Quirks of Nuclear Deterrence," *International Relations* 24:3 (2020): 296.

4. Carl von Clausewitz, *On War,* trans. Col. J. J. Graham, new and rev. ed. (London: Kegan Paul, Trench, Trubner, 1918): 1.

5. Barry Nalebuff, "Brinkmanship and Nuclear Deterrence: The Neutrality of Escalation," *Conflict Management and Peace Science* 9:2 (Spring 1986): 19. Small efforts to achieve an international culture of safety around nuclear weapon management followed the Cold War; see Igor Khripunov, Nikolay Ischenko, and James Holmes, eds., *Nuclear Security Culture: From National Best Practices to International Standards* (Amsterdam: IOS Press, 2007).

6. Even today the apocalyptic game of nuclear cat and mouse is played out by the United States and Russia. In late 2020 Russian military hackers appear to have largely hacked into US nuclear production and deployment systems, showing these technologies have always only been as safe as their owners are sane, and as secure as the most recent of multiple technological system updates. See Natasha Bertrand and Eric Wolff, "Nuclear Weapons Agency Breached Amid Massive Cyber Onslaught," *Politico,* 17 December 2020: https://www.politico.com/news/2020/12/17/nuclear-agency-hacked-officials-inform-congress-447855 (accessed 30 December 2020).

7. Many of those incidents were detailed in Eric Schlosser, *Command and Control: Nuclear Weapons, the Damascus Incident, and the Illusion of Safety* (New York: Penguin Press, 2013).

8. International Atomic Energy Agency, *Safety Culture in Maintenance of Nuclear Power Plants* (Vienna: IAEA, 2005): 2.

9. IAEA's PRIS database: https://pris.iaea.org/PRIS/home.aspx (accessed 3 May 2019).

10. See IAEA, *Safety Culture in Maintenance of Nuclear Power Plants.*

11. John Downer and M. V. Ramana, "Empires Built on Sand: On the Fundamental Implausibility of Reactor Safety Assessments and the Implications for Nuclear Regulation," *Regulation and Governance,* 2020: doi:10.1111/rego.12300. Carolyn Miller analyzes how the bedrock reactor safety study conducted on behalf of the US Nuclear Regulatory Commission in 1975 relies on "expert opinion" in the absence of statistical data to assure public safety. See Carolyn R. Miller, "The Presumptions of Expertise: The Role of Ethos in Risk Analysis," *Configurations* 11:2 (2003): 163–202.

12. Annie Makhijami and Arjun Makhijami, "Radioactive Rivers and Rain: Routine Releases of Tritiated Water from Nuclear Power Plants," *Science for Democratic Action* 16:1 (August 2009): 1–10. See also "Backgrounder on Tritium, Radiation Protection Limits, and Drinking Water Standards," US Nuclear Regulatory Commission: https://www.nrc.gov/reading-rm/doc-collections/fact-sheets/tritium-radiation-fs.html (accessed 3 May 2019). There was a US-built reactor, PM-3A McMurdo Station, in Antarctica in 1961–1962. It was nicknamed "Nukey Poo" because of its propensity to leak. All spent nuclear fuel (and 12,200 tons of contaminated soil) was removed from the site by 1976. See Owen Wilkes and Robert Mann, "The Story of Nukey Poo," *Bulletin of the Atomic Scientists* 34:8 (1978): 32–36.

13. James Rice, "Downwind of the Atomic State: US Continental Atmospheric Testing, Radioactive Fallout, and Organizational Deviance, 1951–1962," *Social Science History* 39 (Winter 2015): 656.

14. Quoted in Harvey Wasserman and Norman Solomon, *Killing Our Own: The Disaster of America's Experience with Atomic Radiation* (New York: Dell Publishing, 1982): 89.

15. Kate Brown, *Manual for Survival: A Chernobyl Guide to the Future* (London: Allen Lane, 2019): 42.

16. Dwight D. Eisenhower, *Public Papers of the Presidents of the United States, Dwight David Eisenhower, 1954* (Washington, DC: Government Printing Office, 1960): 346.

17. Jan Zalasiewicz et al., "When Did the Anthropocene Begin? A Mid-Twentieth Century Boundary Level is Stratigraphically Optimal," *Quaternary International* 383 (2015): 196–203. For an interesting exploration of how geophysical structure emerges from weapon use, see Joseph P. Hupy and Randall J. Schaetzl, "Introducing 'Bomb-turbation,' a Singular Type of Soil Disturbance and Mixing," *Soil Science* 171:11 (2006): 823–826.

18. Caroline Clason et al., "The Widespread Presence of Fallout Radionuclides in Cryoconite: An Anthropogenic Legacy and Emerging Issue," *Proceedings from the European Geosciences Union Conference,* April 2019. See also Giovani Baccolo et al., "Cryoconite as a Temporary Sink for Anthropogenic Species Stored in Glaciers," *Nature: Scientific Reports* 7:9623 (2017): https://www.ncbi.nlm.nih.gov/pmc/articles/PMC5575069/pdf/41598_2017_Article_10220.pdf (accessed 15 June 2020).

19. W. M. Place, F. C. Cobb, and C. G. Defferding, *Palomares Summary Report* (Washington, DC: Defense Nuclear Agency, 1975). See also Teresa Vilarós, "The Lightness of Terror: Palomares, 1966," *Journal of Spanish Cultural Studies* 5:2 (2004): 165–186. See also Barbara Moran, "The Aftermath of the Palomares Nuclear Accident," *Bulletin of the Atomic Scientists* 65:3 (2009): 48–54.

20. J. Magill et al., "Consequences of a Radiological Dispersal Event with Nuclear and Radioactive Sources," *Science & Global Security* 15 (2007): 125–126. For the radiological legacy of the event, see Carlos Sancho and R. Garcia-Tenorio, "Radiological Evaluation of the Transuranic Remaining Contamination in Palomares (Spain): A Historical Review," *Journal of Environmental Radioactivity* 203 (2019): 55–70.

21. This structural incompetence is the reason that, of all of popular culture related to possible nuclear war that was made during the Cold War, *Dr. Strangelove* has resonated long past its release and remains a touchstone of Cold War illogic. The juvenile incompetence of nuclear weapon and war management on both sides in *Strangelove* speaks more powerfully to generations of viewers than contemporary films immersed in "geopolitics." See Mick Broderick, *Reconstructing Strangelove: Inside Stanley Kubrick's "Nightmare Comedy"* (New York: Columbia University Press, 2017).

22. For a full analysis, see Nate Jones, *Able Archer 83: The Secret History of the NATO Exercise That Almost Triggered Nuclear War* (New York: The New Press, 2016). See also the documents embedded in "Able Archer War Scare 'Potentially Disastrous,'" National Security Archive, 17 February 2021: https://nsarchive.gwu.edu/briefing-book /aa83/2021-02-17/able-archer-war-scare-potentially-disastrous?eType=EmailBlastCo ntent&eId=53ea5247-117a-4861-8bfc-7acff0660ebe (accessed 18 February 2021).

23. See Eric Schlosser, *Command and Control: Nuclear Weapons, the Damascus Accident, and the Illusion of Safety* (New York: Penguin Books, 2014).

24. Petrov was awarded the World Citizen Award by the Association of World Citizens in 2004. See Anastasiya Lebedev, "The Man Who Saved the World Finally Recognized," 21 May 2004, archived at: https://web.archive.org/web/20110721000030 /http://www.worldcitizens.org/petrov2.html (accessed 18 February 2021). See also Len Scott, "Intelligence and the Risk of Nuclear War: Able Archer-83 Revisited," *Intelligence and National Security* 26:6 (2011): 759–777.

25. Robert Jacobs, "Born Violent: The Birth of Nuclear Power," *Asian Journal of Peacebuilding* 7:1 (July 2019): 9–29.

26. Eight of nine Hanford plutonium production reactors were online by 1955, all five Savannah River plutonium production reactors were online by 1955, and the first US commercial nuclear power plant online was at Shippingport in 1957. See also Murray W. Rosenthal, *An Account of Oak Ridge National Laboratory's Thirteen Nuclear Reactors* (Oak Ridge, TN: Oak Ridge National Laboratory, 2009). There were additional experimental reactors online prior to 1957 at Argonne and Idaho National Laboratories.

27. The accident was kept secret inside the former Soviet Union, and first became known in the West when dissident scientist Zhores Medvedev published an article in *New Scientist* in 1976, and later a book about the incident. See Zhores Medvedev,

"Two Decades of Dissidence," *New Scientist*, 4 November 1976: 264–267; Zhores Medvedev, *Nuclear Disaster in the Urals* (New York: W. W. Norton, 1980).

28. William J. Broad, "Disasters with Nuclear Subs in Moscow's Fleet Detailed," *New York Times*, 26 February 1993: https://www.nytimes.com/1993/02/26/world /disasters-with-nuclear-subs-in-moscow-s-fleet-detailed.html (accessed 1 May 2019).

29. Joseph Lelieveld, D. Kunkel, and M. G. Lawrence, "Global Risk of Radioactive Fallout After Major Nuclear Reactor Accidents," *Atmospheric Chemistry and Physics* 12 (2012): 4245–4258.

30. "Nuclear Power Plant Accidents: Listed and Ranked Since 1952," DATABlog, *The Guardian*, 2016: https://www.theguardian.com/news/datablog/2011 /mar/14/nuclear-power-plant-accidents-list-rank (accessed 29 January 2020). See also Downer and Ramana, "Empires Built on Sand."

31. Dean Wilkie, "Fukushima Daiichi Decay Heat & Corium Status Report," SimplyInfo. org, 2016: http://www.simplyinfo.org/?page_id=15924 (accessed 5 July 2020).

32. Suvrat Raju, "Estimating the Frequency of Nuclear Accidents," *Science & Global Security* 24 (2016): 40.

33. Charles Perrow, *Normal Accidents: Living with High-Risk Technologies* (New York: Basic Books, 1984): 5. See also the classic work by Ulrich Beck, *Risk Society: Towards a New Modernity* (London: Sage Publications, 1992).

34. S. Islam and K. Lindgren, "How Many Reactor Accidents Will There Be?" *Nature* 322 (1986): 691–692. See also US Nuclear Regulatory Commission, *Reactor Safety Study: An Assessment of Accident Risks in U.S. Commercial Nuclear Power Plants*. Executive summary: main report [PWR and BWR] (NUREG-75/014), 1975: https:// www.osti.gov/servlets/purl/7134131 (accessed 1 May 2019); US Nuclear Regulatory Commission, Severe Accident Risks: An Assessment for Five US Nuclear Power Plants (NUREG-1150), 1990: https://www.nrc.gov/reading-rm/doc-collections/nuregs /staff/sr1150/v3/sr1150v3.pdf (accessed 1 May 2019).

35. Airi Ryu and Najmedin Meshkati, "Onagawa: The Japanese Nuclear Power Plant That Didn't Melt Down on 3/11," *Bulletin of Atomic Scientists*, 10 March 2014: https://thebulletin.org/2014/03/onagawa-the-japanese-nuclear-power-plant-that-didnt-melt-down-on-3-11/ (accessed 12 July 2020). Fukushima Daiichi is owned and operated by the Tokyo Electric Power Company (TEPCO), and the Onagawa plants are owned and operated by the Tōhoku Electric Power Company.

36. John W. Johnson, "Nuclear Power and the Price-Anderson Act: An Overview of a Policy in Transition," *Journal of Policy History* 2:2 (1990): 214.

37. See George T. Mazuzan and J. Samuel Walker, *Controlling the Atom: Nuclear Regulation 1946–1962* (Washington, DC: Nuclear Regulatory Commission, 1997): 93–121; Harold P. Green, "Nuclear Power: Risk, Liability, and Indemnity," *Michigan Law Review* 71:3 (1973): 479–510; Jeffrey A. Dubin and Geoffrey S. Rothwell, "Subsidy to Nuclear Power Through Price-Anderson Liability Limit," *Contemporary Policy Issues* 8:3 (1990): 73–79. Dubin and Rothwell put the value of the subsidy at $60 million per reactor, per reactor year, through the 1982 revision.

38. Peter Custers, *Questioning Globalized Militarism: Nuclear and Military Production and Critical Economic Theory* (Monmouth, UK: Merlin Press, 2008): 26 (original italics).

39. See International Atomic Energy Agency, *Status and Trends in Spent Fuel and Radioactive Waste Management* (Vienna: IAEA, 2018): 35.

40. Office of Scientific and Technical Information, *Categorization of Used Nuclear Fuel Inventory in Support of a Comprehensive National Nuclear Fuel Cycle Strategy* (Oak Ridge, TN: Oak Ridge National Laboratory, 2012).

41. Alvin M. Weinberg, "Social Institutions and Nuclear Energy," *Science* 177:4043 (7 July 1972): 33–34.

42. For a fascinating analysis of uranium decay in situ over deep time, see Gabrielle Hecht, "Interscalar Vehicles for an African Anthropocene: On Waste, Temporality, and Violence," *Cultural Anthropology* 33:1 (2018): 109–141.

43. Harry H. Hess et al., *The Disposal of Radioactive Waste on Land* (Washington, DC: National Academy of Sciences / National Research Council, 1957): 3 (original underlining). This publication was based on a conference held at Princeton University in 1955, and was primarily concerned with the liquid waste stored in tanks at the Hanford plutonium production site.

44. United States Atomic Energy Commission, *Handling Radioactive Wastes in the Atomic Energy Program* (Washington, DC: Government Printing Office, 1949): 7. This same book advised that "In carrying out its program for safe handling of radioactive wastes, the Commission relies heavily upon the individual programs of its prime contractors" (page v).

45. When a radioactive particle undergoes decay, it can transmute into a different isotope, or a "daughter particle." Plutonium is a daughter particle of decaying ^{238}U. See Don Lincoln, "What Is Subatomic Decay," Fermilab at Work, 24 August 2012: https://news.fnal.gov/2012/08/what-is-subatomic-decay/ (accessed 12 June 2020). There is some controversy over using gendered terms like "daughter" for such entities, which are now sometimes referred to as "progeny."

46. "Mixed Oxide Fuel," World Nuclear Association, October 2017: http://www.world-nuclear.org/information-library/nuclear-fuel-cycle/fuel-recycling/mixed-oxide-fuel-mox.aspx (accessed 17 May 2019).

47. Generally this water does not contain any radionuclides, since the fuel is encased, but the dumping of hot water into rivers, lakes, and oceans does affect local aquatic systems.

48. "Gulls Contaminated with Radiation Culled at Sellafield," BBC News, 25 February 2010: http://news.bbc.co.uk/2/hi/uk_news/england/cumbria/8536094.stm (accessed 21 August 2020). Sellafield is the site of the 1957 Windscale Fire discussed earlier in the chapter. See also Ian Burrell, "Sellafield Scare over Radioactive Pigeons," *The Independent*, 11 February 1998: https://www.independent.co.uk/news/sellafield-scare-over-radioactive-pigeons-1144082.html (accessed 21 August 2020).

49. General Accounting Office, *Spent Nuclear Fuel Management: Outreach Needed to Help Gain Public Acceptance for Federal Activities That Address Liability* (Washington,

DC: Government Printing Office, 2014): 7, 14. See also Per Högselius, "The Decay of Communism: Managing Spent Nuclear Fuel in the Soviet Union," *Risk, Hazards & Crisis in Public Policy* 1:4 (2010): 83–109.

50. GAO, *Spent Nuclear Fuel Management:* 12.

51. Fluor Hanford, *FY 1999 Annual Report* (Richland, WA: Department of Energy, 1999): 6.

52. William S. Burroughs, *Naked Lunch* (New York: Grove Press, 1959): 44.

53. See Achim Brunnengräber and Christoph Görg, "Nuclear Waste in the Anthropocene: Uncertainties and Unforeseeable Timescales in the Disposal of Nuclear Waste," *Gaia* 26:2 (2017): 96–99.

54. Office of Civilian Radioactive Waste Management, *A Monitored Retrievable Storage Facility: Technical Background Information* (Washington, DC: Government Printing Office, 1991). There are also plans for private companies to store spent fuel on a temporary basis, in a variety of settings. See Jeff Brady, "As Nuclear Waste Piles Up, Private Companies Pitch New Ways to Store It," NPR, 30 April 2019: https://www.npr.org/2019/04/30/716837443/as-nuclear-waste-piles-up-private-companies-pitch-new-ways-to-store-it (accessed 11 June 2020).

55. Svensk Kärnbränslehantering AB, "Spent Nuclear Fuel for Disposal in the KBS-3 Repository," December 2010: http://www.skb.se/upload/publications/pdf/TR-10-13.pdf (accessed 16 May 2019).

56. Jin-Seop Kim et al., "Geological Storage of High Level Nuclear Waste," *KSCE Journal of Civil Engineering* 15:4 (2011): 721–737.

57. Waste Technology Section, *The Use of Scientific and Technical Results from Underground Research Laboratory Investigations for the Geological Disposal of Radioactive Waste* (Vienna: IAEA, 2001): 56.

58. Posiva Oy, *The Final Disposal Facility for Spent Nuclear Fuel: Environmental Impact Assessment Report* (Helsinki: Posiva Oy, 1999).

59. Claire Corkhill and Neil Hyatt, *Nuclear Waste Management* (Bristol, UK: Institute of Physics Publishing, 2018): 9.

60. Marika Hietala has written a fascinating article on discursive strategies to make this storage system seem "natural" and "safe" in Finland. See Marika Hietala, "Safer-than: Making Nuclear Waste Disposal More Familiar," *Science as Culture* 30:2 (2021): doi:10.1080/09505431.2021.1872520 (accessed 22 January 2021).

61. Kristin Shrader-Frechette, *Environmental Justice: Creating Equity, Reclaiming Democracy* (Oxford: Oxford University Press, 2002): 105.

62. International Association of Scientific Hydrology, *Bulletin* 3:2 (1958): 8.

63. Karl Philberth, "The Disposal of Radioactive Waste in Ice Sheets," *Journal of Glaciology* 19:81 (1977): 607. Both Philberths were also Catholic priests.

64. Radioactive waste was actually buried beneath the ice sheets of Greenland by the United States in 1967 after the abandonment of Camp Century, an underground site where, starting in 1959, the US military studied "the feasibility of deploying ballistic missiles within the ice sheet." A 2016 study of the conditions at the waste site determined that "Net ablation would guarantee the eventual remobilization of physical,

chemical, biological, and radiological wastes abandoned at the site." See William Colgan et al., "The Abandoned Ice Sheet Base at Camp Century, Greenland, in a Warming Climate," *Geophysical Research Letters* 43 (2016): doi:10.1002/2016GL069688 (accessed 22 December 2020).

65. Terrence R. Fehner and F. G. Gosling, *Origins of the Nevada Test Site* (Washington, DC: US Department of Energy, 2002): 20.

66. Leo Kinlin, "Childhood Leukaemia and Ordnance Factories in West Cumbria During the Second World War," *British Journal of Cancer* 95 (2006): 102–103.

67. Robert Chaplow, "The Geology and Hydrogeology of Sellafield: An Overview," *Quarterly Journal of Engineering Geology and Hydrogeology* 29 (1996): 1–12.

68. Oy, *The Final Disposal Facility for Spent Nuclear Fuel.*

69. Presentation by Posiva Oy spokesman at the Onkalo site attended by author, 13 October 2016.

70. Helen Gordon, "Journey Deep into the Finnish Caverns Where Nuclear Waste Will Be Buried for Millennia," *Wired,* 24 April 2017: https://www.wired.co.uk /article/olkiluoto-island-finland-nuclear-waste-onkalo (accessed 17 June 2019).

71. This is not to dismiss the work of those engaged in designing the Onkalo repository, as Vincent Ialenti has written, "I had to extend some guarded trust that their years of work had left them with a more sophisticated understanding of possible future worlds than I could find in people who had not done the same." There is no doubt that this work is being done with earnest and sophisticated efforts, but it may be beyond our capabilities. We must, but perhaps can only, do our best. See Vincent Ialenti, *Deep Time Reckoning: How Future Thinking Can Help Earth Now* (Cambridge, MA: MIT Press, 2020): 5.

72. Maxime Polleri, "Post-Political Uncertainties: Governing Nuclear Controversies in Post-Fukushima Japan," *Social Studies of Science* 50:4 (2019): https://journals. sagepub.com/doi/10.1177/0306312719889405 (accessed 10 January 2020).

73. "METI Maps Out Suitable Nuclear Waste Disposal Sites," *Japan Times,* 28 July 2017: https://www.japantimes.co.jp/news/2017/07/28/national/meti-posts -map-potential-nuclear-waste-disposal-sites/ (accessed 7 May 2019).

74. Quoted in Polleri, "Post-Political Uncertainties."

75. The spent fuel is only part of the 26,820 tons of radioactive waste accrued by the Japanese nuclear power industry and does not include the 47 tons of plutonium that Japan has stockpiled for fuel reprocessing (no reprocessing facility is currently in operation). See "Where to Put All the Radioactive Waste Is Now the Burning Issue," *Asahi Shimbun,* 20 April 2017: http://www.asahi.com/ajw/articles/AJ201704200039. html (accessed 20 April 2017); Kazunari Hanawa and Takashi Tsuji, "Japan's Plutonium Glut Casts a Shadow on Renewed Nuclear Deal," *Nikkei Asian Review,* 14 February 2018: https://asia.nikkei.com/Politics-Economy/International-Relations/Japan-s -plutonium-glut-casts-a-shadow-on-renewed-nuclear-deal (accessed 19 May 2019).

76. Similarly, when planning was underway for the (currently) scuttled Yucca Mountain DGR in the United States, parameters anticipated and accepted different radiation exposures to future generations over time: "Because radioactive leaks will

increase over time, E.P.A. proposes one radiation exposure-limit for the near future (the next 10,000 years) and another limit—2300 percent higher—for the distant future (the period beyond 10,000 years). For the near future, this annual standard is 15 millirems. For the distant future, it is 350 millirems." Thus, "By setting different exposure limits for different time periods, E.P.A.'s first proposal fails to give all citizens equal protection." Twenty-first-century humans act as the arbiters for one thousand future generations. See Kristin Shrader-Frechette, "Mortgaging the Future: Dumping Ethics with Nuclear Waste," *Science and Engineering Ethics* 11:4 (2005): 519. See also Ethan T. Wilding, "Framing Ethical Acceptability: A Problem with Nuclear Waste in Canada," *Science and Engineering Ethics* 18:2 (2012): 301–313.

77. Joseph Masco, *The Nuclear Borderlands: The Manhattan Project in Post-Cold War New Mexico* (Princeton, NJ: Princeton University Press, 2006): 11–12.

78. Thomas Kuhn, *The Structure of Scientific Revolutions* (Chicago: University of Chicago Press, 1965).

79. "High-Level Nuclear Waste Storage Materials Will Likely Degrade Faster Than Previously Thought," *SciTechDaily,* 27 January 2020: https://scitechdaily.com/high-level-nuclear-waste-storage-materials-will-likely-degrade-faster-than-previously-thought/ (accessed 28 January 2020); Xiaolei Guo et al., "Self-Accelerated Corrosion of Nuclear Waste Forms at Material Interfaces," *Nature Materials* 19 (2020): 310–316.

80. Luke T. Townsend et al., "Formation of a U(VI)—Persulfide Complex During Environmentally Relevant Sulfidation of Iron (Oxyhydr)oxides," *Environmental Science and Technology* 54:1 (2020): 129–136.

81. Anna Demming, "New Form of Uranium Found That Could Affect Waste Disposal Plans," *The Guardian,* 20 December 2019: https://www.theguardian.com/environment/2019/dec/20/new-form-of-uranium-found-that-could-affect-nuclear-waste-disposal-plans (accessed 9 January 2020).

82. Frank von Hippel, Masafumi Takubo, and Jungmin Kang, *Plutonium: How Nuclear Power's Dream Fuel Became a Nightmare* (Singapore: Springer, 2019): 127; SKB, *Long-Term Safety for the Final Repository for Spent Nuclear Fuel at Forsmark: Main Report of the SR-Site Project,* vol. 3 (Stockholm: Svensk Kärnbränslehantering, 2011).

83. See Ulrich Beck, *Risk Society: Towards a New Modernity* (London: Sage Publications, 1992); Dexter Masters, *The Accident* (New York: Alfred A. Knopf, 1955); Nassim Nicholas Taleb, *The Black Swan: The Impact of the Highly Improbable* (New York: Random House, 2007).

84. Peter van Wyck, *Signs of Danger: Waste, Trauma, and the Nuclear Threat* (Minneapolis: University of Minnesota Press, 2005): 45.

85. Kathleen M. Trauth, Stephen C. Horal, and Robert V. Guzowski, *Expert Judgment on Markers to Deter Inadvertent Human Intrusion to Waste Isolation Pilot Plant* (Albuquerque, NM: Sandia National Laboratory, 1993): i.

86. Trauth, Horal, and Guzowski, *Expert Judgement on Markers to Deter Inadvertent Human Intrusion:* F-12.

87. Trauth, Horal, and Guzowski, *Expert Judgement on Markers to Deter Inadvertent Human Intrusion:* F-149.

88. Trauth, Horal, and Guzowski, *Expert Judgement on Markers to Deter Inadvertent Human Intrusion:* G-9, 10.

89. *Permanent Markers Implementation Plan,* US Department of Energy, DOE /WIPP 04-3302, 19 August 2004: 16, 19.

90. Van Wyck, *Signs of Danger:* 46.

91. Françoise Bastide and Paolo Fabbri, "Lebende Detektoren und komplementäre Zeichen: Katzen, Augen und Sirenen," Und in alle Ewigkeit: Kommunikation über 10 000 Jahre: Wie sagen wir unsern Kindeskindern wo der Atommüll liegt?: https:// www.semiotik.tu-berlin.de/menue/zeitschrift_fuer_semiotik/zs_hefte/bd_6_ hft_3/#c185968 (accessed 12 June 2019). The "ray cat solution" is currently being developed by Bricobio, which describes itself as an open laboratory in Montreal.

92. Stephen C. Hora, Detlof von Winterfeldt, and Kathleen M. Trauth, *Expert Judgment on Inadvertent Human Intrusion into the Waste Isolation Pilot Plant* (Albuquerque, NM: Sandia National Laboratory, 1991): C-57.

93. Thomas A. Sebeok, *Communication Measures to Bridge Ten Millennia* (Columbus, OH: Battelle Memorial Institute, 1984): 24.

94. Human Interference Task Force, *Reducing the Likelihood of Future Human Activities That Could Affect Geologic High-Level Waste Repositories* (Columbus, OH: Battelle Memorial Institute, Office of Nuclear Waste Isolation, 1984): 11.

95. "Conserver et transmettre la mémoire," Andra website: https://www.andra. fr/nos-expertises/conserver-et-transmettre-la-memoire (accessed 17 June 2019).

96. This is discussed at length in the documentary *Into Eternity,* directed by Michael Madsen (2011; Atmo Medi Network).

97. Oy, *The Final Disposal Facility for Spent Nuclear Fuel:* 136–137.

98. Human Interference Task Force, *Reducing the Likelihood of Future Human Activities:* 6, 9.

99. Van Wyck, *Signs of Danger:* 46.

100. For a fascinating analysis of the people tasked with working on deep geological storage in Finland, see Vincent F. Ialenti, "When Deep Time Becomes Shallow: Knowing Nuclear Waste Risk Ethnographically," Discard Studies, 9 March 2017: https://discardstudies.com/2017/03/09/when-deep-time-becomes-shallow-knowing-nuclear-waste-risk-ethnographically/ (accessed 11 June 2019). See also Sophie Poirot-Delpech and Laurence Raineau, "Nuclear Waste Facing the Test of Time: The Case of the French Deep Geological Repository Project," *Science and Engineering Ethics* 22:6 (2015): doi:101007/s11948-015-9739-9 (accessed 15 May 020).

101. Martin Fackler, "Tsunami Warnings for the Ages, Carved in Stone," *New York Time,* 20 April 2011: A6. See also Danny Lewis, "These Century-Old Stone 'Tsunami Stones' Dot Japan's Coastline," *Smithsonian Magazine,* 31 August 2015: https:// www.smithsonianmag.com/smart-news/century-old-warnings-against-tsunamis-dot -japans-coastline-180956448/ (accessed 23 June 2020).

102. International Nuclear and Radiological Event Scale (INES), IAEA website: https://www.iaea.org/topics/emergency-preparedness-and-response-epr/international -nuclear-radiological-event-scale-ines (accessed 12 June 2019).

103. National Research Council, *The Hanford Tanks: Environmental Impacts and Policy Choices* (Washington, DC: National Academies Press, 1996).

104. Rob Nixon, *Slow Violence and the Environmentalism of the Poor* (Cambridge, MA: Harvard University Press, 2011): 2.

105. John McNeil refers to the fallout from Chernobyl as "easily the most lasting insignia of the twentieth century and the longest lien on the future that any generation of humanity has yet imposed"; see John R. McNeil, *Something New Under the Sun: An Environmental History of the Twentieth-Century World* (New York: W. W. Norton, 2000): 313.

106. Ialenti, *Deep Time Reckoning*: xv.

107. Los Alamos National Laboratory, "Periodic Table of the Elements: Pluto-nium," August 2013: https://periodic.lanl.gov/94.shtml (accessed 18 June 2021).

108. The United States government tested this hypothesis by successfully detonat-ing a nuclear weapon manufactured with "reactor-grade" plutonium at the Nevada Test Site in 1962. See US Department of Energy, "Additional Information Concerning Underground Nuclear Weapon Test of Reactor-Grade Plutonium," June 1994: https://permanent.access.gpo.gov/websites/osti.gov/www.osti.gov/html/osti/opennet/document/press/pc29.html (accessed 18 June 2021). See also Gregory S. Jones, *Reactor-Grade Plutonium and Nuclear Weapons: Exploding the Myths* (Arlington, VA: Nonproliferation Policy Education Center, 2018): 87–100.

109. Text taken from one proposal for a nuclear marker to be placed at the Waste Isolation Pilot Plant in New Mexico. See Trauth, Horal, and Guzowski, *Expert Judgment on Markers to Deter Inadvertent Human Intrusion to Waste Isolation Pilot Plant*: F-49.

110. Van Wyck, *Signs of Danger*: 4 (original italics).

111. "Scenarios for a Nuclear Exchange," from A. Barrie Pittcock et al., *Environ-mental Consequences of Nuclear War*, SCOPE 28 (New York: John Wiley & Sons, 1985): 25–37.

112. David Allan Rosenberg, "'A Smoking Radiating Ruin at the End of Two Hours': Documents on American Plans for Nuclear War with the Soviet Union, 1954–1955," *International Security* 6:3 (Winter 1981/1982): 3–38.

Afterword

1. Arkady and Boris Strugatsky, *Roadside Picnic*, trans. Antonina W. Bouis (New York: MacMillan, 1977). This novel was the basis for the 1979 film *Stalker* directed by Andrei Tarkovsky.

2. Colin N. Waters et al., "Can Nuclear Weapons Fallout Mark the Beginning of the Anthropocene Epoch?" *Bulletin of the Atomic Scientists* 71:3 (2015): 46–57.

3. Joanna Macy has done elegant work on the concept of nuclear guardianship of this legacy. See Joanna Macy, *World as Lover, World as Self: A Guide to Living Fully in Turbulent Times* (Berkeley, CA: Parallax Press, 2003): 216–217, 226–234.

Index

Authorized Representative in the EU: Easy Access System Europe,
Mustamäe tee 50, 10621 Tallinn, Estonia, gpsr.requests@easproject.com

Printed and bound by CPI Group (UK) Ltd, Croydon, CR0 4YY
18/12/2025
02022706-0002